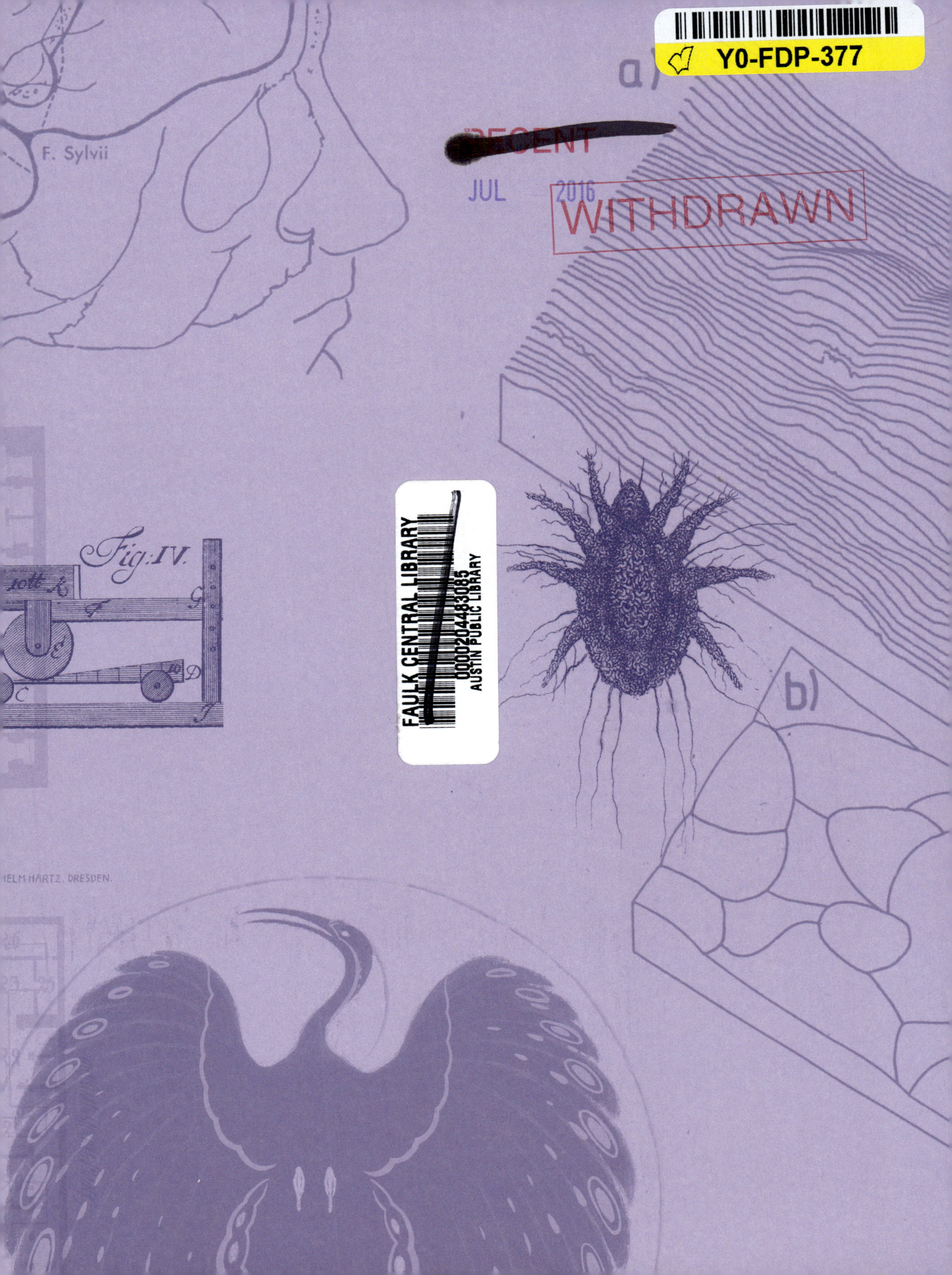
F. Sylvii
a)
Fig:IV.
b)
HELM HARTZ, DRESDEN.

The Technical Image

The Technical Image

A History of Styles in Scientific Imagery

Edited by Horst Bredekamp, Vera Dünkel, and Birgit Schneider

The University of Chicago Press, Chicago and London
Published in association with the Bard Graduate Center, New York City

HORST BREDEKAMP is professor of art history at the Humboldt University of Berlin and a permanent fellow of the Institute for Advanced Study in Berlin.

VERA DÜNKEL is a research affiliate with the "Das Technische Bild" research project.

BIRGIT SCHNEIDER is the postdoctoral fellow of the Fritz Thyssen Foundation at the Institute for Arts and Media, University of Potsdam.

Design: Laura Grey

The University of Chicago Press, Chicago 60637
The University of Chicago Press, Ltd., London

24 23 22 21 20 19 18 17 16 15   1 2 3 4 5

ISBN-13: 978-0-226-25884-3 (cloth)
ISBN-13: 978-0-226-25898-0 (e-book)
DOI: 10.7208/chicago
/9780226258980.001.0001

Library of Congress Cataloging-in-Publication Data

Technische Bild. English.
The technical image : a history of styles in scientific imagery / edited by Horst Bredekamp, Vera Dünkel, and Birgit Schneider.
pages ; cm
Includes bibliographical references.
ISBN 978-0-226-25884-3 (cloth : alk. paper)
ISBN 978-0-226-25898-0 (e-book)
1. Technical illustration. 2. Scientific illustration. 3. Digital images. I. Bredekamp, Horst, 1947– editor. II. Dünkel, Vera, editor. III. Schneider, Birgit, 1972– editor. IV. Title.
T11.8.T4413 2015
604.2—dc23
2014035711

Published in association with the Bard Graduate Center.

Contents

Foreword
Peter N. Miller

The Technical Image: A History of Styles in Scientific Imagery is a landmark book. It embodies a familiar intellectual format—a book—a project that represents a challenge to the meaning of art history. By discovering in an adjacent field, history of science, a whole range of materials amenable to art-historical treatment, this project points to a new horizon for art history. By making "technical illustration" with its reference point in the world of experiment and observation the arena for this gesture, *Das Technische Bild*, as it is called in German, also models a "cultural history of the material world."

In the German intellectual landscape out of which it sprung around the year 2000, *Das Technische Bild* represented an innovative approach to what Horst Bredekamp and Hans Belting had already offered up as *Bildwissenschaft*, or the study of images. This intervention was inspired by Aby Warburg's catholic approach to images in his practice of art history, something that for Warburg constituted a "cultural-historical art history." The study of Warburg, which emerged in Germany in the 1980s, recovered a voice and a practice that had been silenced not only by his death in 1929 but even more by the subsequent relocation of his institute to London in 1933 and the disruption over the next twelve years of his reading audience. Moreover, those followers who continued his work after him, at the Warburg Institute and elsewhere, did not share all of his aims and in any event did not communicate them to the new English-reading public. Indeed, Warburg's own published writings, let alone the vast manuscript materials in the London-based archive, were not translated into English until 1998 (though they were published in Italian in 1968) and still have not been translated into French.

But in Germany, beginning in the 1970s, the study of Warburg's range of work and his place in the intellectual culture of the early twentieth century has become a richly mined subject. It was here that Bredekamp in particular found an intellectual platform from which he could propose a reorientation of art history. Neither formal nor utterly historicizing, Warburg's "cultural-historical art history" opened up whole realms of images that had never before been treated seriously. Even though art historians were humanists trained to study and interpret images, they had neglected much of what was not considered "art." Like the turn to "visual culture" in the Anglophone world, *Bildwissenschaft* represented a healthy imperialistic grab by art historians, reaching out to occupy the abandoned landscape of advertising images, cheap print, digitalia, and, yes, science, among others.

Das Technische Bild was a research project at the Humboldt University in Berlin starting in 2000, chaired by Bredekamp. It drew together professors and students but also postdoctoral fellows. It launched a fascinating journal in 2003, *Bildwelten des Wissens* (Image-Worlds of Knowledge), and created an intellectual market for intensely visual discussions of practice in the natural sciences. In 2008, a representative sampling of the project's ambition and breadth was published by Akademie Verlag in Berlin. Theoretical statements were juxtaposed to case studies, broad treatments with microhistories. The brilliantly illustrated volume is a joy to contemplate. In turn, parallel and related projects have sprung up in Switzerland and elsewhere in Germany.

This is the book we are translating here and bringing to the wider English-reading audience. We are doing this not only because of its intrinsic worth but precisely because it parallels and challenges ongoing work in English. I have already used the term "visual culture" as an analogue for *Bildwissenschaft*. In fact, their aspirations are very similar. In both cases, though on different foundational grounds, there is a desire to broaden the subject matter of art history from the traditional canon and a willingness to confront the three-dimensionality of the image as well. Hence the otherwise ungainly term "visual and material culture."

At the same time, historians of science, especially those working on early modern Europe, have, from their perspective, also been probing the perimeters of art's history. Paying special attention to the ways in which imagery has been deployed in the context of intellectual argument and social practice, these scholars have slowly but surely been effacing any hard boundary between the verbal and the visual, what otherwise could be termed "content" and "form." The work promoted by Lorraine Daston at the Max Planck Institute for History of Science, originally just down the road from the Art History Department of the Humboldt University where *Das Technische Bild* lives (although today it is further away), has been at the center of this development.

What *Das Technische Bild* brought to this new dialogue between science and art is the insistence that visualization is in itself a technology. Imagery is not passive, not simply an illustration of what is, not simply the corollary of the correspondence theory of truth. Rather, the capability of putting some phenomenon into an image is an interpretative intervention. Indeed, were I to use the more contemporary word *imaging*, we would immediately grasp the technological dimension upon which this research agenda insists. It is this productive power of the image that differentiates the notion of "style" used here from the more passive, almost Hegelian notion of style as a general background, milieu, mentality, or paradigm.

It is at this very moment of burgeoning interest in the intersection of the visual and the material, of art and science, that we feel a translation of *Das Technische Bild* is important. It gives an English-reading public access to a German-language initiative that helps makes sense of the future of art history as well as of art history's relationship to its neighboring disciplines. *The Technical Image* represents the Bard Graduate Center's commitment to the idea that in a global scholarly world it is a necessity to know what is going on in other communities. The best would be for everyone to read everyone else's language. Failing that, translations are necessary. We want students who do not know German to know about this project because it is important in itself and as a model.

While its origins lie in an argument about broadening art history, even a casual perusal of this volume shows that many technical "images" are also "objects." This volume, then, assumes importance for anyone interested in the future of those disciplines that study objects. Looked at this way, it should be clear why and how a study of the history of technical imagery could also function like a cultural history of the material world.

For the Bard Graduate Center, the publication of this volume marks the conclusion of a collaboration with the Art History Department of the Humboldt University in Berlin, which has involved hosting a visiting professor and lecture series in 2011 and developing a Focus Gallery exhibit in 2012 (*The Islands of Benoit Mandelbrot: Fractals, Chaos, and the Materiality of Thinking*, with a complementary volume published by the Bard Graduate Center and distributed by Yale University Press). All of these projects together emphasize the vast range of what we can learn about the past from its materialized forms.

—Peter N. Miller, Dean, Bard Graduate Center

INTRODUCTION
The Image—A Cultural Technology: A Research Program for a Critical Analysis of Images
Horst Bredekamp, Vera Dünkel, Birgit Schneider

When the research project "Das Technische Bild" was founded in 2000 as part of the Hermann von Helmholtz-Zentrum für Kulturtechnik at the Humboldt University of Berlin,[1] its focus on the analysis of scientific and technical imagery was a novelty. In the intervening years, several projects and institutions have dedicated themselves to the same field of scientific visual representation;[2] numerous publications document the growing yield of these studies.[3] The project set out from the research hypothesis that the forms of imagery are of no less import than the content and objects they show. Images in the natural sciences are a good example: they do not passively render the findings they serve to represent; they generate and inform them with the distinctive qualities of their own sphere. The transformation of observations, findings, and insights into images partakes actively in the constitution of knowledge. With this active capacity, the production and employment of imagery represent a cultural technique of the first rank. The term "technical" here requires some clarification. For us, it emerged as a fruitful concept because it implies different layers of *techne*. Our objects of study are "technical images" in the sense that they are not artistic, instead primarily originating in the fields of science, technology, and medicine; they are predominantly instrument-based or the results of imaging procedures. On the one hand, "technical" emphasizes the way these specific images are produced (by technical means, apparatuses, instruments, or by hand). On the other hand, images may be thought of as tools or as instruments in their own right. Scientific images thus provide the clearest examples of the technical image's methodological import, but the notion of "the technical image" extends beyond scientific images alone and designates a distinctive and new art-historical approach. The various topics addressed in this volume highlight one or the other reading of the term "technical"; each case study also sheds light on how the different layers of *techne* are interwoven in manifold ways.

From the outset, the project's work has been firmly rooted in the methods of art history; the discipline has built up an unrivaled competence in the analysis of the material features, form, and semantics of images. In this regard our approach differs not only from Anglo-American Visual Culture Studies but also from the collective field of German *Bildwissenschaft*, both of which were developed from a broad field of disciplines.[4] Unlike Visual Culture Studies, we do not first look at the social construction of images but rather at their material form; and unlike the strain of *Bildwissenschaft* rooted in the philosophy of aesthetics,[5] we

1 The project was chaired by Horst Bredekamp until 2012. Its first director was Gabriele Werner, who was succeeded by Matthias Bruhn in 2005, who is also the current chair. The first issue of our yearbook, *Bildwelten des Wissens: Kunsthistorisches Jahrbuch für Bildkritik*, which is published biannually and designed and edited by members of the team, came out in 2003. Having initially relied on funding from the Humboldt University alone, we subsequently received generous third-party funding support from the Getty Foundation, Los Angeles, and the German Research Foundation.

2 In the German-speaking world, we should mention the exemplary work of the Swiss National Center of Competence in Research "Eikones"; the research group "Die Welt als Bild" at the Berlin-Brandenburg Academy of Sciences and Humanities (2005–2008); the Humanities Centre for Advanced Studies "Bildakt und Verkörperung" at the Humboldt University of Berlin; the Research Training Group "Schriftbildlichkeit: Über Materialität, Wahrnehmbarkeit und Operativität von Notationen" at Freie Universität Berlin; the Research Training Group "Sichtbarkeit und Sichtbarmachung: Hybride Formen des Bildwissens" at the University of Potsdam; the Humanities Centre for Advanced Studies "BildEvidenz: Geschichte und Ästhetik" at Freie Universität Berlin; and the Cluster of Excellence "Image Knowledge Gestaltung: An Interdisciplinary Laboratory" at the Humboldt University of Berlin.

3 See, e.g., Bettina Heintz, Jörg Huber, eds., *Mit dem Auge denken: Strategien der Sichtbarmachung in wissenschaftlichen und virtuellen Welten* (Zurich, Vienna, New York: Voldemeer, 2001); David Gugerli, Barbara Orland, eds., *Ganz normale Bilder: Historische Beiträge zur visuellen Herstellung von Selbstverständlichkeit* (Zurich: Chronos, 2002); Martina Heßler, ed., *Konstruierte Sichtbarkeiten: Wissenschafts- und*

follow an inductive and historical approach in the analysis of pictures. *Bildwissenschaft* taken from our perspective originates from art historical traditions.[6] This methodological framework compels the observer to take any visualization seriously as a formal manifestation, delineating its historical context within and without the confines of "art" and going beyond its phenomenological appearance in order to comprehend its modes of operation and specific functions. This approach rests on the assumption that the choice of a particular pictorial form, medium, or type has implications that hide in plain sight and inform the object of study and the manner in which it is studied; Erwin Panofsky used the term "disguised symbolism"[7] to describe this nexus. Moreover, images may trigger new research: the development of genetics as a branch of research, for instance, would have been virtually impossible without the existence of visual representations such as the double helix model and X-ray scattering images. So the ways in which researchers produce and employ images represent more than their conscious intentions, or at least more than their writings and formulas reflect. The project therefore undertakes to investigate the frequently observed constructive agency of images and to determine their intrinsic efficacy.

The goal that "Das Technische Bild" set itself was accordingly to comprehend images not as illustrative representations, but as productive agents and distinctive multi-layered elements of the epistemic process. A particular concern of this investigation is the "principle of disjunction" governing visualization in the natural sciences.[8] This principle seeks to frame the paradoxical observation that a scientific image is often the more thoroughly constructed the more natural its object appears in the rendition.[9] Time and again, technology becomes invisible once its employment becomes a matter of course: by the same token, the artificial character of the image tends to sink into oblivion once researchers begin to work with it.[10] The implication for an understanding of the constructive role scientific images play is that all those conditions that shape the form of an image must be taken into account. The fundamental conviction guiding this critical approach to images is that they must be regarded not as finished products, but with a view to all components of their generation, to the techniques and interventions, the agents who apply them, and the contexts in which they take place: images, that is to say, must be considered in process. The project has accordingly worked from the outset to envision also the traces that have become invisible in the process of rendering something visible. The technologies of image production, in particular, constitute a central element in the study of scientific imagery, as a distinct class of instruments, devices, and tools have been constructed and continually refined that are explicitly designed for the purpose of visualization. This touches upon the enormous significance the act of rendering visible has in the sciences, as well as the technological and media conditions on which knowledge is based.[11] The title of our project refers to this supportive role technological implements play in perception and the generation of images as well as the design and production of images as a more general *techne*.

If scientific images play a constructive role in shaping the findings and insights they illustrate, the representation of an observation in images, however mechanical, however detached from the individual researcher's

Technikbilder seit der Frühen Neuzeit (Munich: Fink, 2006); Inge Hinterwaldner, Markus Buschhaus, *The Picture's Image: Wissenschaftliche Visualisierung als Komposit* (Munich: Fink, 2006); James Elkins, ed., *Visual Practices across the University* (Munich: Fink, 2007).

4 See the bibliography at the end of the book. In the German development of academic disciplines, there is no institutionalized discipline of visual studies or visual culture studies. While there has been a strong tradition of *Bildwissenschaften* in the German-speaking world over the last few decades, today there are still no stand-alone departments for the study of *Bildwissenschaften.*

5 See especially the writings of Klaus Sachs-Hombach, for instance, *Bildwissenschaft zwischen Reflexion und Anwendung* (Cologne: Halem, 2004).

6 Horst Bredekamp, "A Neglected Tradition? Art History as 'Bildwissenschaft,'" *Critical Inquiry* 29, no. 3 (2003): 418–28.

7 Erwin Panofsky, *Early Netherlandish Painting: Its Origins and Character*, 2 vols. (Cambridge, MA: Harvard University Press, 1953). See also Erwin Panofsky, "Probleme der Kunstgeschichte," in *Korrespondenz*, vol. 1: *1910 bis 1936*, ed. Dieter Wuttke (Wiesbaden: Harrassowitz, 2001), 957–64.

8 This concept is borrowed from Erwin Panofsky, who coined the term "principle of disjunction" to describe the disconnect between form and content in the late-medieval reception of antiquity. See Erwin Panofsky, *Renaissance and Renascences in Western Art* (Stockholm: Almqvist & Wiksell, 1960).

9 Horst Bredekamp, Angela Fischel, Birgit Schneider, Gabriele Werner, "Bildwelten des Wissens," in Horst Bredekamp, Gabriele Werner, eds., *Bilder in Prozessen, Bildwelten des Wissens: Kunsthistorisches Jahrbuch für Bildkritik*, vol. 1, no. 1 (Berlin: Akademie, 2003), 15.

10 Hans Blumenberg has proposed this hypothesis, especially in "Lebenswelt und Technisierung unter Aspekten der Phänomenologie," in *Wirklichkeiten, in denen wir leben* (Stuttgart: Reclam, 1993), 7–54. It also underlies the studies contained in the anthologies edited by Heintz, Huber; Gugerli, Orland; and Heßler (see n. 2).

11 Ever since Hans-Jörg Rheinberger and Bruno Latour pointed up this dimen-

choices their appearance may be, likewise becomes an instance of the style of a period, a mindset, a research collective, and a device. Niklas Luhmann's concept of the "medium/form coupling" seeks to capture this process.[12] The reference to "style" in the title of the present book reaffirms the focus on form proposed by Luhmann's argument. It emphasizes that an image not only displays the symptoms and results of a thought style,[13] but also constitutes that style with a quasi-objective power that seems to transcend the work of any individual. The terms "style" and "form" are generally used interchangeably in this volume. We favor "form" as the more neutral and general term of analysis that avoids the historic connotations of "style," which, in the nineteenth century, implied a value judgment. In its wider sense, however, "style" reveals the art historian's assumption that form always implies a symbolic meaning.

The desire for historical order tends to subsume phenomena under concepts, which then become the paradigms of epistemic progress. *Objectivity*, *documentation*, and *evidence* are exemplary concepts that, as the scholarly study of scientific images has thrived, have become so popular and influential that the material has seemed to submit of its own accord to the order they delineate. "Das Technische Bild," however, takes the inverse route; our phenomenological-conceptual spindle always moves upward from the pictorial forms and techniques. This inductive method, which is the standard in archaeology and art history, is particularly apt for demonstrating the extraordinary power that forms of scientific imagery wield. We have found that this perspective provides better insight into how the initial search for suitable visual forms, which is often a process of trial and error, as well as the conventionalization of pictorial forms proceeds. The meticulous study of forms accordingly constitutes the point of departure as well as the final destination of all our analyses of imagery.

Manifesting perceptions, observations, and insights as well as giving them definite shape, images must in a second step be examined in the interdisciplinary perspective that is an indispensable part of art-historical iconology.[14] That is particularly true of the wide range of scientific imagery, and implies the involvement of methods of inquiring into the constructive character of any emerging pictorial form. In consequence, the iconological interpretation of technical images also requires the application of methods and findings from cultural and media studies, anthropology, sociology, and political science; vital, in particular, are the contributions of the history of science and technology studies, with which our endeavor has in recent years built a positively symbiotic relationship.[15] Approaches in aesthetics and the philosophy of consciousness that promise to overcome the Platonism of some analytical philosophy seem to adumbrate another new horizon of collaboration.[16] Based on image-specific analyses, only this polyphony of methods can give an adequate account of the constructedness, mutability, and operability of these images. Accordingly, our team, and the group of contributors to the present volume, includes not only art historians, but also cultural and literary scholars, physicists, philosophers, and historians of science.

In order to trace the evolution of our project since 2000, the present book lays out our findings in conjunction with the full range of our approaches; we hope that it may serve as a textbook of methodology. The

sion, many scholars have emphasized the foundational role that technical and scientific images play by rendering phenomena visible.

12 Niklas Luhmann, *Art as a Social System*, trans. Eva M. Knodt (Stanford, CA: Stanford University Press, 2000), 102–32; and cf. Niklas Luhmann, *Die Gesellschaft der Gesellschaft* (Frankfurt a.M.: Suhrkamp, 1997), 195–202.

13 Ludwik Fleck coined the term "thought style" to describe the system of conventions, agreements, and procedures that, transcending the ideas of any individual researcher, governs the scientific practice of a research collective. See Ludwik Fleck, *Genesis and Development of a Scientific Fact* [1935], trans. Fred Bradley, Thaddeus J. Trenn (Chicago, London: University of Chicago Press, 1979).

14 As early as 1912, Warburg mentioned "iconological analysis" as an "extension of the methodological borders of our study of art, in both material and spatial terms." Aby Warburg, "Italian Art and International Astrology in the Palazzo Schifanoia, Ferrara" [1912], in *The Renewal of Pagan Antiquity: Contributions to the Cultural History of the European Renaissance*, ed. Steven Lindberg (Los Angeles: Getty Publications, 1999), 585.

15 See the exemplary study by Lorraine Daston, *Eine kurze Geschichte der wissenschaftlichen Aufmerksamkeit* (Munich: C. F. von Siemens Stiftung, 2001); Wolfgang Lefèvre, Jürgen Renn, Urs Schoepflin, eds., *The Power of Images in Early Modern Science* (Basel: Birkhäuser, 2003); Wolfgang Lefèvre, ed., *Picturing Machines 1400–1700* (Cambridge, MA: MIT Press, 2004). Many of the contributors to the yearbook *Bildwelten des Wissens* are likewise historians of science.

16 See, e.g., John Michael Krois, "Für Bilder braucht man keine Augen. Zur Verkörperungstheorie des Ikonischen," in John Michael Krois, Norbert Meuter, eds., *Kulturelle Existenz und symbolische Form: Philosophische Essays zu Kultur und Medien* (Berlin: Parerga, 2006), 167–90; Sybille Krämer, "Operationsraum Schrift: Ein Perspektivenwechsel im Schriftverständnis," in Gernot Grube, Werner Kogge, Sybille Krämer, eds., *Schrift: Kulturtechnik zwischen Auge, Hand und Maschine* (Munich: Fink, 2005), 13–32; Frederik Stjernfelt, *Diagrammatology: An Investigation on the Borderlines of Phenomenology, Ontology and Semiotics* (Dordrecht: Springer, 2007); Wolfram Hogrebe, *Echo des*

Nichtwissens (Berlin: Akademie, 2006); Martin Seel, *Ästhetik des Erscheinens* (Munich: Hanser, 2000); Lambert Wiesing, *Artifizielle Präsenz: Studien zur Philosophie des Bildes* (Frankfurt a.M.: Suhrkamp, 2005).

core of the book consists of several historical case studies. They evolved out of individual research projects that have been undertaken at the institute since its inception. Each case study presents a larger constellation of issues in a limited set of specific images. Four methodological articles precede the section of case studies that introduce the reader to the set of questions we study, embedding them in a more general framework. For the English edition, Franziska Brons, Stefan Ditzen, and Reinhard Wendler provided new case studies, while Matthias Bruhn and Gabriele Werner revised and updated their methodological articles. The case studies also form the basis for brief encyclopedia-style definitions interspersed throughout the book. The selection does not aim at lexicological completeness; rather, it highlights specific features and dimensions of scientific and technical imagery and reflects on the particular challenges these features pose. They define, and elaborate on, selected terms that have in recent years become key concepts in the analysis of scientific imagery. For example, the terms "image noise," "observation techniques," and "visuality, visualizing, imaging" bring fields of inquiry and functions of imagery into focus that are of particular significance to scientific images insofar as they allow us to take into account the ways instruments condition them and their embeddedness in generative processes. Terms such as "diagrammatics" and "structuring images as tableaux" refer to species and configurations of images that are of strategic value in the presentation of scientific and technological material. "Comparison as method" and "iconological analysis" represent our attempts to examine the significance and application of art-historical methods in the realm of scientific imagery. These articles hope to provide analytical tools to scholars of scientific images and chart avenues of future inquiry. They muster different concepts from the discipline that may be made fruitful for a further analysis of technical images.

The thematically arranged bibliographical appendix surveys the essential research in the field of scientific imagery. We have sought to adapt the bibliography for an English-speaking audience, but know full well that the result of our efforts may not be comprehensive. In light of the fact that defining and appreciating the intrinsic value of historic materials requires that the present-day interest in them has been rendered visible—and can thus be considered from a distance—the case studies following the four general essays form a sequence moving backward from the present into the historical past. This arrangement parallels the organization of our yearbook, *Bildwelten des Wissens*.[17]

17 The prototype of this organization is Dietrich Mahnke's history of cosmology, which also begins with the present and then proceeds to rewind the history of astronomical models. See Dietrich Mahnke, *Unendliche Sphäre und Allmittelpunkt* (Halle a.S.: Frommann, 1937).

Acknowledgments

The present book could not have been completed without the active support of past and present members of the "Das Technische Bild" team. Former student assistants Violeta Sánchez and Jana August made vital contributions to the articles on key terms. Student assistants Florian Horsthemke and Hanna Felski faced considerable challenges in researching and editing the visual material. Christiane Gaedicke provided valuable administrative assistance.

For the English version of the book, we owe great thanks to Peter N. Miller and Daniel Lee at the Bard Graduate Center in New York City, who initiated this undertaking and supported it with great optimism and patience. The tireless work undertaken by student assistants Theresa Stooß, Rahel Schrohe, Jane Beran, Judith Berganski, Simon Hirsbrunner, and Felix Jäger in clearing more than one hundred image copyrights, researching new images, and providing literature for the English edition was invaluable. We are thankful to Sanford Sanchez for his assistance with the bibliographies. At the University of Chicago Press, we would like to thank Christie Henry for adopting and shepherding the project through the co-publication process and Logan Ryan Smith for his ardent support. We owe especial gratitude to our translator Gerrit Jackson. His witty comments, sense of humor, and knowledge and accuracy of both language and subject matter were constant strengths for us throughout. We also thank Martin Schneider, who scrupulously took care of the copyediting, and Christine Gever, our peerless proofreader. And last, but certainly not least, we owe a debt of gratitude to the designers: Laura Grey, who designed the present edition, and the scientific illustrator Nils Hoff, who designed the German edition on which portions of the English edition are based. Finally, we thank the Bard Graduate Center and the German Research Foundation (DFG) for providing us with the financial support that enabled us to realize this book together with The University of Chicago Press.

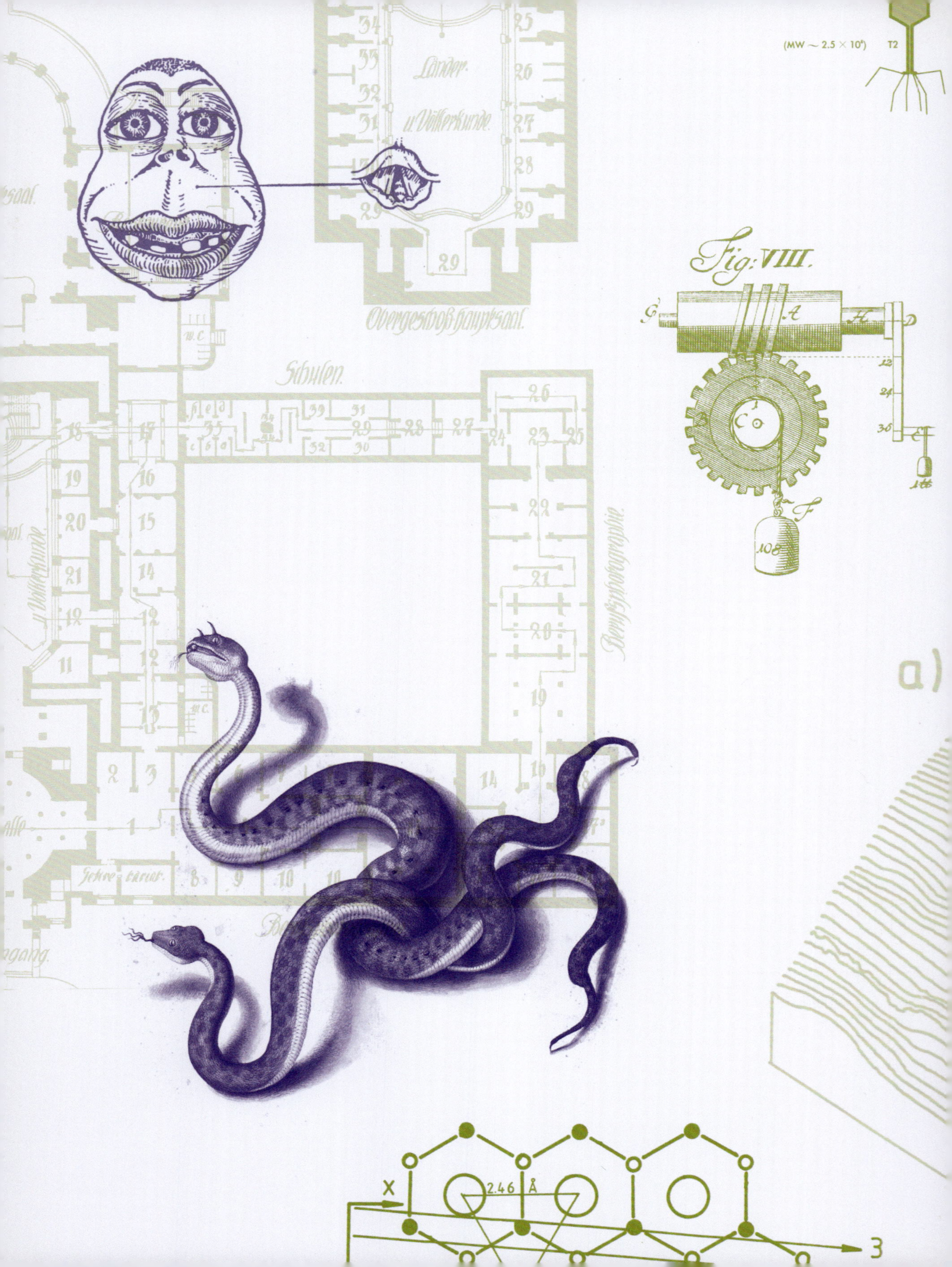
(MW ~ 2.5 × 10⁸)
T2
Fig: VIII.
Länder
u. Völkerkunde
Obergeschoß Hauptsaal.
Schulen.
a)
2.46 Å
X
3

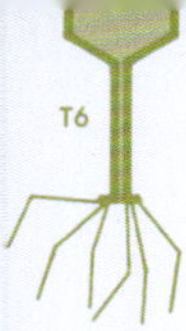
T6

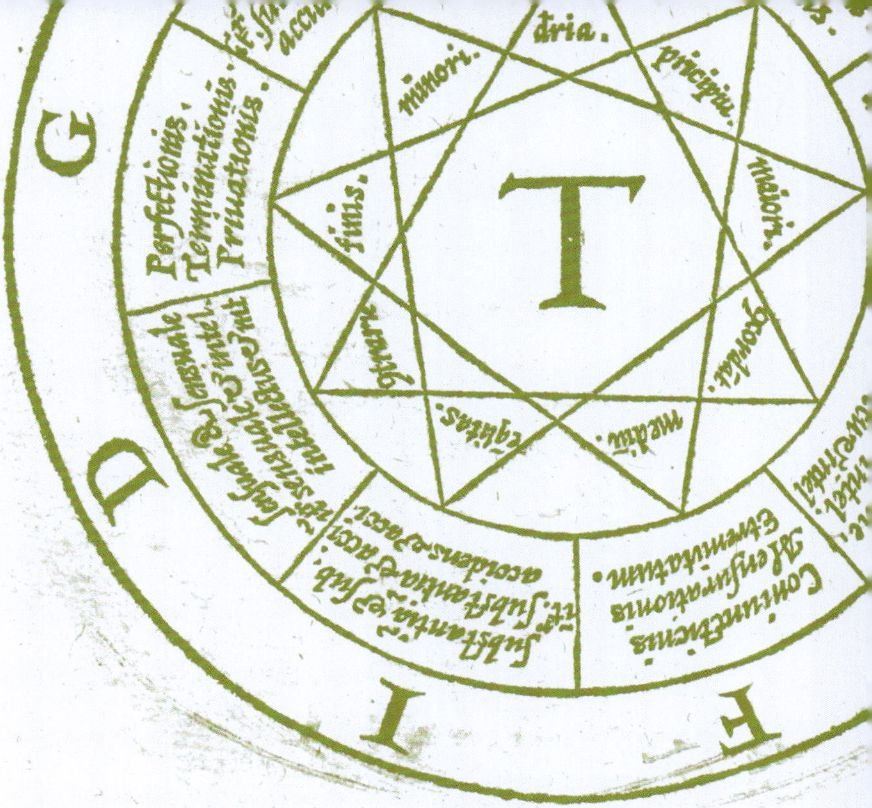
G
D
I
F
T
minori.
dria.
maiori.
finis.
medii.
Perfectionis.
Terminationis.
Priuationis.

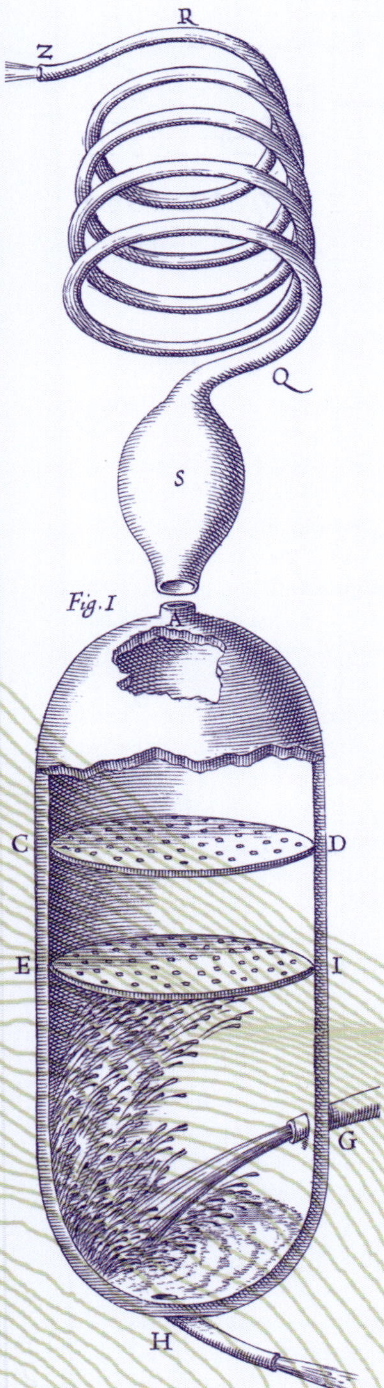
Z
R
Q
S
Fig. I
A
C
D
E
I
G
H

43

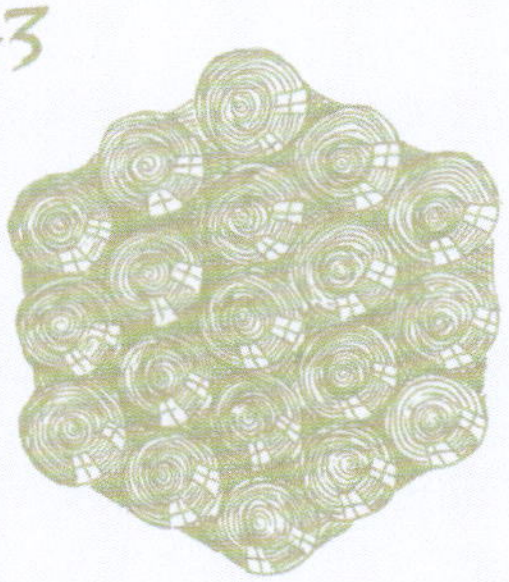

Methods

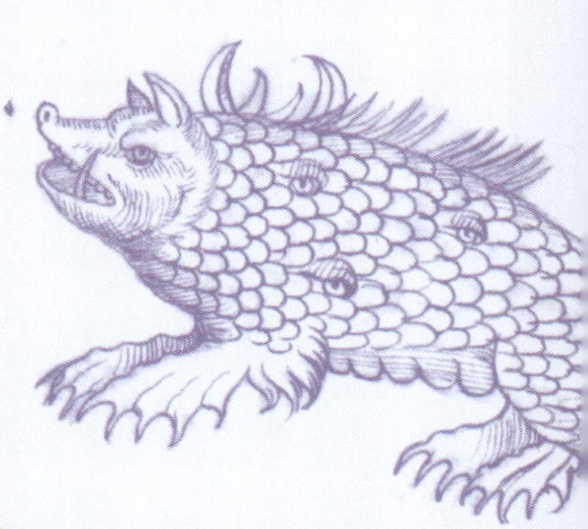

Discourses about Pictures: Considerations on the Particular Challenges Natural-Scientific Pictures Pose for the Theory of the Picture

Gabriele Werner

From Practices to the Picture

The history of the systematic study of pictures from the natural sciences (not the history of the natural-scientific picture as such) begins around 1960, when scholars analyzed various cultural practices and, more particularly, the differences and similarities between artistic and scientific activities. The standpoints were heterogeneous and diverse.[1] Pictures had always been among the material objects that were produced by, or elements of, experimental and laboratory practices.[2] Charting the dimension of congruent intellectual explorations in the sciences and the arts as well as forms of aesthetic influence, scholars now juxtaposed techniques of visualization in the natural sciences, engineering, and technological research with artistic pictures in order to examine their respective epistemological peculiarities.[3] An eminent example of such joint exploration was the *Cybernetic Serendipity* exhibition organized by the curator Jasia Reichardt, which was held at the Institute of Contemporary Arts, London, from August to October 1968. Engineers and information scientists collaborated with artists to show computer-aided work from robotics to drafting machines and textual automata.[4] From that point on and for the next two decades, moreover, it was not yet theories of the picture that were at issue but a critique of representation. Thomas Kuhn's *The Structure of Scientific Revolutions* inaugurated a period of heightened interest in scientific practice, in observation and experimentation,[5] as well as the rehabilitation of skills and craftsmanship as factors that, for all their lack of uniformity and heterogeneity, were of no less significance to the production of scientific insight.[6] Reading the natural sciences as a set of cultural practices with a deep history allowed scholars to turn their attention to the material forms of representation with which researchers work: images, photographs, drawings, charts, statistics, diagrams, preserved specimens, models, computer graphics, animations. Action-theoretical approaches to the generation of knowledge preferred observable epistemic processes to studies that examined the sciences in the perspective of the history of ideas or on a theoretical basis; they enabled scholars to describe recordings, experimental findings, objects of research as agents in the process of scientific insight.[7]

A tradition of scholarship on the scientific picture emerged that made it entirely impossible to see its object as subordinate, a mere illustration. As Michael Lynch wrote, "Visual documents are used at all stages of scientific research. A series of representations of renderings is produced, transferred, and modified as research proceeds from initial observation to final publica-

1 The methodological diversity, but also the shared research interests, are evident in Charles P. Snow, *The Two Cultures and the Scientific Revolution* (Cambridge: Cambridge University Press, 1959); Ernst Gombrich, *Art and Illusion: A Study in the Psychology of Pictorial Representation* (New York: Pantheon Books, 1960); Nelson Goodman, *Languages of Art: An Approach to a Theory of Symbols* (Indianapolis: Bobbs-Merrill, 1968).

2 See, e.g., Caroline A. Jones and Peter Galison, eds., introduction to *Picturing Science, Producing Art* (New York: Routledge, 1998), 1–23; Martina Heßler, "Bilder zwischen Kunst und Wissenschaft: Neue Herausforderungen für die Forschung," *Geschichte und Gesellschaft* 31, no. 2 (2005): 266–92.

3 See Jones and Galison, introduction to *Picturing Science, Producing Art*, 4–5.

4 Jasia Reichardt, ed., *Cybernetic Serendipity: The Computer and the Arts* (special issue, Studio International), 1st ed. (London: *Studio International*, July 1968); 2nd rev. ed. (London: Studio International, September 1968).

5 See Michael Hagner, "Ansichten der Wissenschaftsgeschichte," in *Ansichten der Wissenschaftsgeschichte*, ed. Michael Hagner (Frankfurt am Main: Fischer Taschenbuch, 2001), 19.

6 See Thomas S. Kuhn, *The Structure of Scientific Revolutions* (Chicago: University of Chicago Press, 1962), esp. 23–34; cf. Timothy Lenoir, "Inscription Practices and Materialities of Communication," in *Inscribing Science: Scientific Texts and the Materiality of Communication*, ed. Timothy Lenoir (Palo Alto, CA: Stanford University Press, 1998), 2.

7 See, e.g., Bruno Latour, *Science in Action: How to Follow Scientists and Engineers through Society* (Cambridge, MA: Harvard University Press, 1987), 84, 89.

tion. At any stage in such a production, such representations constitute the physiognomy of the object of the research."[8]

The function of these "visual documents," and hence the kind of information they are intended to convey, depends on the context in which they appear. They may be tools or teaching aids, advertisements or (as historical objects) artifacts; they may document a controversy between researchers or attest to the history of a field of knowledge; they may mediate between a specialized discourse and its popular retelling or have temporary use-value in the circulation of knowledge in laboratories and research institutions. There are durable, medium-term, and volatile media of knowledge, depending on the material, technical, and technological instruments used to generate pictures. When the means of representation change, the way a research object looks changes as well—and the question whether that implies an alteration also of the object itself leads straight to the heart of the problem.

Counter to Lynch's assessment, however, pictures pretend to be more than pure creations of human ingenuity: they are supposed to refer to knowledge or the observation of nature. They are expected to possess a well-founded reference to a theory, an experimental practice, a technique or technology, and a discourse of knowledge, since natural-scientific pictures and objects are assumed to be evidence—they must assert that they have evidentiary value in order to function as objects of knowledge.[9] They show or reveal *something* that both exists outside of them and yet does not come into being without them.

Early on, a critical theory of the picture severed the link between evidence and mere obviousness, embedding the concept in a sociotechnical context according to which evidence can only emerge in a shared informational structure; the latter is not merely a matter of factual knowledge but also rests on opinions, social rules governing attention, and techniques of visualization.[10] Pictures and objects in the natural sciences *figure* something, and that importantly includes not only the *representation* of something but also its *generation*; whence the expectation on the part of their beholders that the information they visualize may be verified and re-created by others and that the gestures, apparatuses, and theoretical principles involved in their genesis allow them to be replicated and reproduced.[11] The pictures and objects of the natural sciences are always technical in this comprehensive sense.

From the Picture to Theories of the Picture

The road from the picture to theories of the scientific picture led through an (ongoing) methodological controversy whose central issue was and still is the critique of what is called logocentrism. Its proponents saw it as a critique of the dominance of language and the comparatively disparaging assessment of pictures, as a category whose eidetic logic was not entirely—in the assessment of some writers, not at all—comparable to the logic of language. The first strike in this controversy was dealt by Richard Rorty's anthology *The Linguistic Turn*, which first came out in 1967 (an expanded edition appeared in 1992); Rorty criticized the conviction in language-based philosophy that its claims to truth trump the articulations of knowledge in the arts and the sciences.[12] Yet it is an inadmissible simplification to read Rorty's critique of the alleged transparency

8 Michael Lynch, "The Externalized Retina: Selection and Mathematization in the Visual Documentation of Objects in the Life Sciences," *Human Studies* 11, nos. 2–3 (April 1988): 201–34; repr. in Michael Lynch and Stephen Woolgar, eds., *Representation in Scientific Practice* (Cambridge, MA: MIT Press, 1990), 153–86, quotation 154.

9 See Karin Knorr-Cetina, *Epistemic Cultures: How the Sciences Make Knowledge* (Cambridge, MA: Harvard University Press, 1999), 81.

10 David Gugerli, "Soziotechnische Evidenzen: Der 'Pictorial Turn' als Chance für die Geschichtswissenschaft," *Traverse: Zeitschrift für Geschichte* 6, no. 3 (1999): 131–59.

11 For this concept of the technical, see Werner Rammert, "Die Form der Technik und die Differenz der Medien: Auf dem Weg zu einer pragmatischen Techniktheorie," in *Technik und Sozialtheorie*, ed. Werner Rammert (Frankfurt am Main: Campus, 1998), 293–326.

12 See Richard M. Rorty, "Introduction: Metaphilosophical Difficulties of Linguistic Philosophy," in *The Linguistic Turn: Essays in Philosophical Method; with Two Retrospective Essays*, ed. Richard M. Rorty (Chicago: University of Chicago Press, 1992), 1–39.

of philosophical language and an epistemology that describes language as the representation of "reality" as rejecting the fundamentally language-based structure of all production of meaning.[13]

Rorty's philosophical skepticism concerning truth was due, rather, to an anti-essentialist stance that regarded any insight as constructed and transmitted by media. Yet language, too, is a medium in this sense, so the linguistic turn in fact made it possible to acknowledge the contingency and ambiguity of verbal utterances—very much including the utterances of the sciences—and the role language played "as an antecedent prerequisite for the sociality of the subject."[14] In the context of this methodological controversy raging in the history of science (but not only there), Timothy Lenoir was being distinctly polemical when he wrote in 1998 that "'the semiotic turn' is upon us"; he was not referring only to a dispute between constructivists and realists that was coming to a head at the time. The project of an "inscribing science" was committed to the endeavor of surveying the complexity of inscriptions leading to the articulation of scientific fact, from the deconstruction of written (theoretical) demonstrations of truth to the materiality of visual resources, from unarticulated skills to the grand cultural narratives and their echoes in museums and popular culture. Writing is examined with a view to its linguistic and semiological techniques of the production of meaning and not as distinct from the picture; whence the picture as a vehicle of meaning becomes susceptible to analysis in the same semiological perspective with a view to its means of making sense.[15]

This approach to pictures differs fundamentally from the positions that have been, and still are, described by the phrases "iconic turn" (Gottfried Boehm) and "pictorial turn" (W. J. T. Mitchell): the latter, Mitchell writes, "is the realization that spectatorship (the look, the gaze, the glance, the practices of observation, surveillance, and visual pleasure) may be as deep a problem as various forms of reading (decipherment, decoding, interpretation, etc.) and that visual experience or 'visual literacy' might not be fully explicable on the model of textuality."[16] The talk of such "turns" triggered a controversy—even today, some contributions to the debate are highly confrontational in tone—over a distinct logic of the picture that is at its core a critique of writing and textuality conducted in writings and texts, a critique of the propositional logic of language and formal determinants of textuality as analogous with pictoriality. Yet regarding the latter, in spontaneous and formulaic fashion, as extralinguistic misses the problem of seeing as seeing-something and of showing as showing-something[17] and fails to appreciate the immensely productive effects of this controversy.

"In its fundamental, first definition, the concept of the picture encompasses any and all forms of visual fashioning"; we should speak of pictures "once natural objects evince even minimal traces of human manipulation."[18] Fashioning and manipulation are, in the perspective of this seminal concept of the picture, not the exclusive privilege of the hand; more precisely, these terms apply also where the hands, using a keyboard, write an algorithm that generates a pictorial, rather than a textual or acoustic, form out of a data set. This anthropological conception of the picture allows us to resolve the arid dispute over whether digital images are pictures even though they lack a physical basis comparable to that of the painted panel. Many other

13 See Susanne Lummerding, *Agency @? Cyber-Diskurse, Subjektkonstituierungen und Handlungsfähigkeiten im Feld des Politischen* (Vienna: Böhlau, 2005). Lummerding offers a series of arguments as to why the recourse to the logic of language excludes a reductionistic contrast between picture and text as isolable entities; the necessity to differentiate between things due to the inherent logic of language underlies all production of meaning. Her arguments are quite extensive; we refer the interested reader to the passages listed in the subject index under "Sprache." For a representative example of the simplistic reading of Rorty's argument, see Gottfried Boehm, "Jenseits der Sprache? Anmerkungen zur Logik der Bilder," in *Iconic Turn: Die neue Macht der Bilder*, ed. Christa Maar and Hubert Burda (Cologne: DuMont, 2004), 28–34.

14 Sigrid Schade and Silke Wenk, *Studien zur visuellen Kultur: Einführung in ein transdisziplinäres Forschungsfeld* (Bielefeld: Transcript, 2011), 42–43.

15 See Lenoir, "Inscription Practices," 4–5.

16 W. J. T. Mitchell, "The Pictorial Turn," in *Picture Theory: Essays on Verbal and Visual Representation*, ed. W. J. T. Mitchell (Chicago: University of Chicago Press, 1994), 16.

17 See Andreas Beyer and Markus Lohoff, "Bildhandeln: Eine Einführung," in *Bild und Erkenntnis: Formen und Funktionen des Bildes in Wissenschaft und Technik*, ed. Andreas Beyer and Markus Lohoff (Munich: Deutscher Kunstverlag, 2005), 12.

18 Horst Bredekamp, *Theorie des Bildakts: Frankfurter Adorno-Vorlesungen 2007* (Berlin: Suhrkamp, 2010), 34–35.

kinds of material research objects are images in this sense as well, as is particularly apparent when the terms used to describe them note the activity: rendering visible, visualization, imaging.

From Theories of the Picture to Practices

Art always establishes insight on the level of form and/or content, but it does not necessarily convey information and knowledge; it moves through series of actions in the form of usually unalterable unique items. Natural-scientific imagery, by contrast, is embedded in far more complex ensembles of events and actions: it allows for the visualization of a scientific achievement that may be reversible or merely preliminary or may be utilized as a fact. In that sense, natural-scientific imagery is instrumental and technical. Its potential intrinsic eidetic value did not become a matter of interest until the 1990s, which then also raised the issue of the extraordinarily wide range of such imagery in the general sense; any discussion henceforth necessarily examined the term as encompassing a plurality of phenomena. Images, photographs, drawings, charts, statistics, diagrams, preserved specimens, models, computer graphics, animations, in other words, came to be seen no longer merely as material elements of a (communicative) practice but also as "epistemic agents."[19] Pictures not only cease to be illustrations; like objects of knowledge, they are also treated as active and operative entities that organize and regulate a knowledge process.[20] Visibility is preceded by the action of rendering a defined something visible,[21] which raises further questions: How is knowledge manifest in the picture? Of what kind is the knowledge being shown? And how can it be seen as such? These questions lead to different answers concerning the knowledge surrounding the picture, the particular quality of the representation of the research—and hence eidetic—content, and the adequate and (ostensibly) formal visual realizations of such content.

1. Eidetic competence

The picture is defined as structurally contingent with regard to its openness to interpretation and its form, and that concerns not only the research object—which is not the picture but what the picture presents information about or serves as empirical evidence of. The contingency of the eidetic also concerns the fact that normativities and conventions are not objective givens but informed, in their genesis, by mental and "inward" images; in other words, they are influenced by cultural pictorial traditions from within as well as without the discipline.

Jochen Hennig describes this conjunction of images in the example of a tunneling microscope image from a DNA study. Its creator, Erich Stoll, uses coloration to transform the image of a recA-DNA complex into the cartography of a setting in the South Seas.[22]

2. Conceptual images

Pictures that illustrate an object of knowledge must possess an acuity of replicable judgment and evidentiary definition that approximates the quality of the concept; examples include atomic models, the DNA double helix, and medical atlases. This approach to the natural-scientific picture presupposes that scientific language is at bottom predicative and propositional and that concepts are unequivocal. That is the yardstick by which

19 Monika Wulz introduced the concept of the "epistemic agent" (*Erkenntnisagent*); see Monika Wulz, *Erkenntnisagenten: Gaston Bachelard und die Reorganisation des Wissens* (Berlin: Kadmos, 2010).

20 See Donna Haraway, "Situated Knowledges: The Science Question in Feminism and the Privilege of Partial Perspective," *Feminist Studies* 14, no. 3 (Fall 1988): 575–99; Bruno Latour, "A Collective of Humans and Nonhumans: Following Daedalus's Labyrinth," in *Pandora's Hope: Essays on the Reality of Science Studies* (Cambridge, MA: Harvard University Press, 1999), 177. See also Bruno Latour, "On Recalling ANT," in *Actor Network Theory and After*, ed. John Law and John Hassard (Oxford: Blackwell, 1999), 15–25.

21 See Karin Knorr-Cetina, "'Viskurse' der Physik: Konsensbildung und visuelle Darstellung," in *Mit dem Auge denken: Strategien der Sichtbarmachung in wissenschaftlichen und virtuellen Welten*, ed. Bettina Heintz and Jörg Huber (Zurich: Edition Voldemeer, 2001), 308–9; see also Soraya de Chadarevian, "Sehen und Aufzeichnen in der Botanik des 19. Jahrhunderts," in *Der Entzug der Bilder: Visuelle Realitäten*, ed. Michael Wetzel and Herta Wolf (Munich: Fink, 1994), 121–44.

22 Jochen Hennig, *Bildpraxis: Visuelle Strategien in der frühen Nanotechnologie* (Bielefeld: Transcript, 2011), 107–8, 177–78.

the evidentiary capacity of the picture is measured; the question is then not whether the picture is capable of illustrating a set of facts in a manner that holds up to scrutiny but solely whether the visual means are the right ones to convey that set of facts in understandable fashion.[23] Pictograms, geometric models, or circuit layouts are offered as evidence that there are forms of imagery that require no textual explanation.[24]

3. The technique of illustration

A different way of contrasting the contingent and the given pits the form of representation against the content being represented. This view refers to the indeterminacy of the technique of "illustration." The implicit ("technical," in the sense outlined above) potentials of the picture are subordinated to a set of facts in order to visualize a principle more clearly. The technical drawing is one such pictorial genre that is "manipulated" in order to be "truthful";[25] so are the pictures widely used in medicine, biology, and even astronomy that are generated from large volumes of data, where the visualization technique presents opportunities to intervene in order to reduce complexity, clarify intentions, and articulate assessments. For an example, we may look to the vicissitudinous history of the representation of the clitoris. The intermittent disappearance and reemergence, the waxing and waning, of the clitoris in the specialized literature on anatomy since the mid-nineteenth century ties imaging techniques to constructions of gender difference. Manipulation or truth, this example suggests, is not a matter of imaging procedures and their technical potential but of the political and discursive practices behind the transmission of pictures.[26]

These different perspectives on what pictures accomplish demonstrate that the analysis of natural-scientific pictures must always also consider their genesis in a discursive and communicative process as well as the conditions of their production: the picture presents a substantially defined object to the eye whose evidentiary value, objectivity, and aptitude for truth are not only liable to historical change but also subject to the process of the scientific verbalization of communicable knowledge. Techniques, pictures, and observers alike are inscribed in this process. The term *technique*—or, more precisely, *visualization technique*—then refers not solely to technology or apparatuses but also, for example, to the operational potential of a patient's manually drawn temperature curve.[27] Processes of normalization and standardization that allow beholders to assess the evidentiary value of a picture are likewise techniques of inscription.

A theory of the picture that takes not only the technique of representation but also the picture's technical quality as discussed above into account qualifies the status of the physical perceptual experience and the body as an instrument of observation and measurement.[28] Far from negating the cognitive subject, this view locates it in the hand that works with and produces things (and pictures) and in the eye that interprets them. As natural-scientific pictures are involved in activities and historical processes, it becomes clear how they refer to other pictures in the sense of an "endless production of traces, where the place of the referent is always already occupied by another trace."[29] What becomes clear is that the reference of these pictures need not have a basis outside the realm of

23 Arguments that pursue this direction engage with the claim that "pictures cannot assert" made in Ernst H. Gombrich, *The Image and the Eye: Further Studies in the Psychology of Pictorial Representation* (Oxford: Phaidon, 1982). See, e.g., Helmut Spinner, "Ein Wort sagt mehr als tausend Bilder? Entwurf einer Wissenstheorie des Bildes," in *Bild—Wissen—Medien: Visuelle Kompetenzen im Medienzeitalter*, ed. Hans Dieter Huber, Bettina Lockemann, and Michael Scheibel (Munich: Kopaed, 2002), 195.

24 See Martin Scholz, *Technologische Bilder: Aspekte visueller Argumentation* (Weimar: VDG, 2000), 138–39. Yet as the international sign indicating arrival gates at airports—to take a simple example—illustrates, the understanding of pictographic information presupposes a major learning process. To prevent the beholder from interpreting them as designating the place where airplanes crash, the downward-facing airplane in the symbol is shown with its landing gear extended.

25 See Anja Zimmermann, *Ästhetik der Objektivität: Genese und Funktion eines wissenschaftlichen und künstlerischen Stils im 19. Jahrhundert* (Bielefeld: Trancript, 2009).

26 See Claudia Reiche, "Zu ähnlich: Neue Klitorisbilder aus Kunst und Wissenschaft," in *GenderMedia Studies: Zum Denken einer neuen Disziplin*, ed. Hedwig Wagner (Weimar: VDG, 2008), 155–67; Nancy Tuana, "Coming to Understand: Orgasm and the Epistemology of Ignorance," in *Agnotology: The Making and Unmaking of Ignorance*, ed. Robert Proctor and Londa Schiebinger (Palo Alto, CA: Stanford University Press, 2008), 108–45.

27 For numerous examples from the fields of oceanography, tourism, urban planning, ecology, statistics, economics, and medicine, see the contributions in David Gugerli and Barbara Orland, eds., *Ganz normale Bilder: Historische Beiträge zur visuellen Herstellung von Selbstverständlichkeit* (Zurich: Chronos, 2002).

28 See Lorraine Daston and Peter Galison, "The Image of Objectivity," in "Seeing Science," special issue, *Representations*, no. 40 (Autumn 1992): 81–128.

29 Hans-Jörg Rheinberger, *Toward a History of Epistemic Things: Synthesizing Proteins in the Test Tube* (Palo Alto, CA: Stanford University Press, 1997), 104.

pictures; it instead results from the specific employment and function of pictures. A theory of the picture founded on its technical quality is apt to show that the different kinds of pictures that everyday scientific practice produces and actuates as "epistemic agents" cannot be comprehended by a single categorically defined conception of the picture; they must be described in detail by analyzing the different ways of interacting with pictures. It is precisely because natural-scientific pictures are about the generation of (temporary) eidetic evidence that theories that interrogate the quality of being evident to the eye with a view to the local conditions that sustain it are valuable.

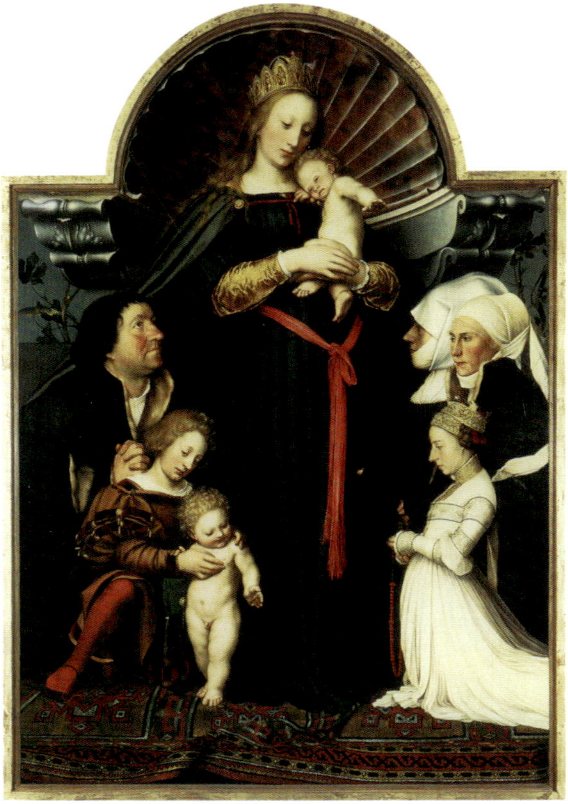

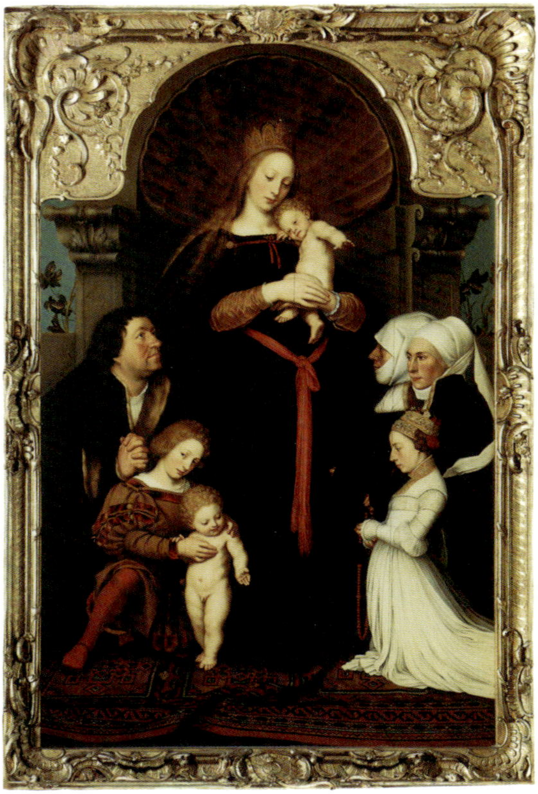

FIGS. 1a, 1b: *Darmstadt Madonna* (*Madonna* of Jakob Meyer zum Hasen) by Hans Holbein, 1526, 146.5 × 102 cm, Städelsches Kunstinstitut Frankfurt am Main, and a copy of Holbein's *Madonna* by Bartholomäus Sarburgh, 1635/1637, 158.9 × 103 cm, Gemäldegalerie Alter Meister, Dresden. For a long time, the painting on display in the Dresden Gemäldegalerie (right) was regarded as a product of the hand of Hans Holbein. In the nineteenth century, a painting with the same motifs and composition was discovered (left), and a dispute about the authenticity of the Dresden version broke out. Public comparison and discussion of the two paintings during an exhibition in 1871 decided the dispute in favor of the second picture. Direct comparison made a strictly formal and precise comparison of styles possible. An empirical procedure was established that based its findings on the examination of details, the individual characteristics of the artist's hand, painterly methods, techniques, and color materials. This procedure has since become an integral element of art-historical methodology. Oskar Bätschmann and Pascal Griener, *Hans Holbein the Younger: Die Darmstädter Madonna* (Frankfurt am Main: Fischer-Taschenbuch-Verlag, 1998). Fig. 1a: © Hans Holbein d. J. Madonna des Bürgermeisters Jacob Meyer zum Hasen, 1525/26 und 1528 (Öl auf Nadelholz, 146.5 × 102 cm), Würth Collection, Inv. 14910, Photographer: Philipp Schönborn, München. Fig. 1b: © Bartholomäus Sarburgh (Kopie nach Hans Holbein d. J.): Die Madonna des Basler Bürgermeisters Jakob Meyer zum Hasen, Gal. Nr. 1892. Gemäldegalerie Alte Meister, Staatliche Kunstsammlungen Dresden, Photographer: Hans-Peter Klut.

COMPARING IMAGES

Generally, comparison can be described as a fundamental method that orients people in their environment so that they can distinguish (differentiate) and categorize (classify) what they see and experience. More specifically, it is an intellectual technique used in various scientific disciplines and performed and evaluated in many different ways (Lutz et al. 2006; Elkins 2007). While texts, verbal imagery, patterns of behavior, styles of thinking, and cultures are compared in cultural and literary studies, art history deals explicitly with the comparison of images. In art history, comparative visual analysis is one of the central methodological paradigms of analysis and argumentation. As such, it serves to order works of art historically into schools and eras, to identify unknown masters, and to identify authentic and inauthentic, genuine and counterfeit works (Wölfflin 1932; Friedländer 1930; Gombrich 1960; figs. 1a and 1b). At the same time, comparative visual analysis plays a fundamental role as a basis for argumentation in presenting such findings in publications and slide lectures.

One reason for the difficulties in theorizing comparative visual analysis seems to lie in the double role

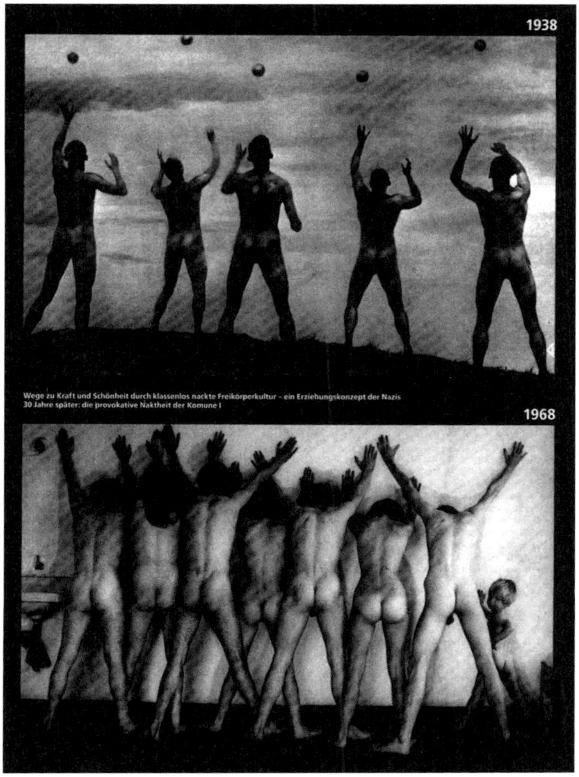

FIG. 2: A page from the March 2008 issue of *Cicero: Magazin für politische Kultur*. These two photographs were compared in the context of an article by the German historian Götz Aly. The essay argues for the existence of direct parallels between the rise of the Nazis in the 1930s and the student revolts of 1968 in Germany and Berlin. According to Aly, both were "young" movements with a "totalitarian language" and a "tendency towards violent activism" that pursued a "takeover of power." The two images are supposed to convey this connection; at first glance, the similarities in their composition and subjects are striking. Yet the question is whether, beyond the formal similarities of the images, the differences do not in fact predominate. The people in the upper image are men involved in a ballgame in the open air, with raised heads and outstretched arms. In the lower photo, by contrast, men, women, and a child pose in a confined space, their heads lowered, arms and legs rigidly spread along a wall; the arrangement would seem to quote the scene of an arrest. Criticizing the *Cicero* article, the daily newspaper *Die Tageszeitung* (*Taz*) was able to show in an article entitled "Nackerte Tatsachen" ("Bare Facts") that the presentation of the pictures obscured or falsified their contexts and the dates on which they were made. The second photograph, of Kommune I, was taken by Thomas Hesterberg after the shah's visit to Germany in 1967; the other is a still from Wilhelm Prager's 1924 film *Wege zu Kraft und Schönheit,* which seeks to portray the rebirth of the body from the spirit of antiquity. The comparison seems evocative and persuasive enough at first glance but on closer inspection turns out to be not only formally but also historically untenable. Götz Aly, "Unser Kampf: 1968," *Cicero,* March 2008, 105 and © akg-images and © Hesterberg, Thomas / Süddeutsche Zeitung Photo.

that comparison plays in this discipline as an instrument of analysis *and* argumentation. Comparative visual analysis was part of the implicit basic know-how or "tacit knowledge" (Michael Polanyi) of art-historical writing long before art history was established as an academic discipline in the nineteenth century (Friedländer 1942). Even today, it forms a self-evident and largely unquestioned part of the discipline as a means of acquiring knowledge, of analyzing and describing images. This may explain why comparative practices have only rarely been investigated as a problem in their own right.

Based on the German art historian Heinrich Dilly's seminal research, art history scholars have more recently carried out studies that have investigated the meaning and scope of comparison as a method in the development of art history as an academic discipline; in particular, they traced the history of the material foundations of comparison, which were laid by the emergence of new reproductive media such as photography and their connections with science (Dilly 1975; Wenk 1999; Nelson 2000; Ratzeburg 2002; Reichle 2002; Bader 2007). At the same time, art history has repeatedly reproached itself for not adequately substantiating the comparisons it makes: with the help of apparently "evident" visual comparisons, relationships between images are more implicitly asserted than verbally or contextually deduced. The objection that there are images that simply cannot be compared implies the accusation that relations are claimed where divergence—indeed, irreconcilable difference, regarded as essential—is the strikingly predominant observation (Geimer 2006). It is a rhetorical quality of comparative presentation that showing two images in juxtaposition tends to suggest relationships more than

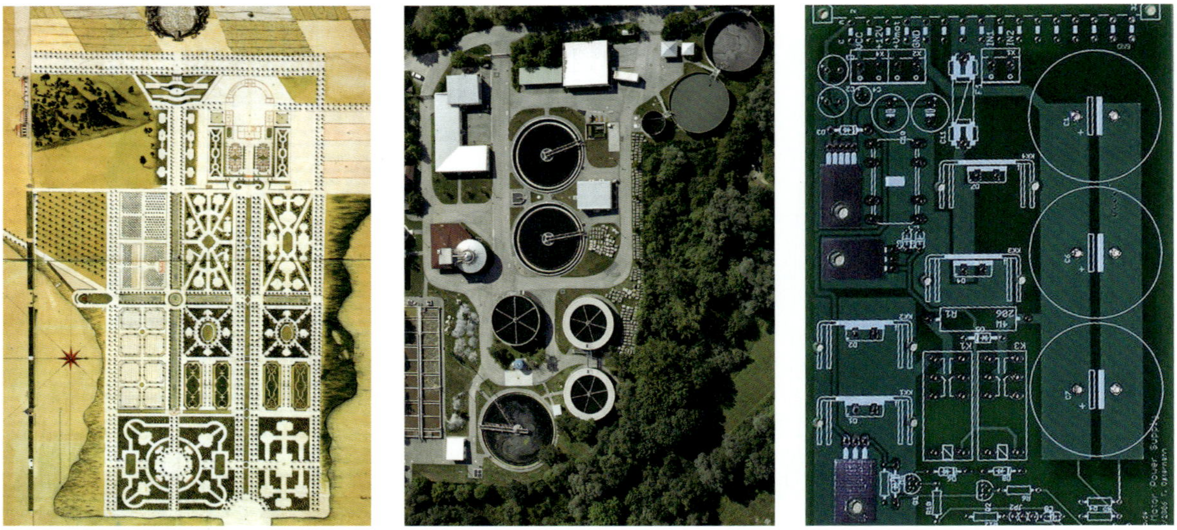

FIGS. 3a–3c: Baroque garden (a), sewage treatment plant (b), and circuit board (c). The *tertium comparationis* for the comparison of these three images lies not in a shared time horizon, nor in the subject or context. Comparability is derived here solely from the formal level of order and structure. In all three images, round, square, and diamond-shaped elements are linked by thin lines as in a labyrinth; lines are laid straight, diagonally or slightly bent across the surface. In all images, the structures seem to follow well-ordered and ornamental laws and logic. This comparability is supported and largely made possible by the bird's-eye perspective common to the three images and their unified size. Only seeing these objects as images allows a microchip to resemble a baroque garden. But can such similarities increase knowledge, and is there a shared basis for these structures? To answer this question, we would have to study the issue of the overarching requirements on the planning of a coherent networked system. In all three cases, planners were faced with the task of arranging several functions rationally and symmetrically on a surface, optimally positioning paths and guidance systems. The formal resemblance is a symbol of functional networking. But while the ornamental structures of the gardens of Versailles reflect the ideal of a tamed nature, the sewage treatment plant and the microchip produce symmetries in the quest for optimum planning in a technical sense. Fig. 3a: Adrian von Buttlar and Nargita Marion Meyer, eds., *Historische Gärten in Schleswig-Holstein* (Heide: Boyens, 1996), 118. © Cecilia Heisser, Nationalmuseum, Stockholm. Fig. 3b: © FOTAG Luftbild München. Fig. 3c: http://www.mechapro.de/shop/images/product_images/popup_images/132_0.jpg (accessed November 2013).

distinctions, just as positioning two or more images on a surface or in a space seems to suggest a certain direction in which to "read" them. In consequence, the danger can arise of generating "accidental similarities" (Geimer 2010; fig. 2).

It should be noted, however, that it is initially a fundamentally open question what a juxtaposition of images—particularly during research—is supposed to express. The visual confrontation of two objects is inherently capable of both revealing relationships and emphasizing difference and specificity more sharply; it can distinguish more clearly or bring closer together, as the art historian Otto Pächt explained in *The Practice of Art History* (Pächt 1999, 87–104). To this must be added the potential of montage to create something new beyond the individual images from which it is composed, as the French art historian Georges Didi-Huberman has pointed out (Didi-Huberman 2010).

Pächt emphasized that what is compared is oriented toward what the comparison is supposed to achieve. The collating of images is carried out based on decisions and in accordance with epistemic interests that may vary widely. For example, in an attempt to initially view and classify visual material, comparison can provide a first orientation and then sharpen the eye, leading to an understanding of the particularity of a work or a group of works. We might say that every comparison serves to elucidate an image by drawing on other images. Following Pächt, the distance between the objects compared in a first step must be as short as possible (Pächt 1999, 87). But it does not follow that comparability simply results from formal congruities. In fact, comparability is constituted by at least one constant, which, as the *tertium comparationis*, forms its basic precondition, and formal similarity is just one such criterion among many. Comparability can also be based on a *tertium comparationis* that is not present at the visible level of the image but is, for example, established by a shared context. Further bases for comparison in this sense might be a shared time or place of origin, a shared theme or subject, or the same design purpose. In such cases, two images may be juxtaposed that show two completely different outcomes even though they were, for example, created to achieve the same end.

With regard to technical images, the spectrum of comparison criteria is expanded by the scientific contexts from which the images come, the functions they serve in these contexts, and the imaging technology used to generate them. A particular challenge is posed by comparisons of artistic and non-artistic images and the investigation of migrating imagery that diffuses through various disciplines. Here, too, comparison must always be made based on an awareness of the epistemic interests and deliberately selected criteria. For instance, when similar images are brought together with a particular interest in the transmission and adoption of forms, such comparison should be undertaken, on the one hand, with an awareness of the possible difference between individual associations, the viewer's own visual memory, and that of the creator of the image. On the other hand, formal proximity must be continually related to other criteria such as function, context, and the conditions under which the images were produced. It is then just as possible for a comparison to fail to produce useful knowledge as it is for it to visibly provide associative, productive cause for thought and initiate new investigations (figs. 3a–3c). Whether a comparison is useful can only be determined in the course of further research and a detailed investigation of the objects brought together as well as in the discussion following a public presentation of comparisons. In this process, the relative significance of equivalences and differences should be subject to ongoing critical revision. The legitimacy of a presented comparison can only be determined by the verbal argumentation that accompanies it. This also means that the respective interests in it must be disclosed and made transparent. —VD

LITERATURE

Bader, Lena. "Imaging Imagery: On the Visuality of Iconic Criticism and the Early Media of Visual Studies in Their Current Meaning." In *Image-Problem? Medienkunst und Performance im Kontext der Bilddiskussion*, ed. Slavko Kacunko and Dawn Leach, 67–86. Berlin: Logos-Verlag, 2007.

Baxandall, Michael. *Patterns of Intention: On the Historical Explanation of Pictures*. New Haven, CT: Yale University Press, 1985.

Didi-Huberman, Georges. "Was zwischen zwei Bildern passiert: Anachronie, Montage, Allegorie, Pathos." In *Vergleichendes Sehen*, ed. Lena Bader, Martin Gaier, and Falk Wolf, 537–72. Munich: Fink, 2010.

Dilly, Heinrich. "Lichtbildprojektion: Prothese der Kunstbetrachtung." In *Kunstwissenschaft und Kunstvermittlung*, ed. Irene Below, 153–72. Giessen: Anabas-Verlag, 1975.

Elkins, James. *Visual Practices across the University*. München: Wilhelm Fink Verlag, 2007.

Friedländer, Max J. *Genuine and Counterfeit: Experiences of a Connoisseur*. New York: A. and C. Boni, 1930.

Friedländer, Max J. *On Art and Connoisseurship*. Boston: Beacon Press, 1942.

Geimer, Peter. "Unwillkürliche Kunstgeschichte: Peter Geimer im Gespräch mit Texte zur Kunst." *Texte zur Kunst*, no. 62 (June 2006): 56–71.

Geimer, Peter. "Vergleichendes Sehen oder Gleichheit aus Versehen? Analogie und Differenz in kunsthistorischen Bildvergleichen." In *Vergleichendes Sehen*, ed. Lena Bader, Martin Gaier, and Falk Wolf, 45–69. Munich: Fink, 2010.

Ginzburg, Carlo. "Morelli, Freud and Sherlock Holmes: Clues and Scientific Method." *History Workshop* (1980): 5–36.

Gombrich, E. H. *Art and Illusion: A Study in the Psychology of Pictorial Representation*. New York: Pantheon Books, 1960.

Klonk, Charlotte, and Michael Hatt. *Art History: A Critical Introduction to Its Methods*. Manchester: Manchester University Press, 2006.

Lutz, Helga, Jan-Friedrich Missfelder, and Tilo Renz, eds. *Äpfel und Birnen: Illegitimes Vergleichen in den Kulturwissenschaften*. Bielefeld: Transcript, 2006.

Nelson, Robert S. "The Slide Lecture, or the Work of Art 'History' in the Age of Mechanical Reproduction." *Critical Inquiry* 26, no. 3 (2000): 414–34.

Pächt, Otto. *The Practice of Art History: Reflections on Method*. New York: Harvey Miller, 1999.

Ratzeburg, Wiebke. "Mediendiskussion im 19. Jahrhundert: Wie die Kunstgeschichte ihre wissenschaftliche Grundlage in der Fotografie fand." *Kritische Berichte*, no. 1 (2002): 22–39.

Reichle, Ingeborg. "Medienbrüche." *Kritische Berichte*, no. 1 (2002): 40–56.

Wenk, Silke. "Zeigen und Schweigen: Der Kunsthistorische Diskurs und die Diaprojektion." In *Konfigurationen: Zwischen Kunst und Medien*, ed. Sigrid Schade and Georg Christoph Tholen, 292–305. Munich: Fink, 1999.

Wölfflin, Heinrich. *Principles of Art History: The Problem of the Development of Style in Later Art* [1915], trans. M. D. Hottinger. New York: Dover Publications, 1932.

A History of Styles of Technical Imagery: Between Description and Interpretation

A Conversation with Horst Bredekamp

DAS TECHNISCHE BILD:[1] In the history of art, the concept of style has been a central category since Giorgio Vasari, helping scholars to organize artifacts. The studies of scientific imagery collected in the present volume likewise rely on the category of style as an important point of reference. Yet the application to scientific images of a concept developed in the discussion of art also generates tensions. The German historian of science Michael Hagner, for example, has cast doubt on the idea of a history of styles of imagery in the natural sciences, arguing that it paints the divergent problems and phenomena of the history of science with a single broad brush. A perspective focused solely on the form of scientific images, he claims, fails to do justice to their complexity. For example, he argues, it's impossible to comprehend the "epistemic agency" of images without broadening the analysis to include experimental procedures, technological presuppositions and conditions imposed by equipment, and local traditions of knowledge and mentalities.[2] The central question is still: can there be a history of styles of technical imagery? But first things first: what do you understand by "style"?

HORST BREDEKAMP: The term *style*, I believe, designates recognizably shared traits of created forms that transcend the individual producer. Two elements need to be involved: at least two designers and no fewer than two works that, though made independently of one another, nonetheless evince sufficient similarity that shared features of their created forms become evident. That is the smallest common denominator of an art-historical definition of style.

DTB: Hasn't this definition, or at least its employment in scholarship, become obsolete by now? If you raise the question of style, you're interested in general and overarching features, in qualities derived from the concrete forms. That was the object of the grand endeavors to write a history of styles in the nineteenth century, like those mounted by Gottfried Semper (1803–1879)[3] and Alois Riegl (1858–1905);[4] the histories of linear development they wrote also contain distinctly evaluative assessments of specific eras and artifacts. One might think that such undertakings move the focus away from the specificity and complexity of individual forms and contexts. By contrast, it sounds much less pretentious to speak of a history of forms. Isn't that enough? What's the benefit in holding on to the concept of style? What do we gain from it?

1 The "Das Technische Bild" team is represented by Jana August, Franziska Brons, Violeta Sánchez, Vera Dünkel, Jochen Hennig, and Birgit Schneider, all of whom took part in this conversation.

2 "Bildunterschätzung—Bildüberschätzung: Ein Gespräch der Bildwelten des Wissens mit Michael Hagner," in *Bilder in Prozessen, Bildwelten des Wissens: Kunsthistorisches Jahrbuch für Bildkritik*, vol. 1, no. 1 (Berlin: Akademie, 2003), 110.

3 Gottfried Semper, *Style in the Technical and Tectonic Arts, Or, Practical Aesthetics*, trans. Harry Francis Mallgrave and Michael Robinson, vol. 1: *Textile Art: Considered in Itself and in Relation to Architecture* [1860] (Los Angeles: Getty Research Institute, 2004), 102–465.

4 Alois Riegl, *Problems of Style: Foundations for a History of Ornament* [1893] (Princeton, NJ: Princeton University Press, 1992).

HB: Defenders of the concept of style have long faced up to these questions. The idea of linear stylistic development was abandoned early on in favor of a complex vision in which styles are interwoven, but also at odds, with each other; it was the art historian Wilhelm Pinder who, in 1926, before he became a rather problematic person, coined the ingenious formula of the "non-simultaneity of the simultaneous."[5] Since then, scholars have worked to develop progressively more nuanced understandings of style, lending new legitimacy to the employment of the concept[6] and once again establishing the specific form as a central criterion of analysis.

DTB: The Polish immunologist and philosopher of science Ludwik Fleck (1896–1961) developed a theory of what he called "thought styles."[7] Any particular thought collective is governed by historically evolved conventions. They define which research questions are asked and which research methods are used to address them, but also what is regarded as true. Fleck diagnosed a collective's thought style primarily in historical texts, but also in images from the history of medicine, which makes his endeavor closely akin to ours. What's the difference between the art-historical concept of style and the concept of style in the history of science?[8]

HB: The history of science that evolved after Fleck has tended to use the term *style* to describe one element of mentality more broadly conceived. Thomas Kuhn replaced the thought style with the concept of the paradigm; imagery slipped out of focus.[9] Fleck's thought style likewise aims at something general that transcends the individual, designating a frame that illuminates and at once delimits the possibilities of thinking. It's about shared questions and perspectives from which the scholar may abstract commonalities for a history of mentalities. Style in the art-historical sense, by contrast, describes a form, a material quality. So the art-historical concept shifts the problem of style from a matter of psychological-mental disposition to form as it has become. That's why form, to us, stands at the beginning of any analysis. And here is the core of the question we raise: is there, in the production of the natural sciences, a style that manifests itself, irrespective of the individual researchers, in the visual objects they design—the prepared objects, the instruments, the images? If such a recognizable style exists, we have offered strict proof that images never—not even and especially not in the natural sciences—illustrate; that, on the contrary, they fuse the visualization of an object with the history of their own application.

DTB: Now, in recent years, historians of science have also begun to pay closer attention to form.[10] Issues of form are central, for example, to the analysis of image noise and contamination effects and the role they play for image-based work in the sciences. In the history of image-generating technology, for example in microphotography, the challenge was to distinguish such artifacts from the research object the photograph was to capture (fig. 1).[11] Are these forms generated by the technical implement already part of a style?

5 Wilhelm Pinder, *Das Problem der Generation in der Kunstgeschichte Europas* (Berlin: Frankfurter Verlags-Anstalt, 1926), 1–12. See Horst Bredekamp, "Wilhelm Pinders 'Ungleichzeitigkeit des Gleichzeitigen,'" in *Rekursionen: Von Faltungen des Wissens*, ed. Ana Ofak and Philipp von Hilgers (Munich: Fink, 2010), 117–24.

6 Robert Suckale, "Die Unbrauchbarkeit der gängigen Stilbegriffe und Entwicklungsvorstellungen," in *Stil und Funktion: Ausgewählte Schriften zur Kunst des Mittelalters*, ed. Peter Schmidt and Gregor Wedekind (Berlin: Deutscher Kunstverlag, 2003), 287–302.

7 Ludwik Fleck, *Genesis and Development of a Scientific Fact* [1935], trans. Fred Bradley, ed. Thaddeus J. Trenn and Robert K. Merton (Chicago: University of Chicago Press, 1979).

8 On the concept of style in the history of science, see also Lorraine Daston and Michael Otte, eds., "Style in Science," special issue, *Science in Context* 4, no. 2 (1991); Ian Hacking, "'Style' for Historians and Philosophers," in *Historical Ontology* (Cambridge, MA: Harvard University Press, 2002), 178–99.

9 Thomas S. Kuhn, *The Structure of Scientific Revolutions* (Chicago: Chicago University Press, 1962).

10 See the bibliography at the end of the book.

11 Peter Geimer, *Bilder aus Versehen: Eine Geschichte fotografischer Erscheinungen* (Hamburg: Philo Fine Arts, 2010).

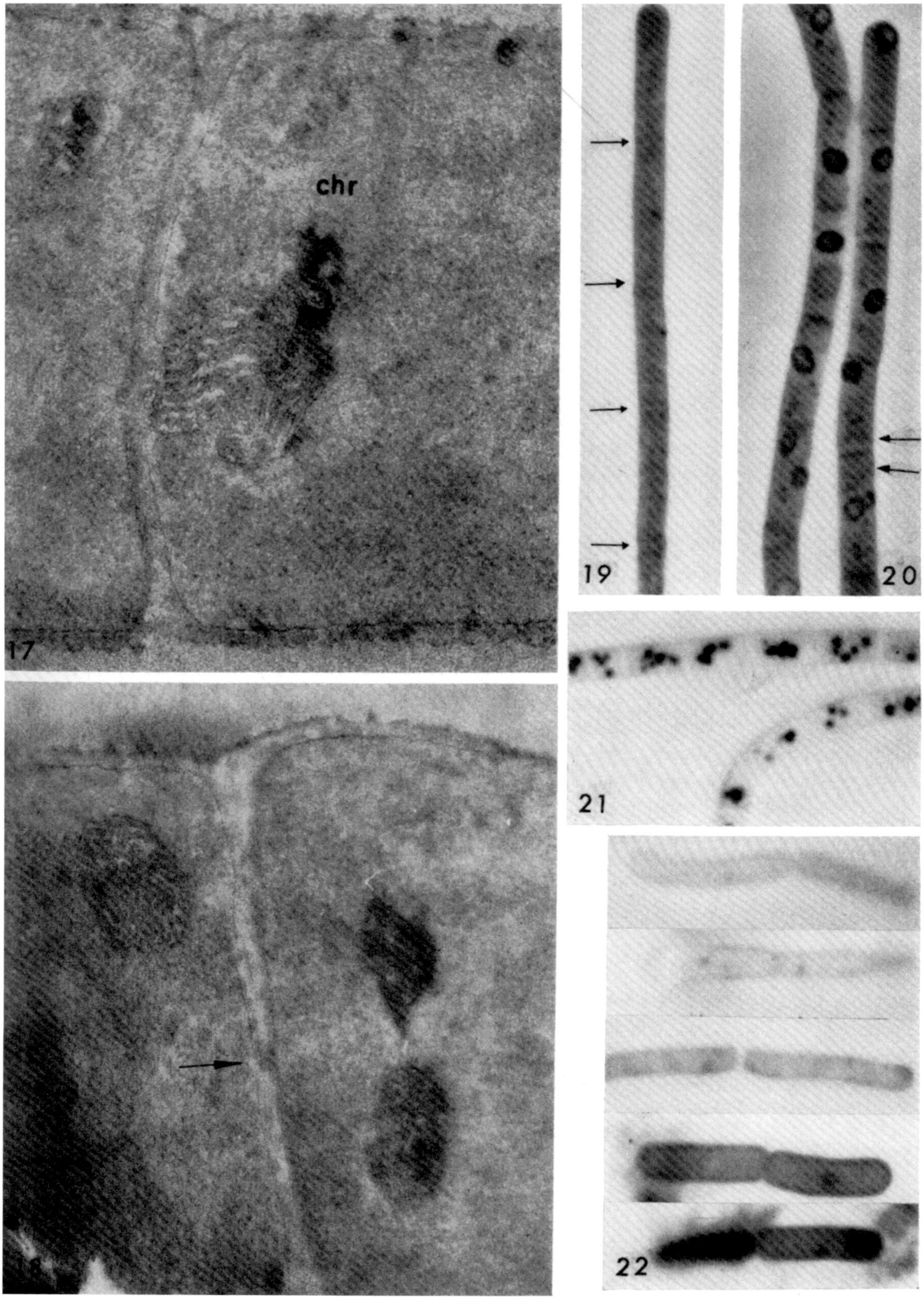

FIG. 1: Microphotographs of thin sections of bacterial cells, 1960. For a long time, the clearly visible dark artifacts in the samples were falsely identified as a special type of cell organelles, the so-called mesosomes. Philip C. Fitz-James, "Participation of the Cytoplasmic Membrane in the Growth and Spore Formation of Bacilli," *Journal of Biophysical and Biochemical Cytology* 8, no. 2 (October 1960): 521, figs. 17–22. ©1960 Rockefeller University Press. Originally published in *Journal of Biophysical and Biochemical Cytology* 8, no. 2 (507–28).

HB: That would displace the element of shared design into the device, which is structurally non-individual and produces non-intended traces. I wouldn't describe such noise that is due to the equipment as directly part of what we mean by style as a potential shared form. But it may stimulate the emergence of a new visual form and thus influence a novel style.

DTB: To explain similar features appearing at the same time and stylistic shifts, scholars have often relied on concepts such as the "common cultural sphere," the "milieu," a shared "spirit of the age," a "mentality," or a "paradigm." These concepts suggest that style is a bit like fate, a framework to which artists and scientists respond in primarily passive ways. Style becomes a symptom of something. Alois Riegl and Wilhelm Worringer, for their part, coined the term "Kunstwollen," or "will to art," in the attempt to conceive style as a product of active intentions, which confers much greater agency on style.[12] Would you say that style is unidirectional, a "symptom" and expression of mentalities or ways of thinking? Is it not conversely also true that the objects of design actively inform these mentalities and ways of thinking?

12 Riegl, *Problems of Style*; Wilhelm Worringer, *Abstraction and Empathy*, trans. Michael Bullock (London: Routledge and Kegan Paul, 1953).

HB: Riegl's and Worringer's concept of "will to art" is significant for our context, as are Wölfflin's "principles" or, say, Warburg's "pathos formula."[13] All these attempts around and after 1900 to define visual design in terms that transcend the individual have lost none of their crucial relevance. I implicitly rely on them when I see a style not as the expression of something behind, above, or next to it, of something that frames it, but instead consider style itself as co-generative of the mentality whose emanation it later appears to be. We cannot emphasize strongly enough that style is not at all the illustration of a mentality or a shared thought system; it's part of a spindle that shuttles back and forth between the creation of form and conscious awareness to create such collectives in the first place. That is what Riegl called "will to art." Without such active determination of style, any "illustration" in the natural sciences would be no more than the inert image of a movement of thought that took place elsewhere.

13 See Heinrich Wölfflin, *Principles of Art History: The Problem of the Development of Style in Later Art* [1915], trans. M. D. Hottinger (New York: Dover Publications, 1932). "Pathos formula" was the label Aby Warburg used to describe the recurrent universal and therefore formulaic depictions of gestures and facial expressions he traced; see Aby Warburg, "Dürer and Italian Antiquity" [1905], in *The Renewal of Pagan Antiquity: Contributions to the Cultural History of the European Renaissance*, trans. David Britt (Los Angeles: Getty Research Institute for the History of Art and the Humanities, 1999), 555.

DTB: The counterargument would be: but doesn't every thought originate in the mind?

HB: Of course it does, but what is the mind without the body? Disembodiment is an idea a misconceived Platonism tries to spread in forever new guises. Against it, we must point out that the design informing the physical material directly contributes to shaping the thought, so that the fingers must be addressed as organs of thinking no less than the brain. That's also why style is not a "symptom" but a moving agent; it expresses a form of thinking it generates and fashions. Therein lies its paradox, which is as irritating as it is fertile. In order to be able to understand the active and formative role the created form plays, I insist on concrete precision in the description of an image and the determination of its material form. These are not external procedures but the descriptive reconstruction of how an idea was conceived.

DTB: The case studies collected in the present volume always start out from the description of an artifact. It's a methodological decision to base the critical analysis of the specificity of an image or field of images on a concrete visual object and its description. Is description, rather than interpretation, what allows us to apprehend a style?

HB: Discerning a style fundamentally requires morphological comparison. We must first compile large numbers of objects on the table, the screen, or another platform so that we may compare them. Only then can we recognize a shared design quality we may call a style. After comparing and grouping the objects, we may ask what this style sustains and which interpretation it is amenable to. Is it an expression of mentality, does it bear similarity to it? Or is the mentality embedded in it? I believe that the latter is the case. Style, and that is the unbridgeable gap that distinguishes it from "mentality," works by indirection through the outside world; it operates in design as a form of thinking, and therein it's of a higher order than mentality. I have tried to show this with the concept of the "picture act."[14]

14 Horst Bredekamp, *Theorie des Bildakts: Frankfurter Adorno-Vorlesungen 2007* (Berlin: Suhrkamp, 2010).

DTB: Why do you place such great emphasis on the aspect of transcending the individual? Might we not also compare several images fashioned by one individual and derive a personal style?

HB: Yes, that would be "style" in the sense the English word usually carries, the manner in which someone fashions things, presents oneself, engages with other people, conducts one's life. Yet such "style" can become a style in our sense only when other people make a comparable behavior a specific phenomenon that transcends any individual, not as an acquired disposition (habitus), but rather by active design.

DTB: Isn't it also the case that this super-individual quality is much more easily discernible in certain kinds of visual artifacts? Semper and Riegl deliberately relied on non-artistic objects—products of artisanship and industry—to develop their histories of styles.[15] Such artifacts would seem to have in common with scientific imagery that the individual (artist) wields considerably less formative influence because intended applications and functions are the defining factors in their production. In holding on to the concept of style in the study of technical imagery, to what extent do we carry on the tradition of an "art history without names"?

HB: That's indeed the foundation laid for the history of science by the three representatives of "art history without names": there's Alois Riegl's "will to art," a concept we would ultimately have to complement with that of a "will to imagery";[16] of course, there's Heinrich Wölfflin's art history as a history of seeing;[17] and last but not least, there's Aby Warburg's concept of the pathos formula.[18]

DTB: Where would you set the standard? At which point can we start to speak of a style—is a set of five images enough, or does it take many more? To put it another way, where does individuality end and style begin?

15 Riegl, *Problems of Style*; Alois Riegl, *Late Roman Art Industry* [1901–1923], trans. Rolf Winkes (Rome: Bretschneider, 1985); Semper, "Textile Art."

16 Jörg Trempler, "Der erste Blick in das Innere eines Atoms oder das Bildwollen," in *Beständig im Wandel: Innovationen, Verwandlungen, Konkretisierungen. Festschrift für Karl Möseneder zum 60. Geburtstag*, ed. Christian Hecht (Berlin: Matthes & Seitz, 2009), 456–64.

17 Wölfflin, *Principles of Art History*.

18 Martin Warnke, "Vier Stichworte: Ikonologie—Pathosformel—Polarität und Ausgleich—Schlagbilder und Bilderfahrzeuge," in Werner Hofmann, Georg Syamken, and Martin Warnke, *Die Menschenrechte des Auges: Über Aby Warburg* (Frankfurt am Main: Europäische Verlagsanstalt, 1980), 53–83.

HB: At what point a style begins and the formative influence of the individual ends is a matter that is tied to the experience of the scholar analyzing the style. There will never be an unimpeachable standard based on which a style may be defined as such. The experiences of archaeology and art history teach us that any style will always be discussed along the margins of its application, in a context defined by the diversity of styles.

DTB: By distinguishing between inclusion and exclusion, Carlo Ginzburg has put the finger on the problem of how to define style without reference to ideologies or authorities.[19] As an organizing category, he argues, style cannot be more than a heuristic instrument. The attempt to define styles always presupposes an interpretive grid that must be present before any particular object is examined. By contrast, you have frequently insisted on the need to approach the contemplation of forms without presuppositions. What is your view of this difficult issue?

HB: That's the crucial question concerning the amenability of the object to scholarly understanding. True, we find ourselves in a hermeneutic circle that's also a vicious cycle. But there's no alternative to breaking out of it. We stake our endeavor on a naïveté that, needless to say, we no longer possess and that we nonetheless need to invoke to be able to begin afresh time and again. We've been trained, we're conscious of the objects we have studied before, and we bring a radically different experiential horizon to the task than, say, a longshoreman in Yokohama would. Given that that's the situation from which we start, it's indispensable that we describe the form in a first approach in the mode of a radically "naïve" phenomenology. To crack the hermeneutic circle, one must first be able to describe with the utmost precision and without adjectives: in the style of New Objectivity, as it were. I don't believe that there's a viable alternative to this form of initial approach.

DTB: If that's how I first examine images, I must take an attitude defined by the assumption that a maximally objective grasp of them is possible. This attitude, too, is already part of an ideology with a history we would have to trace; that's an issue that merits further discussion.[20] But let's go back to the core question of the style of scientific imagery. To which features of such images can we tie the idea of style?

HB: In theory, we might locate style on several levels, beginning with the preserved specimen, which is manufactured, prepared, and made available for examination in certain circumstances and in a manner that safeguards comparability. This is the first super-individual training in how to design the object to be examined. We will be able to recognize styles of handling the primary material, of transforming the material being prepared, in the case of a three-dimensional body, into a sculpture (fig. 2).[21] Hans-Jörg Rheinberger founded this kind of research within the history of science.[22] Technical implements occupy the second level: there are comparable lenses in telescopes or microscopes, comparable technical standards of the cameras used to record the objects, in the elaborately designed microscopes, for example, that allow for a similar epistemic style in the material sense (fig. 3). The third level would be what is usually

19 See Carlo Ginzburg, "Style as Inclusion, Style as Exclusion," in *Picturing Science, Producing Art*, ed. Caroline A. Jones and Peter Galison (New York: Routledge, 1998), 27–54.

20 See Lorraine Daston and Peter Galison, *Objectivity* (New York: Zone, 2007); Caroline A. Jones and Peter Galison, "Introduction: Picturing Science, Producing Art," in *Picturing Science, Producing Art*, ed. Jones and Galison, 8–10.

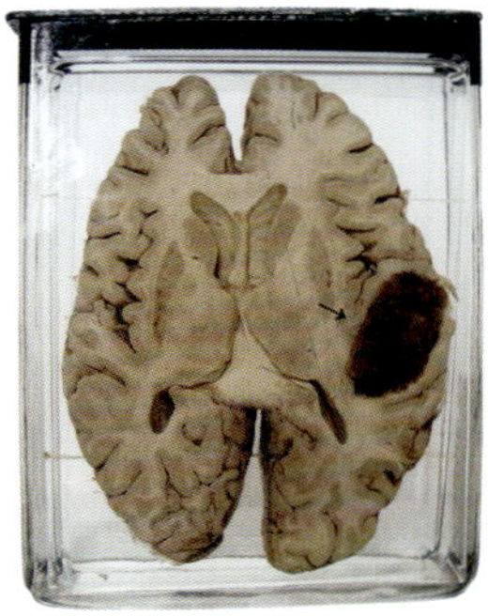

FIG. 2: "Specimen sculpture": cerebral abscess, wet specimen; undated; 21 × 18 × 7.5 cm; Berliner Medizinhistorisches Museum der Charité, Berlin. Sven Dierig and Thomas Schnalke, eds., *Apoll im Labor: Bildung, Experiment, Mechanische Schönheit*, exh. cat., Berliner Medizinhistorisches Museum der Charité, May 13–October 2, 2005 (Berlin: Medizinhistorisches Museum der Charité, 2005), 33, fig. 25. © Berliner Medizinhistorisches Museum der Charité, Photographer: Christa Scholz.

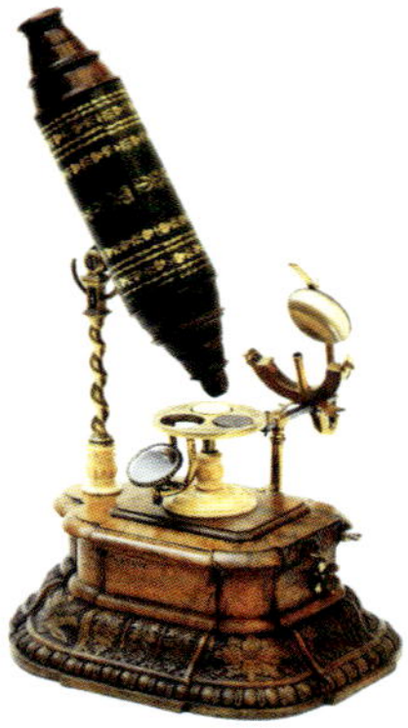

FIG. 3: Hertel's microscope; after 1716; height: 39 cm; Hessisches Landesmuseum, Darmstadt. Kurt Hemmerling and Hanns Feustel, eds., *Historische Mikroskope des physikalischen Kabinetts des Hessischen Landesmuseums Darmstadt*, Kataloge des Hessischen Landesmuseums Darmstadt, vol. 13 (Darmstadt: Hessisches Landesmuseum, 1983), 33, cat. no. 11. © Hessisches Landesmuseum Darmstadt 1983, Photographer: Werner Kumpf.

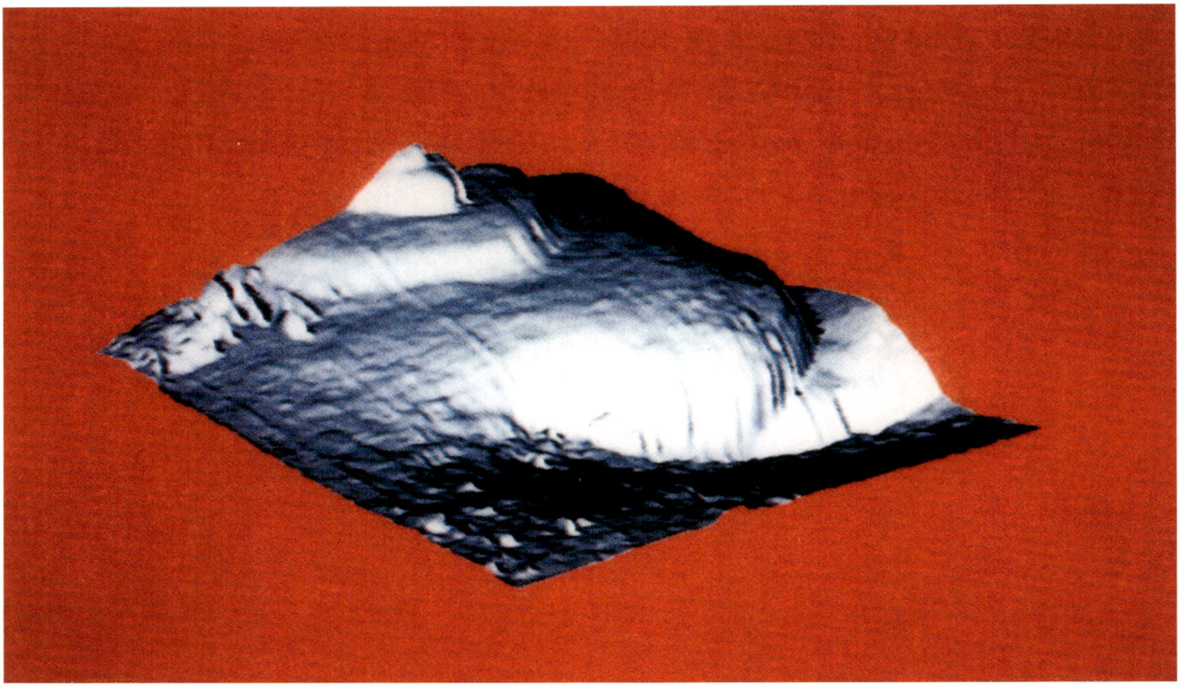

FIG. 4: Scanning tunneling microscope image; late 1980s. Roland Wiesendanger, "Rastertunnelmikroskopie an nichtkristallinen Festkörpern" (PhD diss., University of Basel, 1987), 167, fig. V21. © Roland Wiesendanger.

called "illustration." Within this level, there's another gradation, beginning with the seemingly mechanical and passive depiction in the photograph as well as the drawing. From a distance of a hundred years, scholars will be able to date a depiction produced in nanotechnology research to within two or three years at first glance (fig. 4); the same goes for the images of fractal mathematics.[23] The products of the active hand present a more difficult problem. If we succeed in recognizing a super-individual style in images designed by the hand, we have ascertained that the "illustration" in the natural sciences, far from being a passive derivate of research results, is a distinctive constructive achievement. Medical imagery abounds with such styles of anatomical depiction. My study of Galileo aims at this core. He is obviously the paradigmatic figure who, by means of the specific ability of the hand, implemented a distinctive epistemology of visual-graphical praxis.[24]

DTB: The art historian James Elkins, for example, has spoken of "crystallographic cubism,"[25] adopting a stylistic term from art history and projecting it onto the imagery of crystallography. Can such figurative applications be the goal of a history of styles of technical imagery?

HB: I would add that we would need to develop a specific nomenclature. Elkins addresses a different question: to what extent can the domain of manually produced visualization be related to the artistic tendencies of a time? In that regard, I don't think, to give you an example, that the visualization of the double helix created by Odile Crick, who was an artist, would have been conceivable without Alexander Calder's mobiles;

21 Angela Matyssek, *Rudolf Virchow: Das Pathologische Museum. Geschichte einer wissenschaftlichen Sammlung um 1900*, Schriften aus dem Berliner Medizinhistorischen Museum, vol. 1 (Darmstadt: Steinkopff, 2002); Jutta Helbig, ed., *Präparate, Bildwelten des Wissens: Kunsthistorisches Jahrbuch für Bildkritik*, vol. 9, no. 1 (Berlin: Akademie, 2012).

22 Hans-Jörg Rheinberger, "Präparate: 'Bilder' ihrer selbst. Eine bildtheoretische Glosse," in *Oberflächen der Theorie*, ed. Horst Bredekamp and Gabriele Werner, *Bildwelten des Wissens: Kunsthistorisches Jahrbuch für Bildkritik*, vol. 1, no. 2 (Berlin: Akademie, 2003), 9–19; Hans-Jörg Rheinberger, "Die Evidenz des Präparates," in *Spektakuläre Experimente: Praktiken der Evidenzproduktion im 17. Jahrhundert*, ed. Helmar Schramm, Ludger Schwarte, and Jan Lazardzig (Berlin: de Gruyter, 2007), 1–17.

23 See Nina Samuel, ed., *The Islands of Benoît Mandelbrot: Fractals, Chaos, and the Materiality of Thinking* (New Haven, CT: Yale University Press, 2012); Nina Samuel, *Die Form des Chaos: Bild und Erkenntnis in der komplexen Dynamik und der fraktalen Geometrie* (Munich: Fink, 2013).

consider, in particular, the three-dimensional model.[26] Another example would be Heinz-Otto Peitgen's fractal images (fig. 5). The experience of psychedelia—of the discos and record covers of the 1970s and the post-'68 pop culture more generally—patently plays a role in these designs. You can systematically look for styles from art in this manner. But detecting such formal transfers strikes me as a tad too easy.

DTB: What's the conceptual framework you would propose instead?

HB: One worthwhile goal would be to identify autonomous principles of visualization and concepts in the field of the natural sciences that operate analogously to Wölfflin's conceptual pairs like "open–closed," "linear–painterly," "recessive–planar." Only then would we be truly justified in using the term "style." What we need is the acuity of Worringer's analysis in *Abstraction and Empathy*, which used these two primarily formal concepts to articulate a fundamental critique of his own era. That reaches deeper than, say, the talk of a "cubist style of visualization." We would need to muster a similar cold conceptual rage and the same neutral precision with which Wölfflin developed his bipolar concepts of form to determine super-individual stylistic features in the natural sciences.

24 See Horst Bredekamp, *Galilei der Künstler: Der Mond, die Sonne, die Hand* (Berlin: Akademie, 2007).

25 James Elkins, *The Domain of Images* (Ithaca, NY: Cornell University Press, 1999), 20–27.

26 See James D. Watson and Francis H. C. Crick, "Molecular Structure of Nucleic Acids: A Structure for Deoxyribose Nucleic Acid," *Nature* 171 (1953): 737–38.

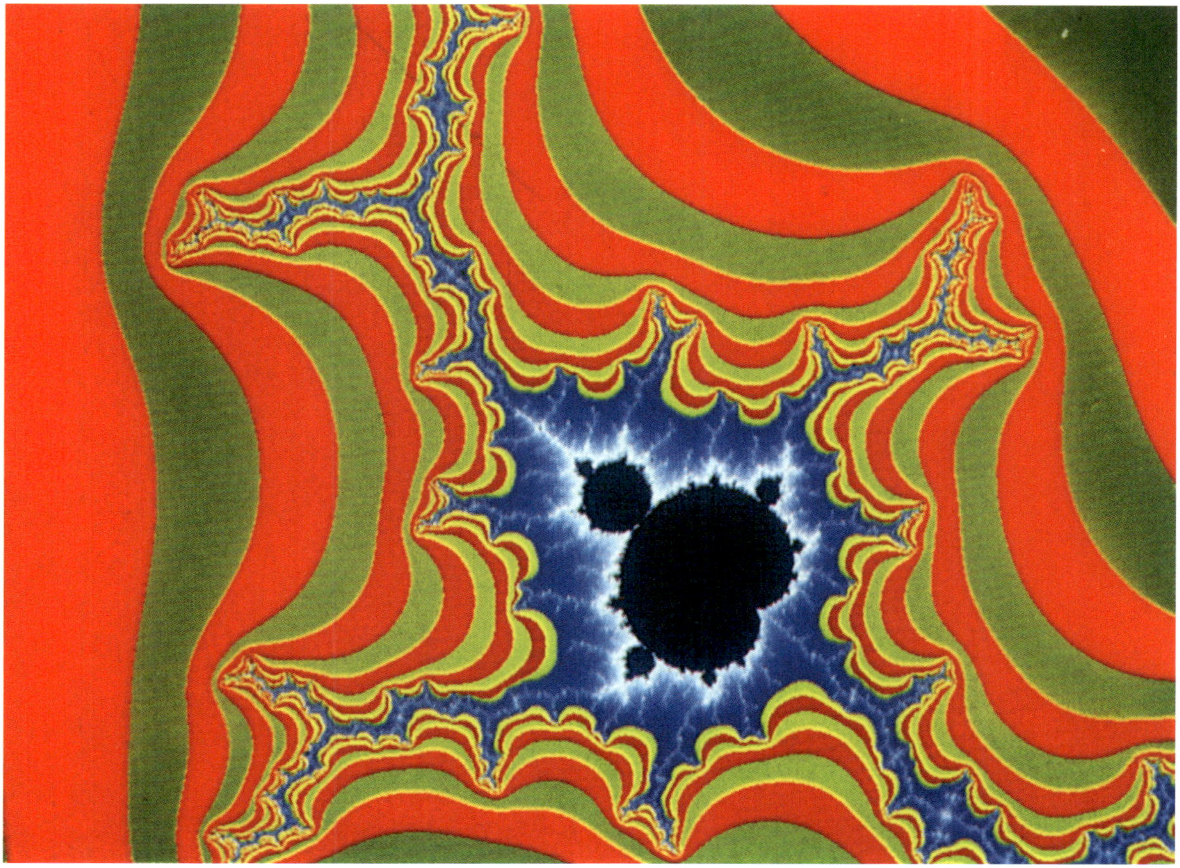

FIG. 5: Fractal image; Heinz-Otto Peitgen; published in 1986. Heinz-Otto Peitgen and Peter H. Richter, *The Beauty of Fractals: Images of Complex Dynamical Systems* (Berlin: Springer, 1986), 87, fig. 52. © Heinz-Otto Peitgen.

DTB: In describing what the project "Das Technische Bild" is about, we are frequently confronted with the fact that our focus on the "intrinsic value of the images" always also implies interest in a sort of "excess." By the latter, we mean, on the one hand, the effects of an image that cannot be controlled, the elements and significations that were not intended. On the other hand, we mean the observation that an image always has aspects that exceed the goal of a depiction conceived as illustration. A difference, which may be described as a form of surplus, intervenes between the intended purposes and objects and their representation in the image—"one plus one equals three or more," as Josef Albers put it.[27] Does an analysis focused on style aim at an "excess" of this sort?

HB: The concept of style indeed seeks to capture the intellectual activity at the core of this excess of design. It aims at the constructed quality of the worlds of the natural sciences. But there's another aspect: the formative power of content. Certain motifs and formal solutions slip into depictions as a surplus throughout the history of imagery. Therein lies the constructive power of the history of semantically invested forms in the iconographical sense. Yet style passes through the content in order to endow it with a higher significance. Not unlike the scholarship of Erwin Panofsky's iconology, studies in style aim at the essence of the work of art.[28] Jochen Hennig's study of the images of early scanning tunneling microscopy strikes at the central point of this concept of style.[29] Their history begins with an observation, which led to a question being asked; the hands-on work of building a model and gluing it together turned that into conviction that finally developed its own gravitational force. Now, with this organic construction—because a second scientist sees the form, finds it exciting, tries something similar, a third scientist joins them, makes comparisons, and develops a logic of requirements with regard to the visualization of nano-worlds—the early nano-images evolved something like a style. And they became the *conditio sine qua non* of thinking about nano-objects.

DTB: Since you brought up the case of several researchers calibrating a visual form by comparing images, does that mean that a study on the question of style must also take the communicative aspect into account? Karin Knorr Cetina, for example, has examined this aspect in the example of physicists discussing images and sought to describe it with her concept of the "viscourse."[30]

HB: Personal contact is usually crucial. But similarities may also appear in the work of people who operate within the same temporal horizon even though they are on different continents. So both exist: specific visual features that are the result of communicative training and exchange, and homogeneity that emerges across different localities.

DTB: How can we explain the observation that a shared style develops in several places at once without any demonstrable connection between the various protagonists?

HB: That's difficult to explain. It's the problem that led C. G. Jung to develop his theory of the collective unconscious—which became a big

27 Josef Albers, "One Plus One Equals Three or More: Factual Facts and Actual Facts," in *Search Versus Re-search* (Hartford, CT: Trinity College Press, 1969), 17–18. Albers's idea of the emergence of a pictorial system was taken up by Edward Tufte, *Envisioning Information* (Cheshire, CT: Graphics Press, 1990). On the matter of the iconic difference, see Gottfried Boehm, "Die Wiederkehr der Bilder," in *Was ist ein Bild?*, ed. Gottfried Boehm (Munich: Fink, 1994), 11–38.

28 Panofsky first presented this view in a lecture at the Kiel section of the German Kant society in 1931; see Erwin Panofsky, "On the Problem of Describing and Interpreting Works of the Visual Arts," trans. Jaś Elsner and Katharina Lorenz, *Critical Inquiry* 38, no. 3 (Spring 2012): 467–82.

29 See Jochen Hennig, *Bildpraxis. Visuelle Strategien in der frühen Nanotechnologie* (Bielefeld: Transcript, 2011).

30 See Karin Knorr Cetina, "'Viskurse' der Physik: Wie visuelle Darstellungen ein Wissenschaftsgebiet ordnen," in *Konstruktionen Sichtbarkeiten: Interventionen*, ed. Jörg Huber and Martin Heller (Zurich: Edition Voldemeer, 1999), 247.

hit. We're not Jungians, but we recognize why he developed his theory: when the same style surfaces in different localities at once, that is one of the most startling observations in our field. Why is it that, around 1400, the same indistinguishable hand is at work between Novgorod and Sicily? The art historian Karl Heinz Clasen has sought to answer this question by reconstructing a single "Master of the Beautiful Madonnas" who completed a full tour of Europe.[31] But there's not a single source that mentions him, no contract with a patron, nothing! The International Style around 1400 was the last time a homogeneous global style ruled the European scene. It reminds me of the 1956 science fiction movie *Invasion of the Body Snatchers*, in which all the invaders share the same brain. What one being thinks, all beings think at the same time. That picture gave narrative form to the problem you've raised.

31 See Karl Heinz Clasen, *Der Meister der schönen Madonnen: Herkunft, Entfaltung und Umkreis* (Berlin: de Gruyter, 1974).

DTB: And the motor that keeps things moving? After all, style is not static, frozen in time; it evolves.

HB: How does this style change? And why? That's what your question is really trying to get at. In art-theoretical terms, I think that stylistic change can be described using the two different concepts of imitation. Human beings have two conflicting drives: the urge to imitate the existing creation (*natura naturata*), and the incessant and curious urge to imitate the creativity of nature (*natura naturans*). Every new style evolves between these two forms of mimesis. That's the dialectic of the profitability of imitation and a loss of originality. Imitation always produces a loss of originality. Originality, for its part, produces a lack of style. And that defines the balance that keeps swinging back and forth throughout the history of humankind: style—innovation—style—innovation, the two principles of imitation.

FIG. 6: Photographs of supersonic projectiles and the phenomena they produce in the surrounding atmosphere. Pictures from Ernst Mach and Peter Salcher's first series of experiments, probably May or June 1886. Mach attached these copies to the special print of the notice in which he reported the first successful attempts to take such photographs. At the time, he could not tell with certainty what exactly caused the various phenomena that appear in the pictures. Diameter of the circular images: ca. 7 mm. Archive of the Deutsches Museum von Meisterwerken der Naturwissenschaft und Technik, Munich, estate of Ernst Mach, NL 174/0166155. Archival photograph by Christoph Hoffmann.

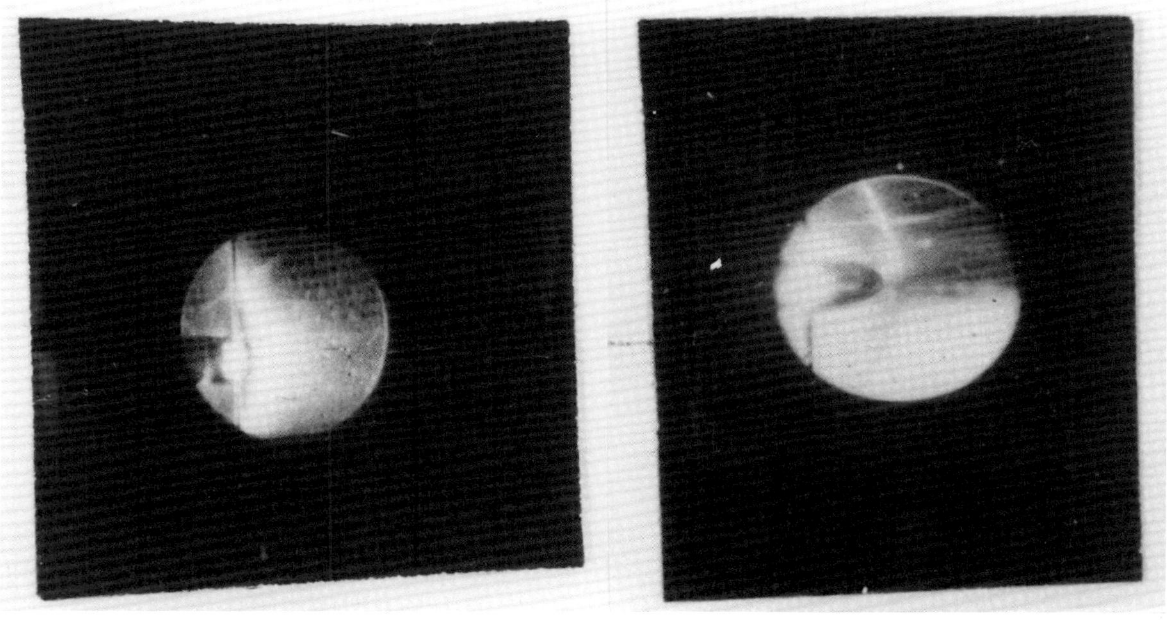

DTB: But there are also images that were entirely without precedent when they came into existence. In the case of Ernst Mach's projectile photographs, for example, stylistic categories are arguably peripheral or even altogether obsolete (fig. 6): when these pictures first appeared, even their makers weren't sure how to categorize and interpret this form.[32] A trace of something surfaced that was not yet clearly identifiable: was it visual noise, the material itself, or the inscription of something else? The picture was without tradition and therefore defied categorization; it wasn't even clear where up and down were. A technical image, we might say, befalls experimenters who are not prepared for it, and then necessitates subsequent actions. How can we conceptualize this characteristic as pertaining to a style of scientific imagery? Isn't the concept of style itself out of place here?

32 See Christoph Hoffmann, "Die Dauer eines Moments: Zu Ernst Machs und Peter Salchers ballistisch-fotografischen Versuchen 1886/87," in *Ordnungen der Sichtbarkeit: Fotografie in Wissenschaft, Kunst und Technologie*, ed. Peter Geimer (Frankfurt am Main: Suhrkamp, 2002), 352; Christoph Hoffmann and Peter Berz, eds., *Über Schall: Ernst Machs und Peter Salchers Geschossfotografien* (Göttingen: Wallstein, 2001).

HB: A "first image" indeed seems to emerge from the void, so unprecedented that the question of its style, which presupposes comparability, is initially pointless. The "first image" has no "style." It becomes associated with a style only when, as an "icon," it acquires super-individual appeal, spurring the production of similar images in various places and infecting other subject areas, which is what happened in the case of projectile photography.

DTB: Which styles of technical imagery can you name? Would you speak, for example, of "the Galilean style"? Would you describe the nano-images we talked about earlier as exemplifying a "nano-style"?

HB: Specific concepts are the eventual goal; but even Wölfflin's categories are already eminently useful. So we might say that the early nano-images evince a painterly style: their surfaces look soft, like cotton wool, transitions are blurred, and they generate an illusion of waves or slopes. With regard to the conflict that pitted Christoph Scheiner against Galileo

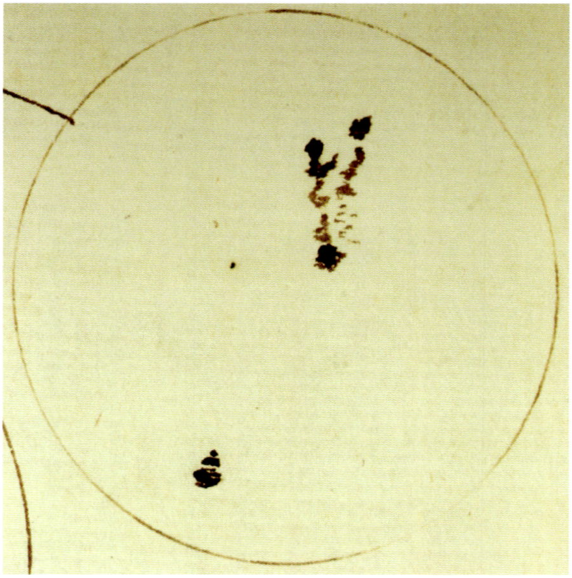

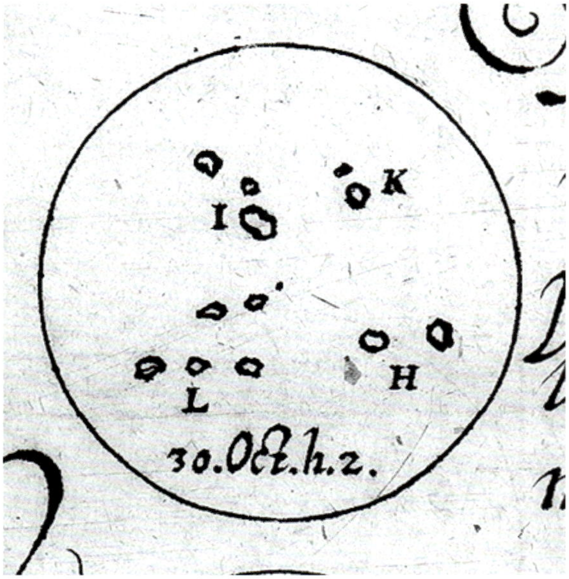

FIGS. 7a, 7b: Sunspots in stylistic comparison: Galileo Galileo's "diffusion-style" (a) is apparent in his sunspot observation dated May 1, 1612, which he renders in subtle gradations and glazes. By contrast, Christoph Scheiner's "consistency-style" in the medium of the copperplate print is characterized by clearly delineated forms. His sunspot observation (b) is dated October 30, 1611; the engraving was made by Alexander Mair. Horst Bredekamp, *Galilei der Künstler: Der Mond, die Sonne, die Hand* (Berlin: Akademie Verlag, 2007), 232, 235.

Galilei and Lodovico Cigoli over their respective sunspot pictures, I would speak of the antagonism between a consistency-style and a diffusion-style.[33] Over the course of thirty years, Scheiner evolves a consistency-style (fig. 7b); he wants to show that the sunspot stars pass freely between the sun and the Earth. Galileo and Cigoli, by contrast, develop a diffusion-style (fig. 7a) designed to show that the sun's spots are located on the sun itself as a haze, as features that are hard to define. The hand is trained in the interplay between seeing, depicting, and discovery. In Galileo, the earliest black spots are very consistent, as though they were solid bodies, but toward the end of the series, he suddenly uses the brush, switching from the quill, which registered precise dots, to a watercolor-like inscription that builds diffuseness.

33 Bredekamp, *Galilei der Künstler*, 217–82.

DTB: But your example also illustrates how the choice of a technique of depiction is a crucial factor in determining a style. Like the specific skills of the hand, the selection of technical implements—from the etching needle to photographic technology to the image editing program—profoundly informs a style. Each of these tools has its specific representational capacities. Before Riegl, Worringer, and Warburg, Gottfried Semper had defined style as something that is crucially influenced by material and technological conditions.[34] What do you see as the significance of the material and technology used in making images? Wouldn't this be one way of designating styles of technical imagery more precisely? Instead of speaking of a "consistency-style" and a "diffusion-style," might you not just as well use the terms "copperplate engraving style" and "watercolor style" or—adopting Wölfflin's dichotomies—"linear" and "painterly"?

34 Semper, "Textile Art."

HB: That's a good point. But "linear" might describe any object, whereas the concept of the consistency-style is already more semantically specific with regard to a particular object. Scheiner designed his style to represent the consistency of matter, and he commissioned the engraver Alexander Mair to translate his findings into sharp-edged lines. Galileo's diffusion-style, by contrast, was precisely meant to visualize the diffuse aspect of the phenomenon. The watercolor was the technique of his choice. Formally speaking, these coincide with "linear" and "painterly"; with the terms I chose, I wanted to sharpen the contours of these two forms of representation with a view to the conflict between the findings they visualized. Because the antagonism between the bodily and consistent aspect of the material phenomenon being represented and its open and hazy quality plays the central part in this conflict. But perhaps my terms are already too suggestive, too tied up with their object.

DTB: That means that concepts of style can be used in two ways: descriptively, to designate a formal specificity, and interpretatively, to indicate the effect or signification of this formal specificity. Based on what you've said, you would prefer stylistic terms chosen with a view to the signification conveyed by a particular style. But to go back to the example you've mentioned, Galileo: does artistic training—which is to say, how good a painter Galileo was—play a role in this context?

HB: Galileo was a trained and excellent draftsman. His friend Cigoli was the Medicis' celebrated court painter. They practiced together and commented on each other's work. But what are we to make of people who had no drawing skills, like Leibniz and Darwin? As I see it, the movements of the hand reveal in any case that the individual's designs are permeated by super-individual influences. In that sense, even Leibniz's or Darwin's doodles have "style."[35]

35 See Horst Bredekamp, *Die Fenster der Monade: Gottfried Wilhelm Leibniz' Theater der Natur und Kunst* (Berlin: Akademie, 2004); Horst Bredekamp, *Darwins Korallen: Frühe Evolutionsmodelle und die Tradition der Naturgeschichte* (Berlin: Wagenbach, 2005).

DTB: Your example illustrates once more that a style becomes manifest even in the absence of virtually all deliberate fashioning or technical skill. Putting it succinctly, we might say that there's no form without style. That is true also of images generated by technical equipment, like photographs. But what do we gain by being able to identify two different styles?

HB: To begin with, a comparative determination of the precision of a design practice is valuable as such. The second level is that Scheiner's consistency-style aims at compact bodies, whereas Galileo envisions clouds and hazes. The different styles produce objective renditions of differences between the objects. On the third plane, to follow the iconological model, these competing modes of representation tie in with what we may positively call worldviews: Scheiner's consistency-style seeks to salvage the traditional cosmology, whereas Galileo's diffusion-style is designed to lift Aristotelianism off its hinges. In his famous letter to Maffeo Barberini, who later became Pope Urban VIII, Galileo described the diffuse features that are his sunspots as the "Last Judgment" on Aristotelianism. A drawing style closes the door on two and a half millennia of cosmology! Style is where the matter of the ancient worldview is decided. That's not a minor episode.

DTB: The category of style as you describe it contradicts the positivist methodological ideal that many in the natural sciences hold dear. In that regard, the search for styles takes on an edge of debunking: it confutes the presuppositionless objectivity also and especially of images that are presented as objective because they are allegedly traces and documentation of events or outcomes of an experiment, and illuminates how they are products of a tradition informed by styles as well. We have personally been met with skepticism by natural scientists, who sometimes perceive our work as nugatory or even hostile. How do you deal with that sort of response?

HB: The rhetoric of debunking is a pitfall we should avoid. What we should emphasize instead is our attempt to raise the epistemological complexity in the domain of the natural sciences. Images constitute a challenge to the positivist conception of what the natural sciences are. Until three decades ago, images tended to be seen as alien to the natural sciences because it was assumed that they contained an admixture of the subjective and were thus inevitably at odds with the imperative of objectivity. "Das Technische Bild" seeks to recognize the strength in this alleged weakness. If there's a history of styles of imagery in the natural sciences, we have demonstrated that images, in rendering an object, also construct it and inform it with the qualities of their own sphere. If that is the case, however, the natural sciences themselves ought to be interested in considering the complexity, beauty, and anarchy of images as particular features that render a distinctive contribution to the sciences. In that regard, I am far more optimistic now than I was even five years ago. We have such an abundance of materials at our disposal that, with all due skepticism, we should over time be able to sketch the outlines of a broad-based style of the natural sciences.

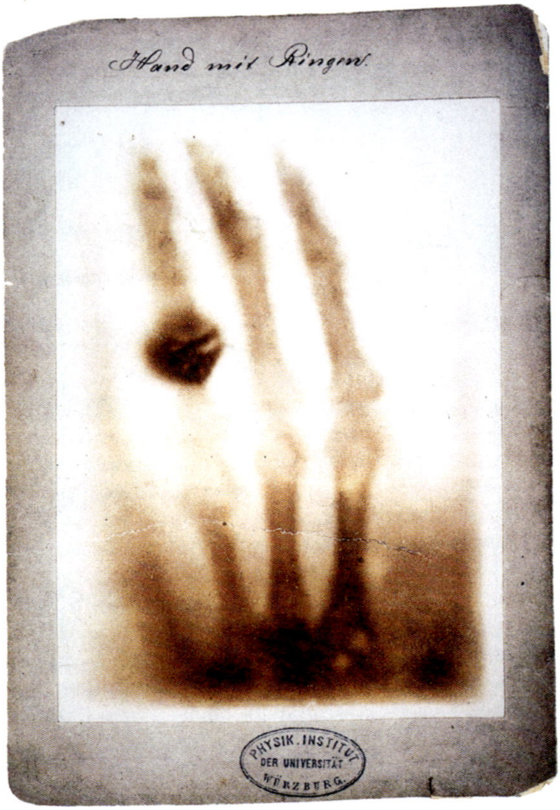

FIG. 1: Wilhelm Conrad Röntgen, X-ray image of his wife Bertha's hand, December 1895. Picture from a series of prints that Röntgen sent to colleagues on January 1, 1896. Archive of the Deutsches Röntgen-Museum Remscheid-Lennep, Germany. © Deutsches Röntgen-Museum, Remscheid-Lennep.

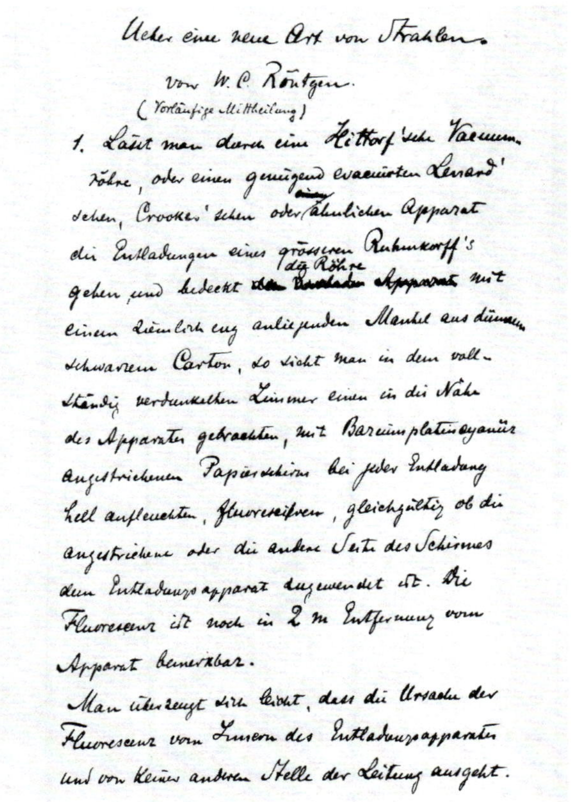
Ueber eine neue Art von Strahlen.
von W. C. Röntgen.
(Vorläufige Mittheilung)
1. Lässt man durch eine Hittorf'sche Vacuumröhre, oder einen genügend evacuirten Lenard'schen, Crookes'schen oder ähnlichen Apparat die Entladungen eines grösseren Ruhmkorff's gehen, und bedeckt die Röhre mit einem ziemlich eng anliegenden Mantel aus dünnem schwarzem Carton, so sieht man in dem vollständig verdunkelten Zimmer einen in die Nähe des Apparates gebrachten, mit Bariumplatincyanür angestrichenen Papierschirm bei jeder Entladung hell aufleuchten, fluoresciren, gleichgültig ob die angestrichene oder die andere Seite des Schirmes dem Entladungsapparat zugewendet ist. Die Fluorescenz ist noch in 2 m Entfernung vom Apparat bemerkbar.
Man überzeugt sich leicht, dass die Ursache der Fluorescenz vom Innern des Entladungsapparates und von keiner anderen Stelle der Leitung ausgeht.

FIG. 2: Röntgen's manuscript "Über eine neue Art von Strahlen" ("On a New Kind of Rays"), sent to the publisher in late 1895, in which he describes his experimental set-up for the first time. Archive of the Deutsches Röntgen-Museum Remscheid-Lennep, Germany. © Deutsches Röntgen-Museum, Remscheid-Lennep.

ICONOLOGICAL ANALYSIS

The term *iconological analysis* designates the art-historical method that seeks to complement the method of iconography, which identifies and interprets subjects and motifs. Iconological analysis synthesizes the precise description of a work with studies of its contexts. This synthesis guides the interpretation of the work (Warnke 1980; Schmidt 1993; Warburg 1999). Although the method is central to the work of the German art historian Aby Warburg (1866–1929), his student Erwin Panofsky was the first to develop and publish a three-stage model of iconological analysis; he called the third stage of this model "iconological interpretation" (Panofsky 1970 and 1981; Holly 1984). Since then, iconological analysis has been taken to refer to the step-by-step interpretation of pictorial artifacts within their cultural and historical contexts (Elsner et al. 2012; Mitchell 1986).

According to Panofsky's model, interpretation begins with a formal description (first stage), proceeds to an iconographic analysis of content (second stage), and then determines the meaning of the work of art (third stage). This last phase analyzes the period in which the work was created and the prevailing social, political, philosophical, and religious attitudes of the era or nation that influenced its creation. The work as a product of these attitudes thus appears as paradigmatic or symptomatic of an epoch or, in the context of the history of ideas, as a historical document of ideas, opinions, and views. Beyond such symptomatic qualities, the extent to which the work itself actively participated in forming these ideas, opinions, and views requires scrutiny as well.

Iconological analysis seeks to reconstruct traditions and reveal layers of meaning by critically investigating literary sources of various provenances, including documents from everyday culture and superstition as well as

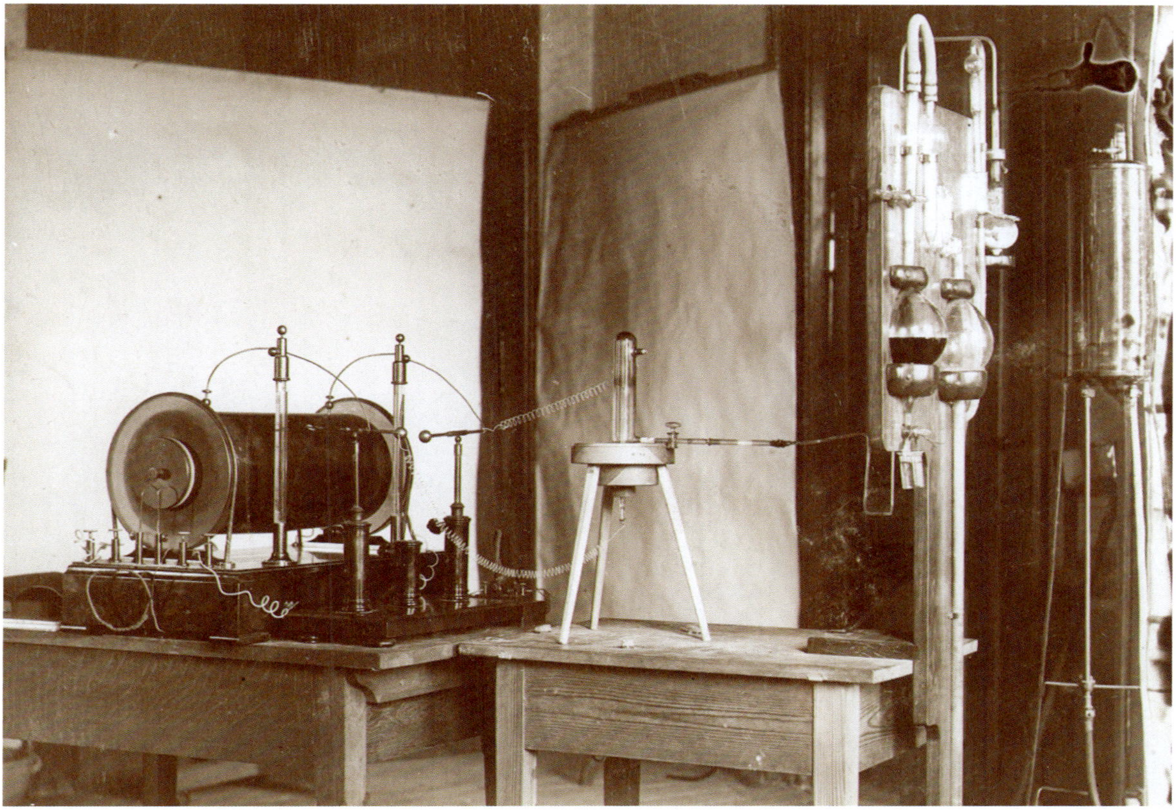

FIG. 3: The laboratory and devices Röntgen used to research the rays; photo taken in Würzburg in May 1923, showing the spark inductor, X-ray tube, and vacuum pump (from left). Archive of the Deutsches Röntgen-Museum Remscheid-Lennep, Germany. © Deutsches Röntgen-Museum, Remscheid-Lennep.

other artifacts related to the work of art. Such meaning may not be apparent to the observer at first glance, nor was it necessarily intended by the producer or commissioner of the work. Iconology as practiced by Panofsky, especially in the concluding phase of the interpretation, seeks to place a work in a wider context of meaning. The interpreter's profound and broad knowledge of the cultural and historical context of the work serves as a corrective to his "personal psychology" or "worldview."

In 1912, Aby Warburg described iconological analysis in the conclusion to his lecture on the frescoes in the Palazzo Schifanoia, Ferrara, as a method "that can range freely, with no fear of border guards, and can treat the ancient, medieval and modern worlds as a coherent historical unity." He proposed that art historians examine "the purest and the most utilitarian of arts as equivalent documents of expression" (Warburg 1999), overcome evaluative categories such as "high" and "low," transcend disciplinary boundaries, and proceed by covering various periods, occasionally anachronistically, when searching for correlations among traditional forms of representation and motifs (Beyer 1992).

The underlying assumption that every form is a historical phenomenon and has a history of its own must therefore also apply to the interpretation of technical and scientific images and be the starting point of their iconological analysis. The latter takes into account first of all the image's specific scientific context; in so doing, it relies on knowledge from other disciplines. To decode the ways in which scientific and technical images create meaning, iconological analysis examines their functions within productive and epistemic processes and attempts to identify the formal properties of images by examining the interplay of technological conditions,

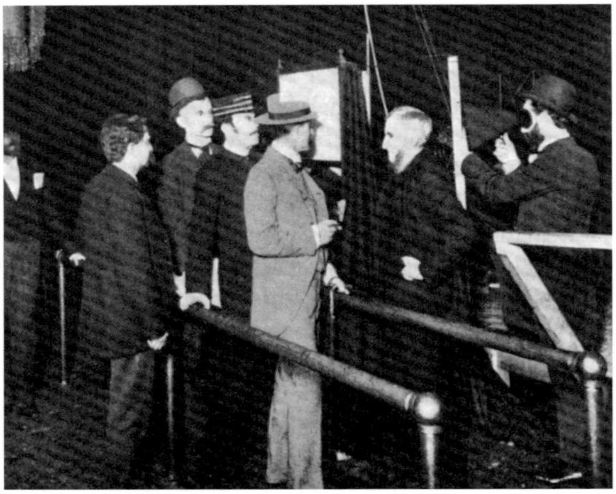

FIG. 4 (left): Demonstration of radiographic screening during the special exhibition organized by Thomas Alva Edison about X-rays as part of the *Electric Light Exhibition* in New York, May 1896. One after another, visitors were invited to hold their hands behind the fluorescent screen. Otto Glasser, *Wilhelm Conrad Röntgen und die Geschichte der Röntgenstrahlen* (Berlin, Göttingen, Heidelberg: Springer, 1959 [1931]), 204. Edward P. Thompson, *Roentgen Rays and Phenomena of the Anode and Cathode,* published in 1896 by the D. Van Nostrand company in New York.

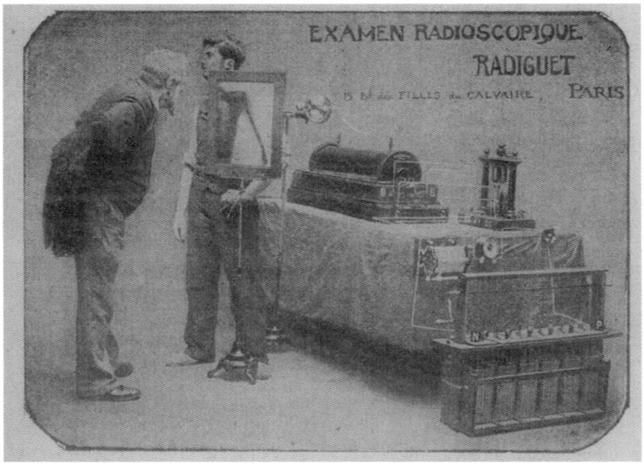

FIG. 5 (bottom left): Medical examination using X-rays, around 1900. Advertising for the Parisian X-ray laboratory of the instrument maker Arthur Radiguet, who made his machines available to doctors. *La Nature. Science et Progrès* (Paris 1897).

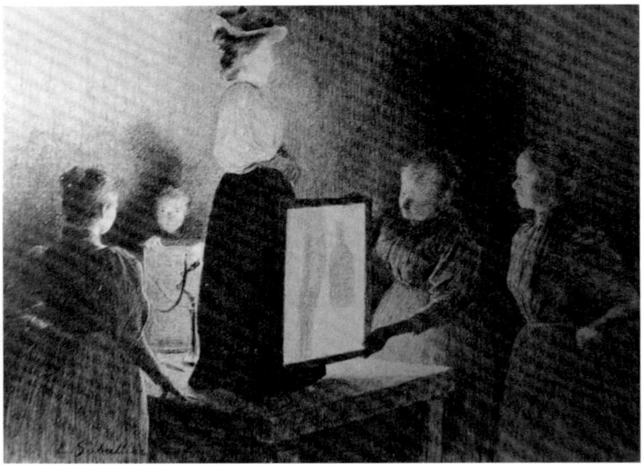

FIG. 6 (bottom right): Smuggler caught by customs officials with the help of X-rays. Popular illustration, 1897. *L'Illustration,* no. 2836, July 3, 1897, 7.

design interventions, and scientific convictions. In light of the ideas proposed by the Polish philosopher of science Ludwik Fleck, then, the aim is to reveal the extent to which social and psychological aspects influence the creation of scientific images that develop in what Fleck called a "communication of ideas" among the scientists. He investigated this process under the headings "thought style" and "thought collective" (Fleck 1979). The method of iconological analysis also focuses on the productive interventions of scientists and apparatuses and their impact on the aesthetics and visual effects of the images.

In order to examine the diverse purposes and claims of scientific and technical images, iconological analysis also relies on contemporary texts such as scientific reports and papers. It inquires into possible references to scientific discourses and into models and traditions of representation within and outside specific disciplines. The transfer of images to the public sphere, their popular reception, and the repercussions in science must also be part of such an analysis, as must the attempt to position scientific and technical images in an overall history of forms of visual representation. Such classification creates a cultural context that does not efface the distinction between the scientific and the popular, between art and non-art, but provides insight into the exchanges, interactions, and interconnectedness of these areas. The challenge of an iconology of scientific imagery, then, is to establish a balance between this integration into a broader history of images and the attention to specific scientific and technological contexts.

Early X-ray images, which, apart from their medical purposes, made it possible actually to see through a wide range of objects, are an exemplary field for such iconological analysis. Once described and historically classified, they may be interpreted as an expression of the voyeuristic curiosity and widespread mania for translucency typical of fin de siècle culture. That culture

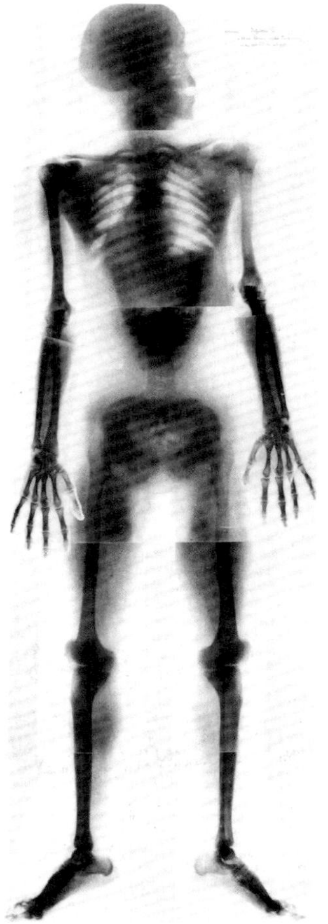

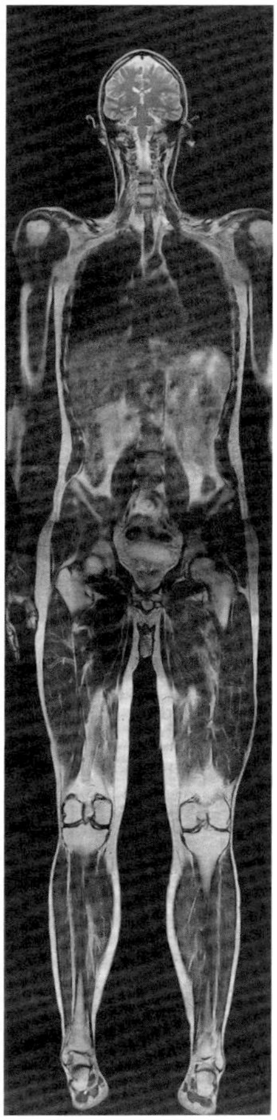

FIG. 7 (left): X-ray image of the human body, a collage of six images of three men, made by the physicist Ludwig Zehnder and photographer Karl Ernst Kempke in the summer of 1896. Height: approximately 1.84 m. Michel Frizot, ed., *Neue Geschichte der Fotografie* (Cologne: Könemann 1998), 281. © Foto Deutsches Museum, BN01755.

FIG. 8 (right): Magnetic resonance tomography image from an article in *Die Zeit*, 2004. The popular reception of new imaging techniques such as computer tomography (CT) and magnetic resonance tomography (MRT) shows a fascination with the technology similar to that elicited by early X-ray images; note, for instance, that—as the article reports—people in the United States even undergo full-body imaging in shopping malls as a precaution. The medical benefits, meanwhile, are of secondary importance, and the debate over the widespread application of this image sectioning procedure is controversial. Some argue that full-body MRT scanning offers preventative health benefits, while others point to the possibility of erroneous results and the difficulties in interpreting the images. *Die Zeit*, January 15, 2004, 28. © Tobias Beck.

expanded around 1900 into diverse scientific and popular areas and still affects the visualization strategies of today's medical imaging processes.

Collections of drawings by sixteenth-century natural philosophers in which traditional representations of monsters and mythical creatures are presented on an equal footing next to images documenting individual observations are another example. This juxtaposition is virtually incomprehensible in the perspective of today's understanding of taxonomy. Yet the iconological method reveals, for example, the pictorial traditions inherent in the discourse on natural philosophy in Aldrovandi's picture collection and the concept of nature that is expressed in the composition of such a collection.
—VS/VD

LITERATURE

Beyer, Andreas, ed. *Die Lesbarkeit der Kunst: Zur Geistes-Gegenwart der Ikonologie*. Berlin: Wagenbach, 1992.

Elsner, Jaś, and Katharina Lorenz. "The Genesis of Iconology." *Critical Inquiry* 38, no. 3 (2012): 483–512.

Fleck, Ludwik. *Genesis and Development of a Scientific Fact* [1935]. Chicago: University of Chicago Press, 1979.

Holly, Michael Ann. *Panofsky and the Foundations of Art History*. Ithaca, NY: Cornell University Press, 1984.

Jordanova, L. J. *Sexual Visions: Images of Gender in Science and Medicine between the Eighteenth and Twentieth Centuries*. Madison: University of Wisconsin Press, 1989.

Kaemmerling, Ekkehard, ed. *Ikonographie und Ikonologie: Theorien, Entwicklung, Probleme*. Cologne: DuMont, 1979.

Mitchell, W. J. T. *Iconology: Image, Text, Ideology*. Chicago: University of Chicago Press, 1986.

Panofsky, Erwin, "The Concept of Artistic Volition" [1920], trans. Kenneth J. Northcott, and Joel Snyder. *Critical Inquiry* 8, no. 1 (1981): 17–33.

Panofsky, Erwin. "Introduction: The History of Art as a Humanistic Discipline" and "Iconography and Iconology: An Introduction to the Study of Renaissance Art." In *Meaning in the Visual Arts*, 23–50 and 51–81. Harmondsworth: Penguin Books, 1970.

Panofsky, Erwin. *Studies in Iconology: Humanistic Themes in the Art of the Renaissance*. New York: Oxford University Press, 1939.

Schmidt, Peter. *Warburg und die Ikonologie: Mit einem Auszug unbekannter Quellen zur Geschichte der Internationalen Gesellschaft für Ikonographische Studien von Dieter Wuttke*. Wiesbaden: Harrassowitz, 1993.

Warburg, Aby. "Italian Art and International Astrology in the Palazzo Schifanoia, Ferrara (1912)." In *The Renewal of Pagan Antiquity: Contributions to the Cultural History of the European Renaissance*, 563–91. Los Angeles: Getty Research Institute for the History of Art and the Humanities, 1999.

Warnke, Martin. "Ikonologie." In *Die Menschenrechte des Auges: Über Aby Warburg*, ed. Werner Hoffmann, 55–61. Frankfurt am Main: Europäische Verlagsanstalt, 1980.

Beyond the Icons of Knowledge: Artistic Styles and the Art History of Scientific Imagery

Matthias Bruhn

Visual Knowledge

Knowledge is communicated by images, but what does it look like? Which context or arrangement makes the line of a chart meaningful or the photographic snapshot valuable as a piece of intelligence or evidence? Illustrations from science or technology may evince recurring patterns and elements such as legends, abstract structures, and mathematical formulae, yet it hardly seems that there is a grammar or aesthetic behind them that makes them definitively scientific. On the contrary, the deeper historians dig into the past, the more examples they find of how differently scientists, humanists, or engineers have employed notes, drawings, photographs, and other media to understand the world and to circulate information. As late as the age of industrialization, an impressive variety of hybrid (chronological and topological, diagrammatic, sequential) forms of illustration made it possible to envision mental processes, geological deep time, or long-term evolutionary development.[1] This creative and sometimes reckless activity led to the gigantic nineteenth-century photographic archives and the standardization of certain formats, modes, and terminologies and to radically new techniques of visualization and display, some of which have been so successful that one might forget they are actually modern inventions.

Visual knowledge, then, is by no means self-evident or self-explanatory. It is a cultural and historical product shaped by social expectations and the views of specialists as well as by the design tools applied; they are constraints of conceptualization. Visual knowledge is not identical with knowledge made visible; as Brian S. Baigrie put it in 1996, scientific images are not translations of a given meaning, nor visual appetizers to make some epistemic entrée more appealing, but a complex of insights that emerges from, and during, the very process of observation and modeling.[2]

For the same reason, their evolution is as telling as the insights they seem to offer. One possible name for this evolution is the "history of styles," at least in its modern sense as established by artists such as the Italian Giorgio Vasari (1511–1574), who distinguished individual and local forms of art production as different *maniere* and evaluated them according to a model of developmental cycles in cultural history (rise, peak, decay). The eighteenth-century archaeologist Johann Joachim Winckelmann (1717–1768) applied this model to the historical phases of classical Greek art, interpreting them as products of a collective and

1 On the visual history of psychology, see Pietro Corsi, ed., *The Enchanted Loom: Chapters in the History of Neuroscience* (Oxford: Oxford University Press, 1991); on the visualization of long periods, see Martin J. S. Rudwick, *Scenes from Deep Time: Early Pictorial Representations of the Prehistoric World* (Chicago: University of Chicago Press, 1992); on evolutionary schemes, see Horst Bredekamp, *Darwins Korallen: Die frühen Evolutionsdiagramme und die Tradition der Naturgeschichte* (Berlin: Wagenbach, 2006); Julia Voss, *Darwin's Pictures: Views of Evolutionary Theory, 1837–1874* (New Haven, CT: Yale University Press, 2010).

2 Brian S. Baigrie, ed., *Picturing Knowledge: Historical and Philosophical Problems concerning the Use of Art in Science* (Toronto: University of Toronto Press, 1996), xix. For literature on aesthetics in science—compare the title of Judith Wechsler, ed., *On Aesthetics in Science* (Cambridge, MA: MIT Press, 1988)—see the bibliography at the end of this volume.

thus as an expression of putative political, environmental, or technical conditions.[3]

In light of the growing collections of antiquarian and natural objects, this new adaptation of the notion of "style" could be seen as implying an autonomous, evolutionary *art history* of man-made artifacts corresponding to what had hitherto been known as *natural history*.[4] Winckelmann's concept had tremendous resonance in and after the second half of the eighteenth century, as it went beyond the narrow territories of the lives of artists and classical masterpieces and expanded the realm of objects to be considered; the underlying assumption was that products of industry (such as merchandise), reflected technical, social, or economic conditions as clearly as the exclusive works of art because they were made in accordance with principles of practicability and tradition and thus "evolved" slowly and in a manner that seemed comparable to the development of a biological body.

Today, the history of styles suffers from a less than stellar reputation; it has been more or less demoted to the status of a tool for the chronological identification of objects. Scholars have objected to the tendency implicit in the concept of "style" to single out certain phenomena as "symptomatic" of a period or society; rather than merely describing changes of forms, they have argued, it has been offered as an explanation for them. In any case, so many factors are involved in shaping products of human creativity that it will remain technically impossible to identify some of them as crucial.

Yet the term still has value, particularly in the area of scientific imagery and visualization. First, it not only circumscribes the obvious phenomenon that images from different epochs or contexts can look different even if they represent identical objects (e.g., a human body), it also suggests that one basic device (e.g., a photographic camera) may result in entirely different depictions, depending on the handling, the choice of position, and other accidental factors. Second, if the common modes of visualization widely used in contemporary science and technology are "applied" forms of representation, with an alleged proximity to their objects of observation, they may reveal aesthetic principles, visualization strategies, or the expectations of the beholders as clearly as (or even more clearly than) the idiosyncratic work of avant-garde art or any high-end design solution.

It was not by accident that the argument over style arose in an age of political and scientific revolution, of changing tastes and scattered art treasures, of new media and visual sensations. Now, the issues scholars and critics confronted around and after 1800 sometimes resemble today's debates concerning the value and virtue of scientific images. The discipline of "art history" that emerged at European universities two centuries ago may in a way be seen as an early response to intellectual and material questions of quantity and quality that remain unresolved.

The following case study examines the work of a largely forgotten German illustrator of around the year 1800 in order to describe the bandwidth of the concept of styles with regard to scientific illustrations and the importance of art-historical methodologies for their interpretation. The goal is to demonstrate the extent to which artistic practices defined an interface between different disciplines in a phase of accelerated

3 On the meaning and etymology of the term, see Willibald Sauerländer, "From Stylus to Style: Reflections on the Fate of a Notion," *Art History* 6 (1983): 253–70; Ulrich Pfisterer, *Donatello und die Entdeckung der Stile: 1430–1445* (Munich: Hirmer, 2002); Jonathan Gilmore: *The Life of a Style: Beginnings and Endings in the Narrative History of Art* (Ithaca, NY: Cornell University Press, 2000).

4 See Wolf Lepenies, *Das Ende der Naturgeschichte: Wandel kultureller Selbstverständlichkeiten in den Wissenschaften des 18. und 19. Jahrhunderts* (Munich: Hanser, 1976); Alain Schnapp: *The Discovery of the Past: The Origins of Archaeology* (London: British Museum Press, 1999), 221–73.

specialization and the emergence of new specialists, audiences, and aesthetic theories facing novel challenges and technicalities of visualization.

Nature Seen as a Piece of Art

> *There are but few Artificial things that are worth observing with a Microscope; and therefore I shall speak but briefly concerning them. For the Productions of art are such rude mis-shapen things, that when view'd with a Microscope, there is little else observable, but their deformity.* —Robert Hooke, *Micrographia* (1665)

With the rise of the handbill and movable-type printing in the early modern era, illustrations had already been discovered both as a medium of popular education and as a sales factor. In the heyday of highly elaborate anatomical atlases around 1800, experienced artists were in great demand in a growing number of fields. While some draftsmen and printers aimed at careers within the fine arts system, defined by public commissions and court art, academies and salons, many of them began to make their living as technicians in the service of book publishers and universities. They became essential to the entire process of scientific communication, and even though some scientists were capable draftsmen themselves,[5] institutions across Europe began to compete for skilled illustrators, seeking to train them or to employ them permanently.[6] Illustrations and illustrators thus acquired a key role as communicators across disciplines.

In 1803, the Royal Society of Göttingen invited scholars to participate in its annual competition by solving a problem in plant anatomy. By the turn of the century, botanists had already identified a considerable number of structural elements in plants; meanwhile, progress in lens-making raised further questions concerning the generation, growth, and composition of life. The Göttingen society now wished to shed more light on this field of research, in particular on the structure and growth of sap-conducting "vessels." Three essays reached the society, by the botanist Heinrich Link and the physicians Carl Rudolphi and Ludolf Treviranus. Historians have deplored that the jury accepted all three papers even though they contradicted each other.

This was partly due to the fact that at the time the competition was announced, a comprehensive theory of the cell as the basic unit of organisms was still far off (it would not be established until after 1830), and the era's terminology was confusing. The *cells*, *boxes*, or *utriculi* that the English microscopist Robert Hooke and the Italian Marcello Malpighi had referenced in their monumental seventeenth-century works were only rough circumscriptions of spaces that reminded them of "tissues," "honeycombs," or "sponges" and were hardly visible under the microscope. As early as the mid-eighteenth century, Caspar Friedrich Wolff had begun to speculate that life-forms in general might arise from foam-like aggregations of globules.

In this physiological debate, illustrations drew growing attention as a point of reference for scholarly communication. When the Göttingen manuscripts were published in separate volumes, they all included copperplate prints executed by a local illustrator, Christian Andreas Besemann (1760–1818). The first of the illustrations in the appendix

5 The famous example being Galileo Galilei. See Horst Bredekamp, *Gazing Hands and Blind Spots: Galileo as Draftsman* (Cambridge: Cambridge University Press, 2000).

6 Elke Schulze, "*Nulla dies sine linea.*" *Universitärer Zeichenunterricht—eine problemgeschichtliche Studie* (Stuttgart: Steiner, 2004).

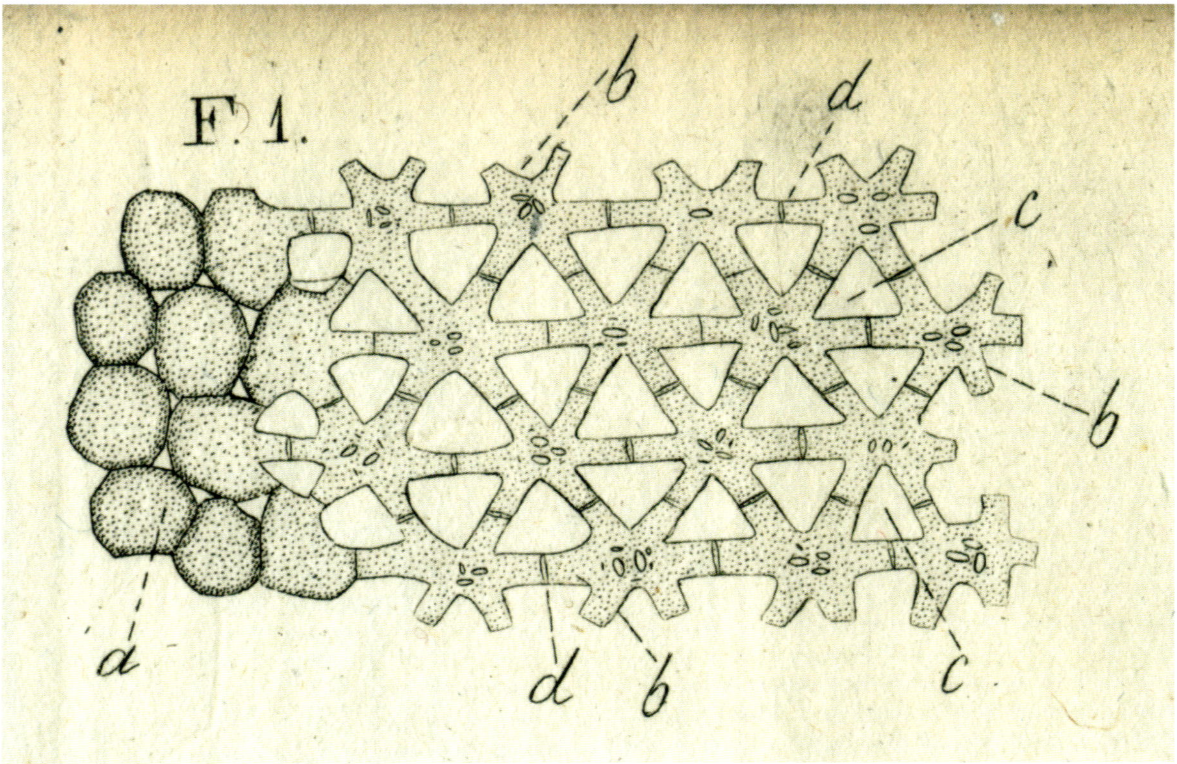

FIG. 1: Christian Besemann, etching from Ludolph Christian Treviranus, *Vom inwendigen Bau der Gewächse [. . .]* (Göttingen, 1806). Detail of plate 1. Photograph by the author.

to Ludolf Treviranus's essay reproduced the microscopic view of a plant parenchyma (fig. 1). It is symmetrical, like a crystal lattice, but changes slightly from the left to the right, presumably showing a part of the tissue undergoing transformation. Treviranus had observed that the symmetrical patterns are not solid tectonic structures but kernels or units that seem to expand slowly. Consequently, he assumed that there had to be spaces *between* those units that were progressively filled in the process of growth.

Heinrich Link, his competitor, rejected Treviranus's final conclusion. Nevertheless, he referred to the other scholar's figures as if they represented the original tissue: "At the point where the cells touch, one may often see double strokes (see fig. 1), an interstice of sorts between the cells."[7] The choice of the words "doppelte Striche" or "double strokes" is remarkable in this context. A drawing is regarded as a physical reality, as if the cell's image had been painted by nature, as if the depiction were itself a piece of evidence. The third competitor, Carl Rudolphi, similarly adopted a terminology of "lines," "hatchings," and "dots" to describe not only the printed illustrations but also what he saw through the microscope.

7 Heinrich F. Link, *Grundlehren der Anatomie und Physiologie der Pflanzen* (Göttingen: Danckwerts, 1807), 13.

Clear-Cut and Distinct: A Classicist Vision of Nature

When the microscopic instrument offers exclusive access to an object, vision is tentative. Judgment concerning the question whether the aspect or object is "in" or "out" of focus is relative and based on previous familiarity with shapes and surfaces. Moreover, metaphors and analogies are inevitably tropes of the imagination, as can be seen in the entire language of cell biology, where words are borrowed from architecture (*cell, stair, wall, chamber*), textiles (*tissue, membrane*), and mechanics and hydraulics (*vessel, vase, pressure*);[8] even now, terminologies maintain the idea of "architectures," "walls," or "channels" to understand the invisible structures of life on the nuclear scale as well as for advanced forms of engineering ("bionics") that emulate nature.

Yet it is impossible to communicate observations verbally when meanings change or standardized nomenclatures are not available. In 1803, for instance, exactly which phenomena terms such as *cell, vesicle, kernel, transpiration*, or *growth* referred to was not definitively settled, especially since there was no theory of aggregate states or cell division; a word like *membrane* might designate skin-like structures on completely different levels of observation. So authors could only hope that their readers would have the same entity in mind when reading a verbal description. On the other hand, metaphors distorted the description of an observation as long as elements called "walls" or "membranes" might be imagined as solid macroscopic structures.

Pictures therefore came into play as a seemingly neutral means of documentation. Draftsmen like Besemann were sometimes explicitly employed as technicians *not* trained in a particular scientific field so that they would furnish a mechanically neutral and "innocent" translation of drawings into prints without any intellectual intervention.[9] But this strategy did not circumvent the technical constraints of representation.

8 On the role of metaphor in science, see Uwe Pörksen, "Zur Metaphorik der naturwissenschaftlichen Sprache: Dargestellt am Beispiel Goethes, Darwins und Freuds," *Neue Rundschau* 89, no. 1 (1978): 64–82; Sabine Maasen and Peter Weingart, eds., *Metaphors and the Dynamics of Knowledge* (New York: Routledge, 2000); Armin Burkhardt and Brigitte Nerlich, eds., *Tropical Truth(s): The Epistemology of Metaphor and Other Tropes* (Berlin: de Gruyter, 2010). Cf. the metaphorology of George Lakoff and Mark Johnson, *Metaphors We Live By* (Chicago: University of Chicago Press, 2011).

9 Lorraine Daston and Peter Galison, *Objectivity*, rev. ed. (New York: Zone Books, 2010); Olaf Breidbach, "Representation of the Microcosm: The Claim for Objectivity in 19th-Century Scientific Microphotography," *Journal of the History of Biology* 35, no. 2 (Summer 2002): 223.

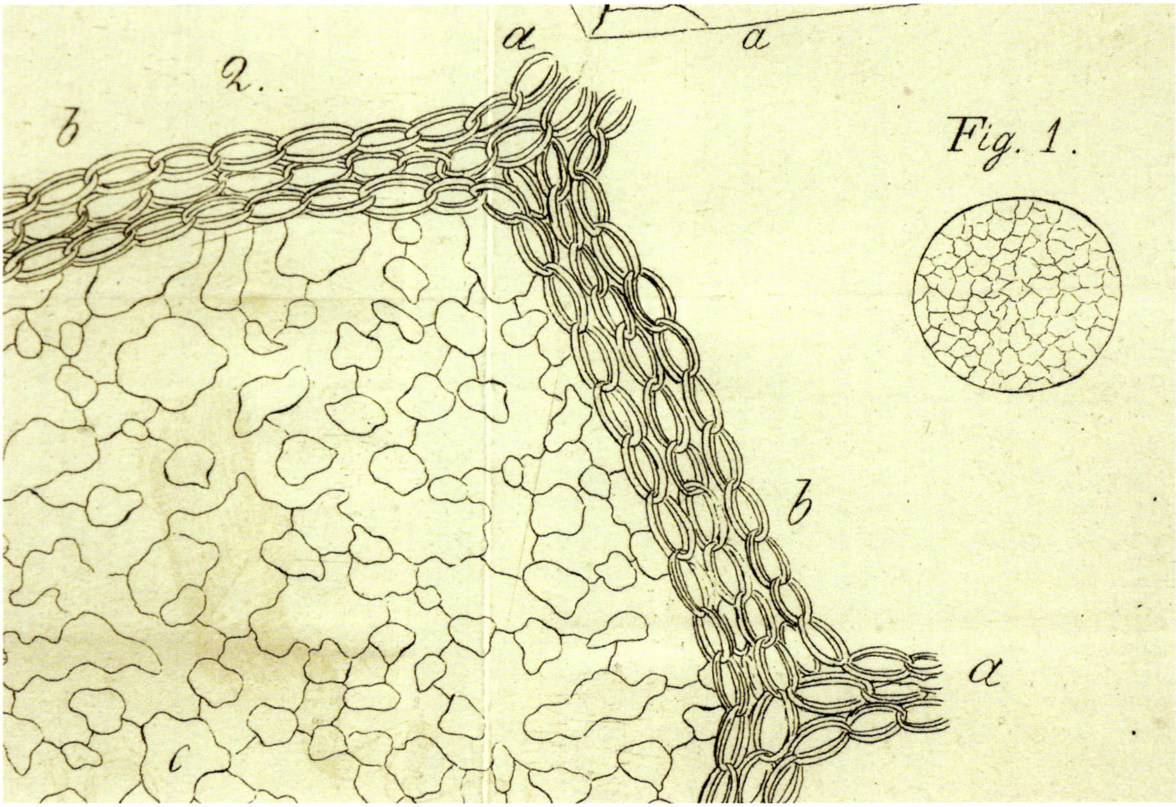

FIG. 2: Christian Besemann, etching from Carl Asmund Rudolphi, *Anatomie der Pflanzen* (Berlin, 1807). Detail of plate 4. Photograph by Barbara Herrenkind.

First, Besemann's workshop was far removed from the cities where his employers lived and worked, so he depended on additional information, such as letters accompanying the observers' drawings; in one instance, the effect was that he interpreted a row of cells, perhaps because it was verbally described as "chain-like," as an actual chain of interlocking rings (fig. 2).

Second, the choice of an artist always implied the choice of a technique or manner of representation. Illustrators like Besemann had to "read" complex pencil drawings and descriptions with the eyes of an expert in copperplate printing. At the time of the essay competition, techniques such as mezzotint, aquatint, and lithography would have been available, and they would have made it possible to translate the fuzzy parts of a microscopic image into a print that would imitate soft gradients or tender pencil sketches. But lithographs were too expensive for small print runs, and an expert like Besemann did not use them—nor did he experiment with the diverse possibilities of etching, which would have been affordable. We only know that he studied with Johann Ludwig Aberli, a Swiss who had introduced a method of printing in outlines to be subsequently colorized by hand.

Besemann thus brought a particular training, attitude, and taste to his work. He was a meticulous and scrupulous worker who handled the copperplate almost like the microtome itself. When asked by Rudolphi to depict the diagonal cut of a twig on the plate next to its magnified image, the purpose being simply to show the original size of the sample, he did not make it a mere circle or a scale but drew delicate details that exceed the grain or "resolution" of the paper fiber. The detail shown in figure 3 measures 1.5 millimeters and is almost invisible without a magnifying glass; it is a symbol of craftsmanship.

This manner of rendering objects may also have been influenced by a classicist doctrine of clarity and purity, an ideal taught in academies across Europe since the sixteenth century that postulated that the *disegno* can bring out the essential characteristics of an object and uncover its abstract structures. In combination with anatomical training, the "linear style" had an impact on scientific perception in that it avoided any distorted or ambiguous parts and rendered observations clear-cut and distinct, as is recognizable in the mid-seventeenth-century engravings after the anatomist Marcello Malpighi's subtle drawings (fig. 4).

Besemann was deeply rooted in this "copperplate era" of anatomical illustration, as the medical historian Goldschmid once called it,[10] and the style in which he worked may be read both as a personal *maniera* and as an expression of contemporary intellectual and material demands and circumstances. Today, the artist's name may be forgotten—it is hardly mentioned in the literature on scientific illustrators—yet in his day he was a sought-after expert.[11] The decision to hire someone like him may well have depended primarily on simple, practical factors such as availability, reliability, and price, but it had consequences for anatomical knowledge in a wider sense, as the accuracy and consistency of his illustrations influenced the subsequent discussion of cells and their walls, membranes, and communications—and may even have promoted the idea that cells are defined by continuous, impermeable borders.

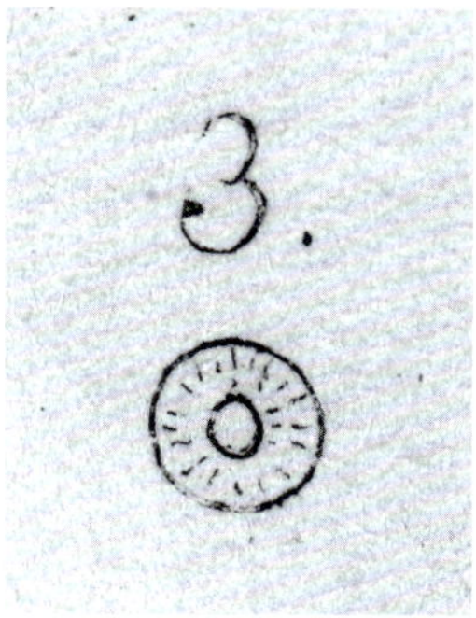

FIG. 3: Christian Besemann, etching from Carl Asmund Rudolphi, *Anatomie der Pflanzen* (Berlin, 1807). Detail of plate 3. Photograph by Barbara Herrenkind.

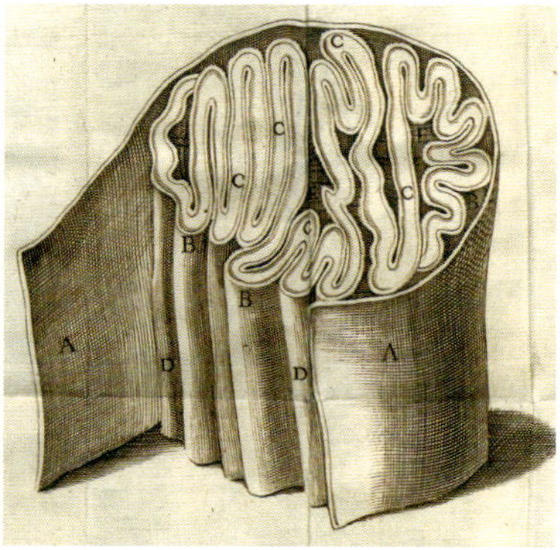

FIG. 4: Marcello Malpighi with Carlo Fracassati, *Tetras anatomicarum epistolarum de lingua, et cerebro* (Bologna, 1665). Illustration accompanying page 18 with microscopic cross section of a nerve fascicle. Gottfried Wilhelm Leibniz Bibliothek–Niedersächsische Landesbibliothek, N-A 1338, Pag. 46.

10 The German term is *Kupferstichzeit* (i.e., the era of Jakob Christof Le Blon and Jacques Fabian Gautier d'Agoty); see Edgar Goldschmid, *Entwicklung und Bibliographie der pathologisch-anatomischen Abbildung* (Leipzig: Hiersemann, 1925), 10.

11 He was offered a staff position at the University of Göttingen in 1806 to prevent him from accepting a similar offer from Russia. See Thomas Appel, "Biographische Ergänzungen zu dem Göttinger Zeichner und Kupferstecher Christian Andreas Besemann (1760–1818)," *Göttinger Jahrbuch* 51 (2003): 27–48.

Multiple Choice: Diversification and Reflection of Imaging Techniques

While art collections and curiosity cabinets underwent progressive transformation over the course of the eighteenth century into historical and typological arrangements—some were also published in the form of printed catalogues—the growing number of fragile items gathered during global expeditions, which were usually prone to decay and fading of their colors, demanded new efforts in conservation and visual documentation. At the same time, physiologists, astronomers, and artists experimented with all kinds of lenses, chemicals, colloids, and printing procedures to record natural phenomena or to observe and preserve compounds and specimens.[12] The phase of industrialization and European expansion thus became a breeding ground for new imaging technologies.

This coexistence of alternative techniques and strategies of representation might have prompted scholars to reconsider the epistemological importance of their own drawings or etchings, especially when the media proved ambivalent or dysfunctional. The growing selection of alternatives underlined that images have an epistemic potential not only when they are (un)true to nature but also when they differ *from the media previously used* and that visual tools not only illustrate certain concepts such as life, development, growth, or motion but in fact generate them, as questions to be answered visually within the frames of the medium a scholar employed.[13] Similarly, in mid-nineteenth-century historism, when a wide palette of styles (Romanesque, Gothic, etc.) forced architects to choose between different building languages, the presence of photography or lithography not only was a new visual experience, it also made it easier to understand that the hatching in a woodcut may be interpreted both as a representation of a shadow and as a fluted surface (fig. 5).

In the context of printing technologies, representation was thus being discussed as a means of translation, for instance in the prefaces of manuals and instruction leaflets designed to promote new printing processes, such as Johann Rudolf Schellenberg's *Short Treatise on the Art of Etching* (1795) and Alois Senefelder's 1809 textbook of lithography.[14] In light of such discussions, when photography became available to scientists in the course of the nineteenth century, drawings and woodcuts remained the preferred medium in many areas, used for emphasizing typical elements of an object or characteristics of a biological species[15]—hence the idea that both *τέχνη* and history play a role for the fabrication of nature.

Based on the widely held opinion that artists and biologists needed education in anatomy and that scholarship required basic skills in drawing and observation, the physician and drawing teacher Eduard Joseph d'Alton (1772–1840) was appointed "Professor of Natural and Art History" at the University of Bonn; his compatriot, the physician Johann Ludwig Choulant, edited the first collection of medical images in 1822,[16] followed by his *History and Bibliography of Anatomic Illustration as They Relate to the Science of Anatomy and the Visual Arts* (1852) and other works;[17] in 1925, the abovementioned Edgar Goldschmid (a collector of caricatures from Hogarth's to his own times) surveyed the genre in a history of the pathological illustration. Even Ludolf Treviranus, one entrant of the Göttingen contest, resolved, in the mid-1850s, to write the first history of the woodcut illustration in botany. While the gap between the artistic and the artificial, between arts and crafts, between art and science

12 See Larry J. Schaaf, *Out of the Shadows: Herschel, Talbot, and the Invention of Photography* (New Haven, CT: Yale University Press, 1992); Geoffrey Batchen, *Burning with Desire: The Conception of Photography* (Cambridge, MA: MIT Press, 1999); Kelley Wilder, *Photography and Science* (London: Reaktion Books, 2009); Peter Geimer, *Bilder aus Versehen: Eine Geschichte fotografischer Erscheinungen* (Hamburg: Philo Fine Arts, 2010); Chitra Ramalingam, "Fixing Transience: Photography and Other Images of Time in 1830s London," in *Time and Photography*, ed. Jan Baetens, Alexander Streitberger, and Hilde Van Gelder (Leuven: Leuven University Press, 2010), 3–26; Steffen Siegel, ed., "Fotografische Experimente," *Fotogeschichte* 31, no. 122 (Winter 2011).

13 Nick Hopwood, Simon Schaffer, and Jim Secord, eds., *Seriality and Scientific Objects in the Nineteenth Century* (Cambridge: Science History Publications, 2010); Janina Wellmann, *Die Form des Werdens: Eine Kulturgeschichte der Embryologie, 1760–1830* (Göttingen: Wallstein, 2010).

14 Johann Rudolf Schellenberg, *Kurze Abhandlung über die Aetzkunst* (Winterthur: Steinerische Buchhandlung, 1795); Alois Senefelder, *Vollständiges Lehrbuch der Steindruckerey [...]* (Munich: Thienemann; Vienna: Gerold, 1818).

15 Cf. Stephen Bann, *Parallel Lines: Printmakers, Painters and Photographers in Nineteenth-Century France* (New Haven, CT: Yale University Press, 2001).

16 Ludwig Choulant, *Tafeln zur Geschichte der Medicin nach Ordnung ihrer Doctrinen: Von den aeltesten Zeiten bis zum Schlusse des 18. Jahrhunderts* (Leipzig: Voss, 1822).

17 Ludwig Choulant, *Geschichte und Bibliographie der anatomischen Abbildung, nach ihrer Beziehung auf anatomische Wissenschaft und bildende Kunst [...]* (Leipzig: Weigel, 1852), published in English as *History and Bibliography of Anatomic Illustration*, trans. Mortimer Frank (Chicago: University of Chicago Press, 1920).

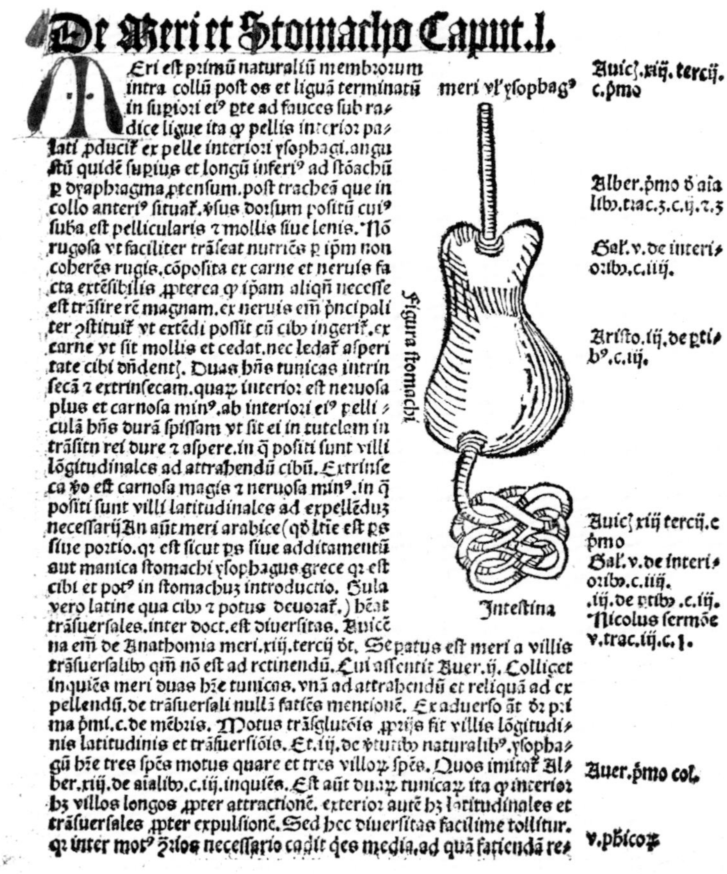

FIG. 5: Magnus Hundt, *Antropologium de hominis dignitate, natura et proprietatibus, de elementis, partibus et membris humani corporis* (Leipzig, 1501). Woodcut illustration of the digestive system, page 157. "Dream Anatomy," an online exhibition of the National Library of Medicine, http://www.nlm.nih.gov/exhibition/historicalanatomies/Images/1200_pixels/hundt_p157.jpg (accessed November 2013).

grew wider in the course of industrialization, scientific illustrations thus maintained the mutual relationship between areas like natural history and emerging disciplines like "art history."

Beyond the Masterpieces of Knowledge

Today, however, the names mentioned in the preceding pages are virtually unknown to the art-historical community, even though they illustrate how many connections once linked the discipline to other fields.
Art historians have discussed the development of forms as a process of adaptation and addressed the role of beauty or creativity in nature, on occasion even inspiring thinking in biology (the relationship has not been a one-way street, as is generally believed). Art history as a field has accompanied and integrated the growing research in color theory and optics, physiology and psychology; experienced the rise of illustrated mass media and the shockwaves of modern art; and produced a multifaceted theoretical literature, from Karl Rosenkranz's *Aesthetics of Ugliness* (1853) to Conrad Fiedler's and Heinrich Wölfflin's theories of vision, Aby Warburg's emotive iconology, Henri Focillon's and George Kubler's

histories of form, Rudolf Arnheim's concept of visual thinking, and integrative design concepts from art nouveau to the Bauhaus. The comparative approach as exemplified by Choulant or Goldschmid was also retransferred to the sciences by microbiologist Ludwik Fleck (1896–1961), who asserted in 1935 that "in science, just as in art and in life, only that which is true to culture is true to nature."[18] According to Fleck, changing forms of observation or representation were an effect of research mentalities he called "thought styles." Once again, "style" implied a political dimension.

18 Ludwik Fleck, *Genesis and Development of a Scientific Fact* [1935], trans. Fred Bradley, ed. Thaddeus J. Trenn and Robert K. Merton (Chicago: University of Chicago Press, 1979), 35. See also Anna Wessely, "Transposing 'Style' from the History of Art to the History of Science," *Science in Context* 4, no. 2 (September 1991): 265–78.

All these approaches took pictures into consideration that ranged far beyond the narrow realm of gallery art and expanded the scope of scholarship tremendously. Due to their own methodology, art historians—as live witnesses to the creative process of "art"—were thus confronted with the problem of quantity. Under the influence of the history of styles, they sought to blend archaeological methods of identification and connoisseurship with modern aesthetics and questions of social taste or national heritage with philosophies of history in order to clear paths through the abundance of human productivity.

Considered in this light, recent claims that modernity is helpless in the face of an overwhelming visual productivity are a retread of a long tradition. If issues pictorial or visual affect every field of science and technology, then umbrella terms like "visuality" might be too general and unspecific. It may be impossible to compile a comprehensive survey of the differentiated and complex imagery that has already evolved in the sciences and technology (let alone in the arts and entertainment).[19] Given the visual output consisting of millions of illustrations published per year or the immense computerized image production in physics or medicine, it will not do to historicize a few paramount icons of science and technology (such as the diagram of the DNA, Lennart Nilsson's photographs of the human embryo, or the pictures taken during NASA's Apollo missions) to explain how the sciences use images or what their impact on society is, especially since, behind each of these icons, there are innumerable test pictures, recordings, and visualizations that are unpublished, unreadable, or lost in oblivion.[20] The field of *imaging science*, for instance, an early type of *Bildwissenschaft*, emerged in the 1980s as an endeavor to find ways to handle the digital data behind those "images" in the face of ephemeral visualizations and constantly changing file formats and operating systems. It is a new variant of "archaeology" or "art history," even if the research is done by trained physicists and information scientists.

19 See, e.g., the surveys in Andreas Beyer and Markus Lohoff, eds., *Bild und Erkenntnis: Formen und Funktionen des Bildes in Wissenschaft und Technik* (Munich: Deutscher Kunstverlag, 2005); James Elkins, ed., *Visual Practices across the University* (Munich: Fink, 2007). On the history of technical depictions, see Hans Holländer, ed., *Erkenntnis, Erfindung, Konstruktion: Studien zur Bildgeschichte von Naturwissenschaften und Technik vom 16. bis zum 19. Jahrhundert* (Berlin: Mann, 2000).

20 A question already implicitly raised in James Elkins, *The Domain of Images* (Ithaca, NY: Cornell University Press, 1999); Matthias Bruhn, ed., *Bilder ohne Betrachter, Bildwelten des Wissens, Kunsthistorisches Jahrbuch für Bildkritik*, vol. 4, no. 2 (Berlin: Akademie, 2006).

In the nineteenth and early twentieth centuries, art history no doubt made numerous disciplines more aware of the virtues and relativity of their own visual tools; today, it might remind them that visual knowledge can also emerge from a historical mining process that recollects, rediscovers, and preserves visual tools and materials once considered "worthless" because of the progressive specialization of the arts and sciences. Formal comparison will remain a fundamental aspect of this mining work, which begins with the identification and classification of recurring forms, formats, and genres of visual information and continues with the description of techniques, contexts, and groups of application.

Following Caroline Jones and Peter Galison's fundamental book

Picturing Science, Producing Art (1998), which devotes a special section to a discussion of the issue of style, we might thus say that today's imaging technologies and technical drawings do not need to be defined as another form of high-brow art in order to be taken as seriously as high art is.[21] Rather, the question would be what "taking seriously" means and how it can be achieved. This was a question once raised by scholars who wished to trace the histories of artifacts in order to learn more about the principles of their design and in so doing constantly returned to the discussion of what their object of investigation actually was. The boundaries keep shifting.

21 Caroline A. Jones and Peter Galison, *Picturing Science, Producing Art*, with contributions by Irene Winter and Carlo Ginzburg (New York: Routledge, 1998); see also Anja Zimmermann, *Ästhetik der Objektivität: Genese und Funktion eines wissenschaftlichen und künstlerischen Stils im 19. Jahrhundert* (Bielefeld: Transcript, 2009).

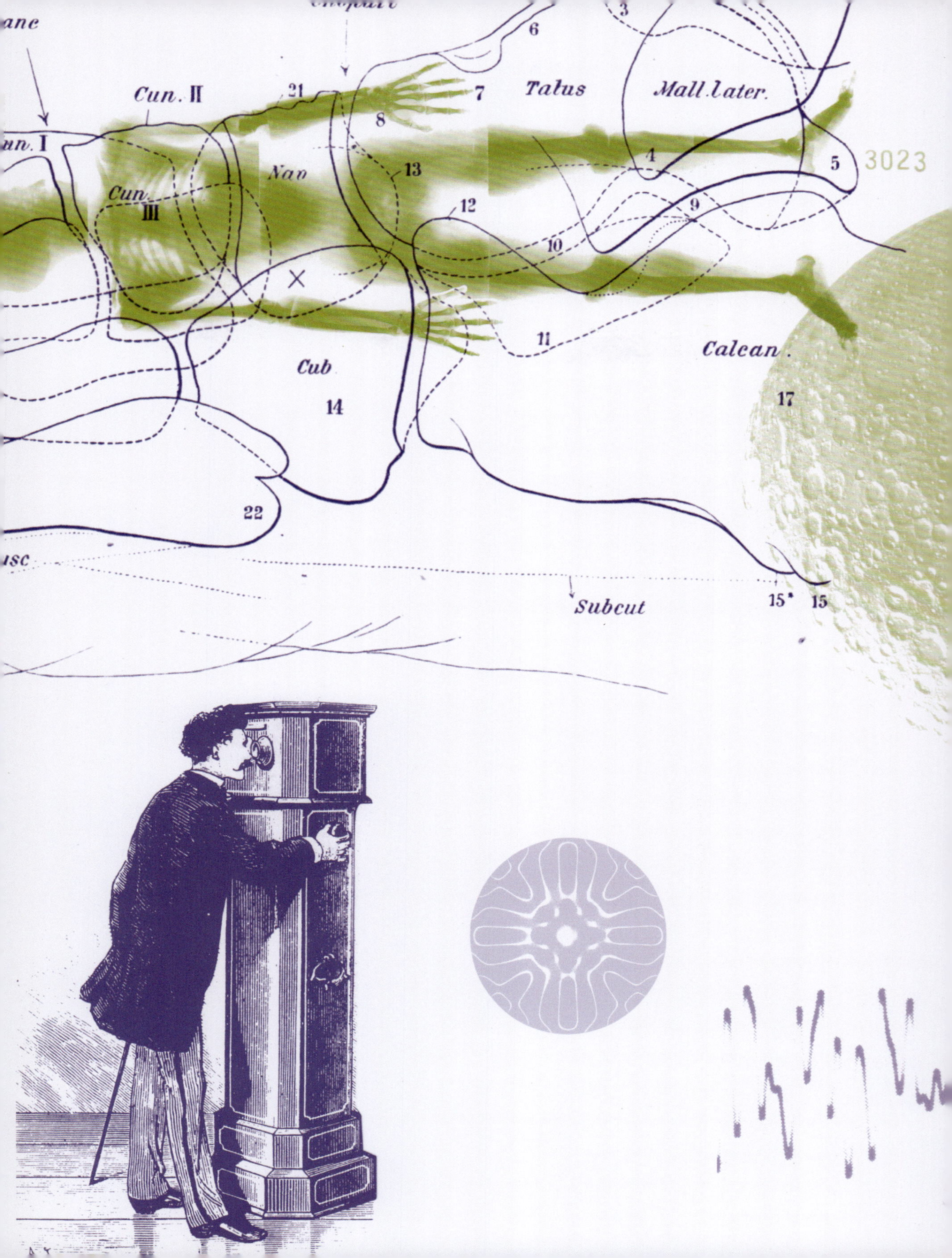

anc
Cun. II
21
un. I
Cun
III
Nan
13
12
6
3
7
8
Talus
Mall. later.
4
5
3023
9
10
11
Calcan.
Cub
14
17
22
isc
Subcut
15*
15

Case Studies

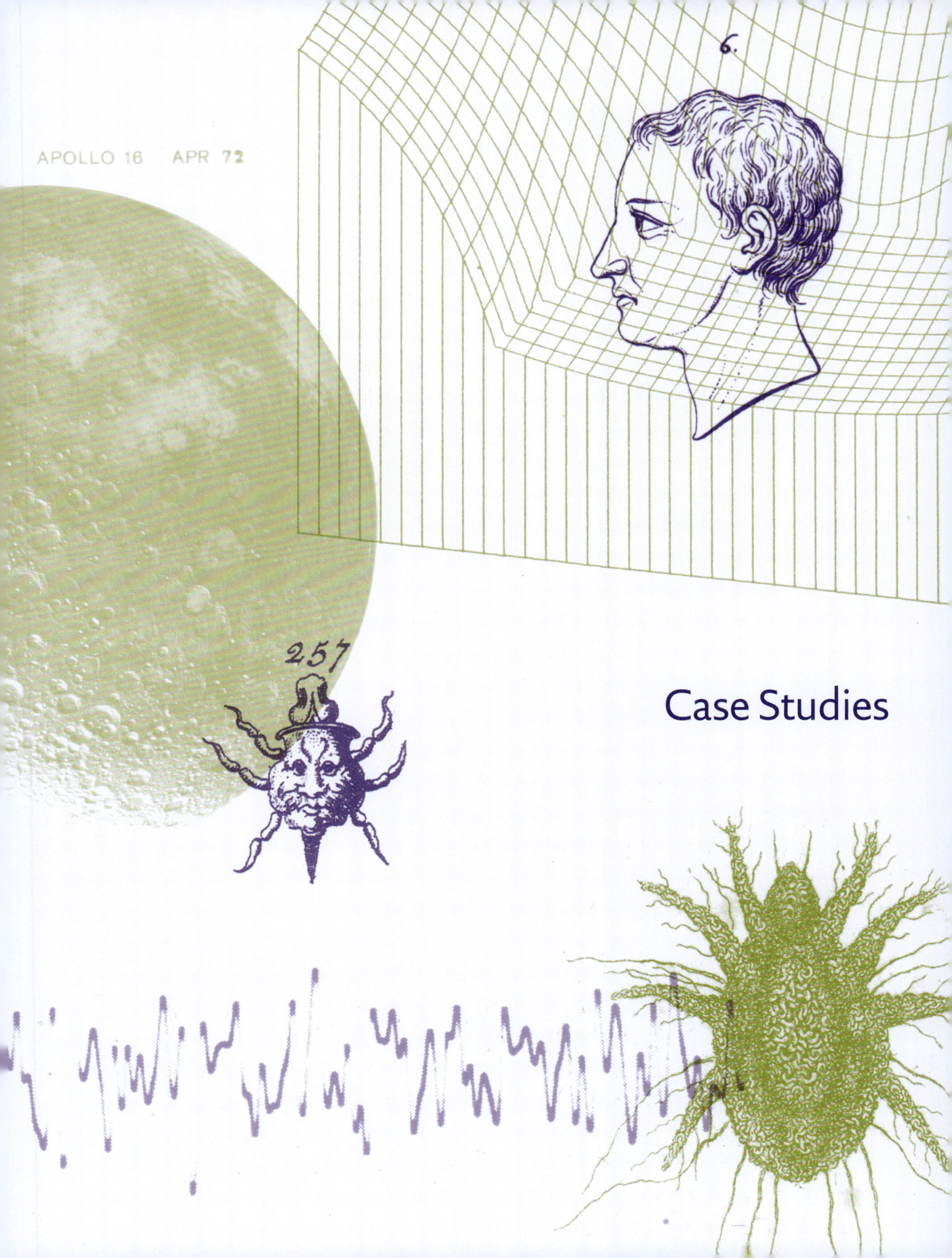

Interacting with Images—Toward a History of the Digital Image: The Case of Graphical User Interfaces

Margarete Pratschke

The development of the graphical user interface marks a turning point in the history of the computer: what had been a machine for the use of experts became a mass medium that anyone can operate intuitively. The graphical user interface replaced the programmer's knowledge and command-line control with graphical interaction, which is to say, with a technology-specific iconic competence. Since the early 1980s, the user's communication with the machine has increasingly moved into pictorial structures that appear on the screen as the interface of interaction. This shift from text to image as the medium of interaction triggered a massive rise in the use of computers. Underlying this technological success story is the idea that the image's advantage over text and complex language consists in its particular ability to convey facts with greater ease and clarity, to be legible intuitively and at a glance.

The standardized iconic structures of graphical user interfaces, which profoundly inform almost all facets of our everyday digital lives, exemplify the basic structure of the digital image and its historical evolution. As the most widespread implementation of computer graphics, graphical user interfaces are, moreover, examples of digital imagery that are not generated by digitizing pictures—the formal appearance of such images, a form of data output, may always be altered—but genuinely engendered by the computer as a medium and displayed on the screen in a firmly defined form. Individual elements of this visual organization of the pictorial interface of the screen, such as overlapping windows, have given rise to a specific style and evolved, distinct formal-aesthetic features whose visual logic is not explained solely by their intended purpose.[1]

The Display as an Interactive Image

When Apple announced the launch of its Macintosh computer in an eighteen-page advertisement in *Newsweek* in late 1983, the copy read: "Of the 235 million people in America, only a fraction can use a computer. Introducing Macintosh. For the rest of us" (figs. 1a and 1b).[2] The slogan "For the rest of us" made it plain: this computer, unlike earlier machines, was not meant for technology experts alone. The feature designed to attract new users, "the rest of us," was a specific form of interaction: the graphical organization of the user interface, i.e., the screen, as an interactive *iconic* surface (fig. 2).[3] The Macintosh's user interface presented the following structural pictorial elements predefined by the operating system: within the confines of the monitor, the user saw a menu bar along the top edge of the display as well as, before a gray backdrop, individual icons, a black arrow indicat-

1 For a comprehensive history of graphical user interfaces, see Margarete Pratschke, *Windows als Tableau: Zur Bildgeschichte grafischer Benutzeroberflächen* (Zurich: Diaphanes, 2014).

2 On the Apple Macintosh, see Steven Levy, *Insanely Great: The Life and Times of Macintosh, the Computer That Changed Everything* (New York: Viking, 1994).

3 The illustrations show screenshots of the respective systems, which inevitably eliminate the interactive aspect of the interfaces. On the problem of visual sources and the reconstruction and preservation of digital imagery, see the debate over the conservation of software: Doron Swade, "Preserving Software in an Object-Centered Culture," in *History and Electronic Artifacts*, ed. Edward Higgs (Oxford: Clarendon, 1998), 195–206; Martin Campbell-Kelly, "Think Piece: Software Preservation; Accumulation and Simulation," *Annals of the History of Computing* 24, no. 1 (January–March 2002): 95–96; John G. Zabolitzky, "Preserving Software: Why and How," *Iterations* 1 (2002): 1–8.

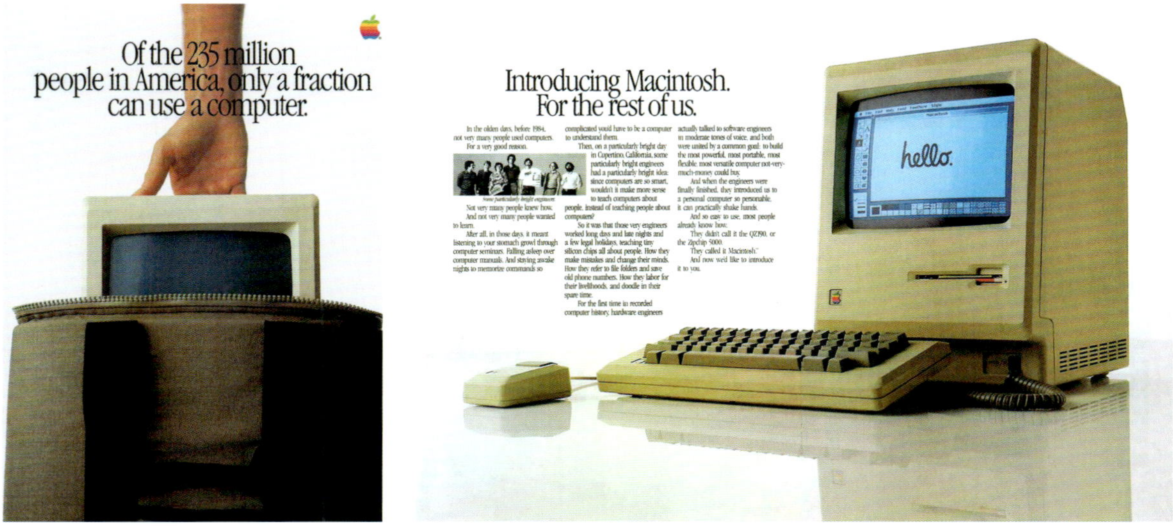

ing the position of the mouse, and two distinct rectangular white image fields—one overlapping or clipping the other—with a rich internal visual and typographic structure. The layering of these fields and the drop shadows visible along the bottom and right edges of their frames generated an illusion of spatial depth. The display became an image, an iconic space—an extraordinarily successful image or iconic structure whose overall appearance was defined by the arrangement of overlapping windows.

FIGS. 1a, 1b: Cover and spread from a twenty-page insert in *Newsweek* advertising the Apple Macintosh, Fall 1984. http://www.macmothership.com/gallery/Newsweek/p001.jpg, 002.jpg, and 003.jpg (accessed November 2013).

Before the 1980s were over, several other window-based graphical user interfaces hit the market.[4] The growing similarities between the basic structures of these visual systems led Apple to wage a legal battle against Microsoft for several years; the lawsuit concerned in particular the feature of overlapping windows, which Microsoft had introduced with its operating system Windows 2.0 in 1987 (fig. 3). In 1992, a California court dismissed Apple's suit against Microsoft, finding that the graphical user interface and the disputed individual elements constituted a general

4 See Martin Campbell-Kelly, *From Airline Reservations to Sonic the Hedgehog: A History of the Software Industry* (Cambridge, MA: MIT Press, 2003), esp. 231–66. Microsoft did not singlehandedly determine the evolution of window-based operating systems; see Martin Campbell-Kelly, "Not Only Microsoft: The Maturing of the Computer Software Industry, 1982–1995," *Business History Review* 75, no. 1 (Spring 2001): 103–45.

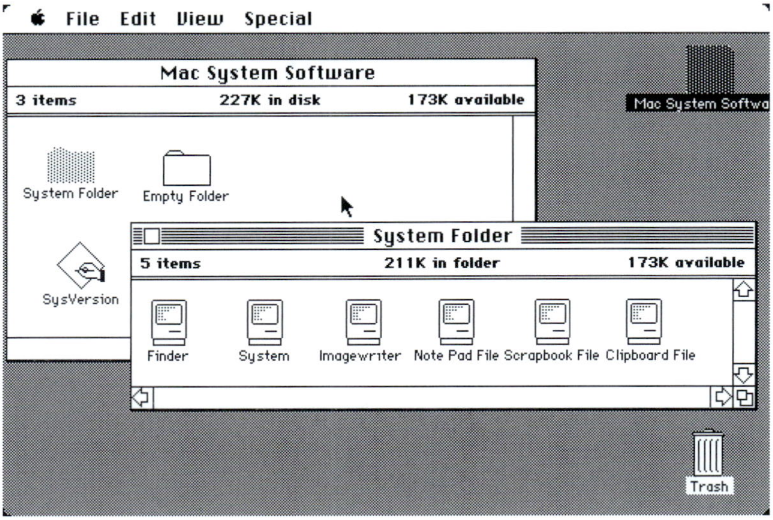

FIG. 2: Screenshot of the Apple Macintosh graphical user interface, 1984. http://toastytech.com/guis/macos1.html (accessed November 2013). Image courtesy of Nathan Lineback, www.toastytech.com.

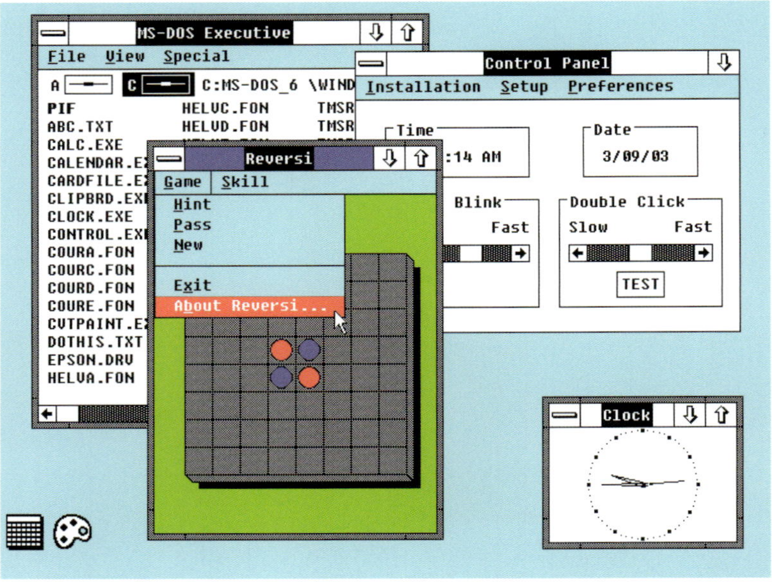

FIG. 3: Screenshot of the Microsoft Windows 2.0 graphical user interface, 1987. http://www.guidebookgallery.org/screenshots/win203 (accessed November 2013). Used with permission from Microsoft.

principle that had become the industry standard and was therefore not entitled to copyright protection.[5] The visual "look and feel" of the interfaces, their formal structures, had evolved into a veritable genre of digital imagery that enabled users not only to interact with the operating system but also to run all applications and perform a wide range of tasks such as writing documents or editing digital photographs.[6] The foundations for this form of interactive imagery, which forms a sort of visual infrastructural system uniting diverse applications, were laid at the Xerox PARC research laboratory in the 1970s.

Technological Competence through/as Iconic Competence

In 1970, Xerox, a manufacturer of photocopiers and office equipment, had set up the Palo Alto Research Center (PARC) in order to explore the potential of a computer-aided "office of the future." The central driving force among the scientists employed at PARC that created new ways of human-computer interaction was Alan Kay, head of the Learning Research Group. This team established the principles of visual interaction with the computer and the elements of the graphical user interface in connection with the development of the programming language Smalltalk, which was designed as a learning environment.[7]

At the time, the so-called time-sharing systems, large-capacity computers shared among many users, were still the standard model; working with computers was the exclusive province of engineers and programmers.[8] Interaction with these mainframes and the earliest microcomputers, which were becoming viable competitors since the 1970s thanks to rising processor performance, took place via terminals, using complex programming languages and command-line input.[9] One early milestone of the turn from textual and programming-language-based to visual interaction, however, had been achieved in 1963: Ivan Sutherland's Sketchpad demonstrated basic features of a form of graphical interaction conducted by drawing.[10]

5 *Apple Computer, Inc. v. Microsoft Corp.*, 799 F. Supp. 1006 (N.D. Cal. 1992), opinion by Judge Vaughn R. Walker.

6 On the transformation of different tasks into structurally similar images and the associated synchronization of cultural techniques in graphical user interfaces, see Margarete Pratschke, "Jockeying Windows: Die bildräumlichen Strukturen grafischer Benutzeroberflächen als visuelle Grundlage von Multitasking," in *Multitasking: Synchronität als kulturelle Praxis*, exhibition catalogue (Berlin: Neue Gesellschaft für Bildende Kunst, 2007), 16–24.

7 For an overview of the work of the Learning Research Group and the development of the programming language Smalltalk, see Alan Kay and Adele Goldberg, "Personal Dynamic Media," *IEEE Computer* 10, no. 3 (March 1977): 31–41; on the history of Alan Kay's work, see his own conspectus in Alan Kay, "The Early History of Smalltalk," *ACM SIGPLAN Notices* 28, no. 3 (March 1993): 69–95; cf. Michael Friedewald, *Der Computer als Werkzeug und Medium: Die geistigen und technischen Wurzeln des Personal Computers* (Diepholz: GNT-Verlag, 1999), 249–61, 311–35.

8 For a general introduction to the history of computers, see Paul E. Ceruzzi, *A History of Modern Computing*, 2nd ed. (Cambridge, MA: MIT Press, 2003); on the history of the personal computer, see the excellent discussion in Friedewald, *Der Computer als Werkzeug und Medium*.

9 For a history of ideas on human-computer interaction, see Jörg Pflüger, "Konversation, Manipulation, Delegation: Zur Ideengeschichte der Interaktivität," in *Geschichten der Informatik: Visionen, Paradigmen, Leitmotive*, ed. Hans Dieter Hellige (Berlin: Springer, 2004), 367–408.

10 Ivan Edward Sutherland, "Sketchpad: A Man-Machine Graphical Communication System" (PhD diss., MIT, 1963), also available online with a new preface by Alan Blackwell and Kerry Rodden as University of Cambridge Computer Laboratory Technical Report, vol. 574, www.cl.cam.ac.uk/techreports/UCAM-CL-TR-574.html; cf. Frieder Nake, "The Display as a Looking Glass: Zu Ivan E. Sutherlands frühen Visionen

Starting in 1972, Xerox PARC developed the Alto, a powerful computer for single-user operation (a personal computer) to be run on a local network. The Alto was equipped with a keyboard, a mouse, and a (bit-mapped) graphic display screen and served the Learning Research Group in its various research projects and in the development of Smalltalk. The screenshot of an early Smalltalk interface running on the Alto around the year 1975 (fig. 4) shows four white rectangles of different sizes both framed and internally subdivided by simple dark lines; the individual image fields are superimposed and seem to clip or overlap one another. In order to make up for the fairly limited size of the Alto's screen—so-called tiled windows that divided the display into fields did not resolve this problem satisfactorily[11]—Kay developed the analogy between the individual document or application display areas and sheets of paper one might layer or stack on a table: the so-called overlapping windows. The basic underlying idea is that the screen is like a desktop in an office environment where papers are laid out or filed away. The visual principle over which Apple and Microsoft fought in the courts in the 1980s, that is to say, originated with neither of the two parties; its roots are in Kay's conception of graphical interaction for untrained users and in Xerox PARC's work on Smalltalk.

As early as 1969, Alan Kay had sketched his vision not only of what computers should be able to do but also, and more importantly, of who

grafischer Datenverarbeitung," in *Geschichten der Informatik*, ed. Hellige, 339–65. See also André Reifenrath, *Die Geschichte der Simulation: Geschichte des Computers von den Anfängen bis zur Gegenwart unter besonderer Berücksichtigung des Themas der Visualisierung und Simulation durch den Computer* (PhD diss., Humboldt-Universität zu Berlin, 2000).

11 The partitioning of the screen for different applications in "windows" appearing side by side was pioneered by Douglas Engelbart's NLS System (ca. 1968). On Engelbart, see Thierry Bardini, *Bootstrapping: Douglas Engelbart, Coevolution, and the Origins of Personal Computing* (Palo Alto, CA: Stanford University Press, 2000); cf. Friedewald, *Der Computer als Werkzeug und Medium*, 191–220.

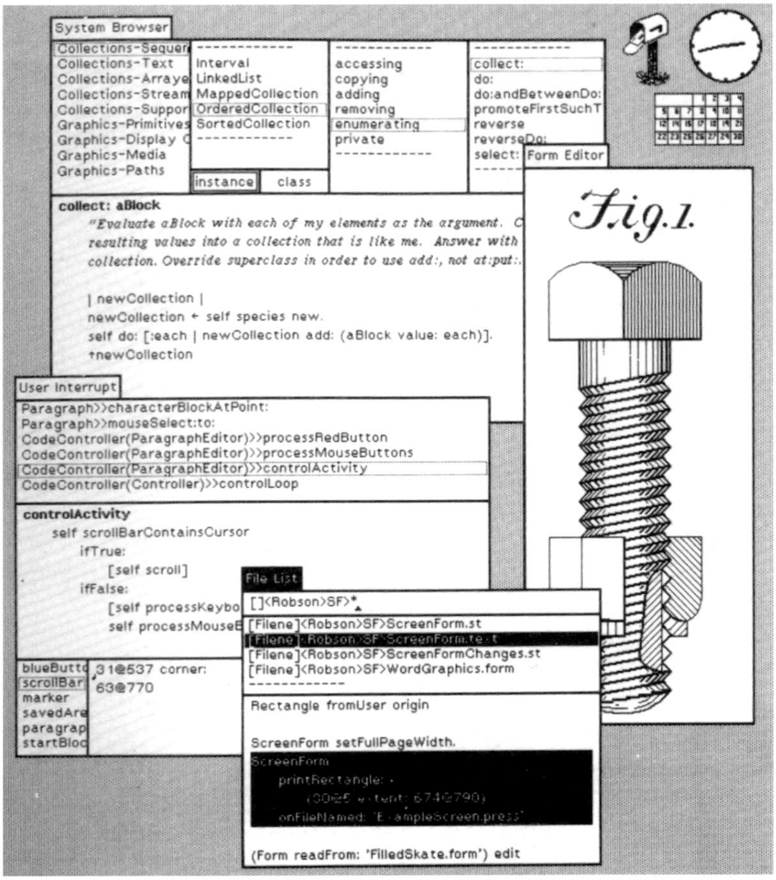

FIG. 4: Screenshot of the Smalltalk graphical user interface running on the Alto, ca. 1975. Butler Lampson, "Personal Distributed Computing: The Alto and Ethernet Software," in Adele Goldberg, ed., *A History of Personal Workstations* (New York: ACM, 1988), 317, fig. 1. Photo courtesy of PARC, a Xerox company.

should be their future users: the computer, he wrote, "must be simple enough so that one does not have to become a systems programmer (one who understands the arcane rites) to use it."[12] The goal of leaving those arcane rites of programming specialists behind also motivated the work of Kay's Learning Research Group at Xerox PARC; the programming language Smalltalk was designed to simplify interaction, opening up computer technology to new audiences.[13] In focusing on the use of Smalltalk by children, who were frequently involved in experimental trial interactions, the Learning Research Group drew heavily on pedagogical theories of learning and developmental psychology in the work of people such as Jean Piaget, Jerome Bruner, Maria Montessori, and John Dewey. Kay wrote, "Millions of potential users meant that the user interface would have to become a learning environment along the lines of Montessori and Bruner." The means to this end were "large scope, reduction in complexity, and end-user literacy."[14]

Based on Jerome Bruner's and Jean Piaget's stage models of the child's cognitive development, which distinguished between an enactive (or sensorimotor), an iconic (or pre-operational), and finally, a symbolic developmental stage, Kay's concept of how the medium would be learned and used intuitively grew out of the endeavor to design interaction in such a way that it would address not only the user's symbolic capability but also his sensorimotor and especially his iconic cognitive faculties, his "iconic mentality."[15] In order to make the computer accessible to broader user groups, Kay and his collaborators conceived a far-ranging program of image-based interaction including, in particular, what they called "iconic programming," which was designed to reduce the complexity of programming and facilitate engagement with the medium.[16]

In their initial reflections on the interaction between "novices" or "non-experts" and computers, concepts such as the desktop metaphor seem to have played a subordinate role; more important were issues of representation and illusion—concepts, that is to say, that are central to the theory of the image. This focus is also evident in the Learning Research Group's interest in theories of perception and gestalt theory, most prominently the works of the art historians Rudolf Arnheim and Ernst H. Gombrich.[17] An important expression of the influence of these concepts of visual thinking is the motto—it almost sounds like a phenomenological aphorism—that "what you see is what you get" (WYSIWYG),[18] which Kay described as a guiding idea in the design of a graphical interface: "Perhaps the most important principle is WYSIWYG: the image on the screen is always a faithful representation of the user's illusion. Manipulating the image in a certain way immediately does something predictable to the state of the machine (as the user imagines the state)."[19] Yet Kay not only places emphasis on the generation of an "illusion" by means of the "image on the screen" but goes further by calling for a "new aesthetic"[20] as well as "magical moments" that are to be an integral feature of the interface—and these moments, he believes, are precisely not brought into existence by metaphors: "One of the most compelling snares is the use of the term *metaphor* to describe a correspondence between what the users see on the screen and how they should think about what they are manipulating. My main complaint is that *metaphor* is a poor metaphor for what needs to be done. At PARC we coined the phrase *user illusion* to describe

12 Alan Kay, "The Reactive Engine" (PhD diss., University of Utah, 1969), 75.

13 See Adele J. Goldberg, "The Community of Smalltalk," in *The Handbook of Programming Languages*, vol. 1: *Object-Oriented Languages*, ed. Peter H. Salus (Indianapolis: Macmillan Technical Publications, 1998), 51–94, esp. 59.

14 Kay, "Early History of Smalltalk," 69. See also Seymour Papert's experiments at MIT and especially his programming language Logo, which had a crucial influence on the work of the Learning Research Group; on Logo, see Seymour Papert, *Mindstorms: Children, Computers, and Powerful Ideas*, 2nd ed. (New York: Basic Books, 1980).

15 On Alan Kay's reading of Jerome Bruner's "Toward a Theory of Instruction" (1966), which gave rise to the motto "Doing with images makes symbols," see Alan Kay, "User Interface: A Personal View," in *The Art of Human-Computer Interface Design*, ed. Brenda Laurel (Reading, MA: Addison-Wesley, 1990), 191–207.

16 On the principle of "iconic programming," see David C. Smith, *Pygmalion: A Computer Program to Model and Stimulate Creative Thought* (Basel: Birkhäuser, 1977).

17 Margarete Pratschke, "Die grafische Benutzeroberfläche als Bild: Zur Rezeption von Rudolf Arnheim und Ernst Gombrich in der Computer Science der 1970er Jahre," in "Nicht-künstlerische Bilder," ed. Christiane Kruse and Sabine Kampmann, *Kritische Berichte* 37, no. 4 (2009): 54–63.

18 The conceptual blueprint underlying the Apple Macintosh in particular shows the strong influence of concepts from the theory of perception and gestalt theory. Apple invoked its "look and feel" and explicitly referred to "gestalt theory" not only in marketing the computer but also in the abovementioned lawsuit against Microsoft, where they figured in the argument that the Macintosh's interface met the standard of originality. For a more extensive discussion of the influence of gestalt theory on designs for graphical human-machine interaction, see Pratschke, *Windows als Tableau*, ch. 2.

19 Alan Kay, "Computer Software," *Scientific American* 251, no. 3 (September 1984): 42.

20 Ibid., 43.

what we were about when designing user interface. [...] It is the magical part that is all-important and that must be most strongly attended to in the user interface design."[21]

By calling for a fundamental aesthetic as well as a "magical illusion" while also stipulating that a correspondence must tie what the user sees to its cognitive referent, Kay is implicitly alluding to other fundamental qualities associated with the image: pictorial magic and representationality, or analogy formations.

Pictorial magic is the term art historians use to address a certain peculiar intrinsic logic or semantic surplus in the image[22] that would seem at odds with the concrete and instrumental purposes of simplified human-machine interaction and the idea that learning to engage in such interaction is the clear-cut acquisition of a technological competence; yet theories of mimetic representation are no less critical to the question of learning with images. The turn from text to image in the interaction with the computer is based on the assumption that images are easier to become conversant with than text owing to the relation of similarity that links them to a specific referent or fact they "portray," such that they may be understood intuitively, at a glance.[23] Such fundamental qualities of images cited by Kay indicate that the graphical user interface is an "image"—a "visual form," not a "metaphorical desktop." *Pace* those media scholars who claim that there is no such thing as a digital pictorial register due to the ephemeral and mutable nature of digital data,[24] examining this digital phenomenon with the art historian's eye reveals that such a register exists and is the subject of theoretical reflection: there is a specific digital form. Graphical user interfaces constitute a genre of digital imagery that, since its beginnings in the 1970s, has evolved stable formal features. The heuristic purchase of this argument may be buttressed by comparing the form of digital imagery to conventional analog images.

Disorderly Image Arrangements

From their earliest days at Xerox PARC to the present, the form of graphical user interfaces has been defined by the principle of overlapping windows: rectangular image fields are arranged within the pictorial format delimited by the screen's frame (fig. 5).[25] The formal appearance of these superimpositions and overlaps involving a large number of rectangular "windows" on the display creates an impression of pictorial depth, although the windows or image fields are not rendered in perspective; there is no diminution or foreshortening. Instead, the individual windows seem to float in front of and behind one another without spatially interacting, moving on planes at different depths whose definition in space remains vague. On the one hand, the overall spatial impression of this graphical interface is fragmented and collage-like; on the other hand, the arrangement is fairly orderly, dominated by orthogonal lines—the rectangular shapes are not dissolved or irreparably disintegrated, as in (analog) physical collages in the visual arts. The partially collaged overall form may be undone by a digital operation; at any time, the user's interaction may reestablish visual order and a focus on a particular window or arrange several intact windows in neat juxtaposition. In this regard, the visual form of the user interface owes less to the collage in general than to issues involving the principles governing the arrangement of pictures

21 Kay, "User Interface," 199.

22 On "pictorial magic" as an implicit critique of a worldview committed to rationalism and terminological specificity, see the summary discussion in Gerhard Wolf, "Bildmagie," in *Metzler Lexikon Kunstwissenschaft: Ideen, Methoden, Begriffe*, ed. Ulrich Pfisterer (Stuttgart: Metzler, 2003), 46–48. For the semantic-aesthetic surplus of the image over language and manifoldness, equivocation, sensory concreteness, and polyvalence as qualities of the image, see the summary discussion in Gottfried Boehm, "Jenseits der Sprache? Anmerkungen zur Logik der Bilder," in *Iconic Turn: Die neue Macht der Bilder*, ed. Christa Maar and Hubert Burda (Cologne: DuMont, 2004), 28–43.

23 For a critique of the theory of resemblance, see Oliver Scholz, *Bild, Darstellung, Zeichen: Philosophische Theorien bildlicher Darstellung*, 2nd ed. (Frankfurt am Main: Klostermann, 2004). On knowledge acquisition through images more generally, see Bernd Weidenmann, ed., *Wissenserwerb mit Bildern: Instruktionale Bilder in Printmedien, Film/Video und Computerprogrammen* (Bern: H. Huber, 1993); Steffen-Peter Ballstaedt, "Bildverstehen, Bildverständlichkeit: Ein Forschungsüberblick unter Anwendungsperspektive," in *Wissenschaftliche Grundlagen der technischen Kommunikation*, ed. Hans Peter Krings (Tübingen: Gunter Narr, 1996), 191–233. See also the fundamental discussion in Karsten Heck, ed., *Bildendes Sehen, Bildwelten des Wissens: Kunsthistorisches Jahrbuch für Bildkritik*, vol. 7, no. 1 (Berlin: Akademie, 2009).

24 See Claus Pias, "Das digitale Bild gibt es nicht: Über das (Nicht-)Wissen der Bilder und die informatische Illusion," *zeitenblicke* 2, no. 1 (2003), www.zeitenblicke.de/2003/01/pias/.

25 This technology, which is associated with multitasking on comparatively small screens, was universal until the arrival of so-called mobile devices such as smartphones and tablets. This development seems to herald the end of the era defined by the interface operating on the basis of windows, icons, menus, and pointers and the dawn of a new form of tactile interaction that is supported through distinct ways of graphic design of the display.

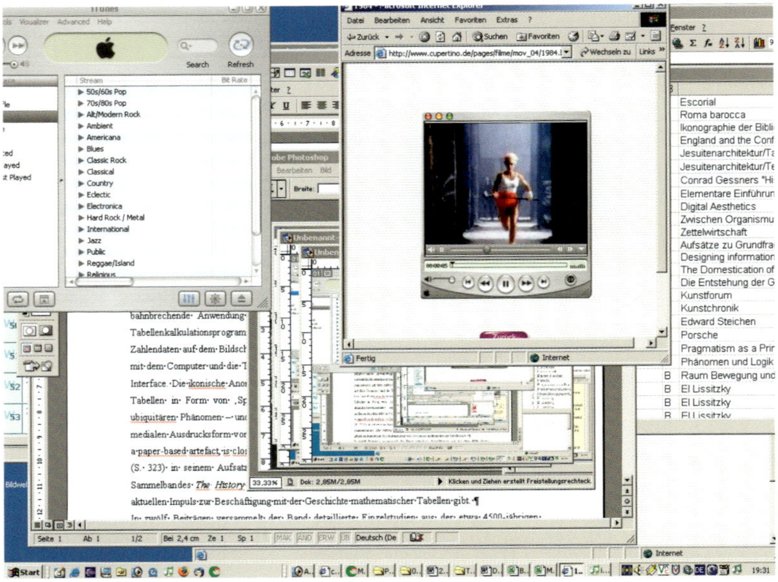

FIG. 5: Screenshot of the Microsoft Windows 2000 graphical user interface with multiple overlapping windows, 2006. Screenshot by the author. Used with permission from Microsoft.

and the representational principle of tableaux for purposes of survey or comparison, particularly if we also consider the predecessor model to the overlapping windows, the juxtaposed or tiled windows, which, like the classical tableau, presented individual pictorial elements side by side.[26]

In formal as well as theoretical terms, moments of interaction between users-beholders and tableaux were a focus of interest in the classical avant-garde of the twentieth century's first three decades. As the pictures themselves are set in motion or the beholder alters their arrangement by moving between vantage points and perspectives, the impression of the pictorial space oscillates between detailed views and synopses, between two and three dimensions. Experimental methods of arranging pictures in time and space play with the beholder's disorientation, spurring him to engage in more vigorous interaction.

The architect Ludwig Mies van der Rohe, rather than relying solely on sketches and models as he designed views of his future buildings, created collages that highlighted the iconic staging of his architecture.[27] In 1939, for instance, he made a collage for Resor House, a vacation home in Wyoming, that visualized the view of a river valley from the building. The architecture itself is only hinted at by the parts of the picture left blank; the sketch focuses on the spatial impression created by three intersecting rectangles. A strip of a nature photograph appears in the background; on the left, a color reproduction of a painting by Paul Klee cuts across it, in front of which, further to the right, a flat, wide rectangle of wood laminate is set as an element of interior architecture. The view from the window, the painting, and the partition are shown as free-floating panels of equal importance; their position in the architectural space remains vague. A play of distance and perspective unfolds, perceived by the beholder as a picture whose illusionistic potential oscillates between two and three dimensions. Mies van der Rohe's architecture itself is designed to foster the perception of staggered layers by a beholder in motion, who may always revise the choices of perspective implicit in the collage.

26 With the principle of tiled windows, unlike with overlapping windows, the individual windows may only be arranged side by side. See Sara A. Bly and Jarrett K. Rosenberg, "A Comparison of Tiled and Overlapping Windows," in *Proceedings of the SIGCHI Conference on Human Factors in Computing Systems*, March 15–17, Gaithersburg, Maryland, United States (New York: ACM Press, 1986), 101–6; Brad Myers, "Window Interfaces: A Taxonomy of Window Manager User Interfaces," *IEEE Computer Graphics and Applications* 8, no. 5 (September 1988): 65–84.

27 On Mies van der Rohe, see Wolf Tegethoff, "Zur Entwicklung der Raumauffassung im Werk Mies van der Rohes," *Daidalos* 13 (1984): 114–23; Neil Levine, "Die Bedeutung der Tatsachen: Mies' Collagen aus nächster Nähe," *Arch+* 146 (April 1999): 59–75. See also Ulrich Müller, *Raum, Bewegung und Zeit im Werk von Walter Gropius und Ludwig Mies van der Rohe* (Berlin: Akademie, 2004).

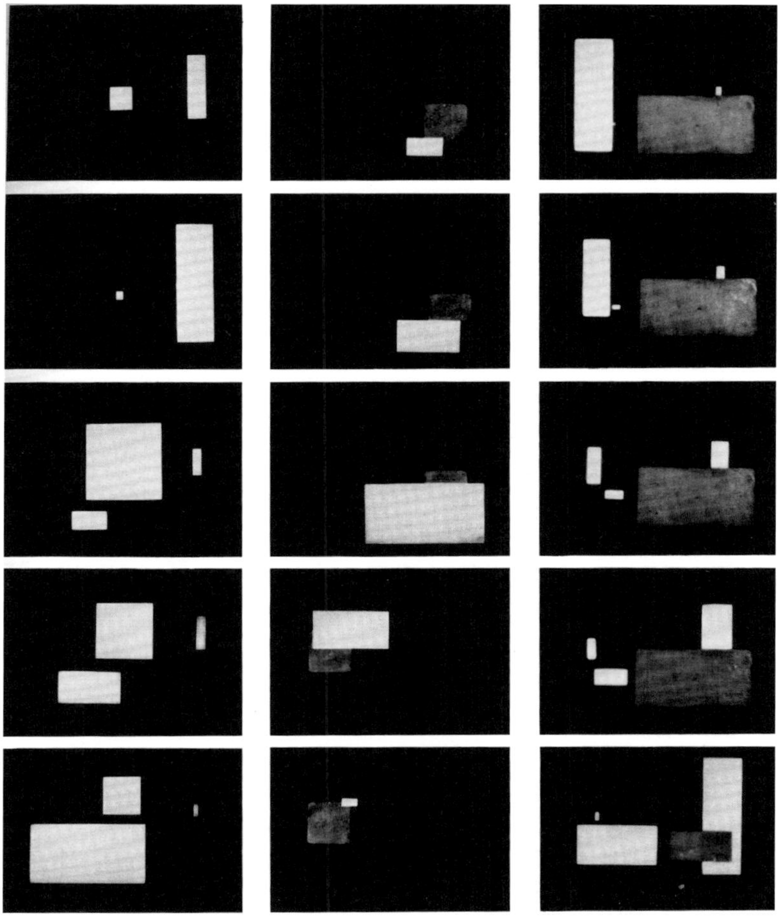

FIG. 6: Hans Richter: "Rhythmus 21," film stills, 1921. *Hans Richter,* with an introduction by Sir Herbert Read (Neuchâtel: Griffon, 1965), 31. © Richter Estate c/o Art acQuest.

Experimental filmmakers likewise explored the possibilities opened up by infusing the motif of floating or swaying rectangles, and hence the impression of pictorial space, with dynamism. As early as the 1920s, the painter Hans Richter began experimenting with film, creating, among other works, the abstract animated sequence "Rhythmus 21" (1921; fig. 6).[28] The entire film, which runs for about three minutes, shows rectangular shapes in shades ranging from white to black appearing in different numbers and formations and moving across the screen at varying velocities. They sway toward the viewer and recede again, fuse, or split. Set before a black or white ground, such expansion and disappearance generates strong illusions of pictorial depth. The ambivalence of the spatial situation in which the different rectangular shapes move builds steadily over the duration of the film; the viewer is invariably occupied with observing which shapes relate to one another in which ways, which figures or configurations constitute the foreground and background. The extraordinarily complex impression made by the picture as a whole makes it virtually impossible to get a firm grasp of the spatial structure of what is seen. To Richter, the crucial point was that film made it possible to create a "motion art" that rendered the "simple juxtaposition of shapes in and of themselves meaningless."[29] He was interested not only in the

28 Jeanpaul Goergen, ed., *Hans Richter: Film ist Rhythmus* (Berlin: Freunde der Deutschen Kinemathek, 2003); see also Stephen C. Foster, ed., *Hans Richter: Activism, Modernism, and the Avant-Garde* (Cambridge, MA: MIT Press, 1998).

29 Hans Richter, "Prinzipielles zur Bewegungskunst," *De Stijl* 4, no. 7 (1921): 109–12, repr. in Goergen, ed., *Hans Richter: Film ist Rhythmus*, 18–20.

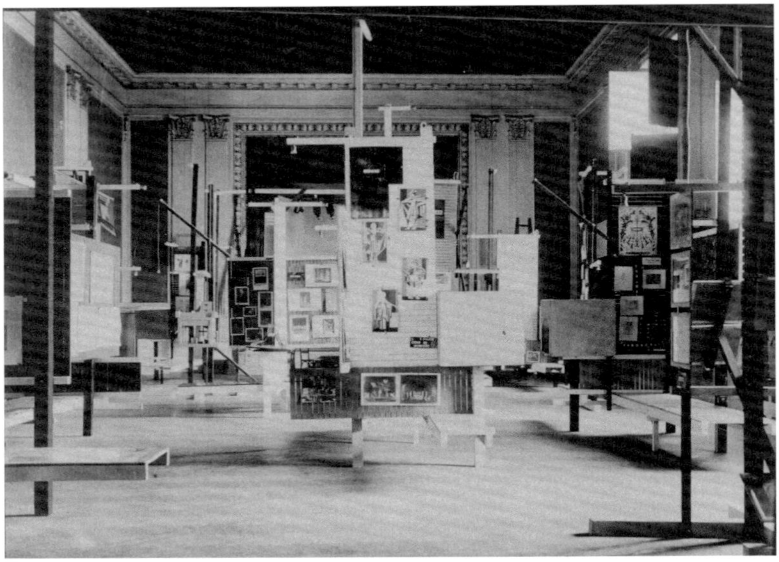

FIG. 7: Frederick Kiesler, *International Exhibition of New Theater Technique*, Vienna, 1924, installation view. © 2014 Austrian Frederick and Lillian Kiesler Private Foundation, Vienna.

simultaneity and synthesis of visual impressions but also, more generally, in creating a "universal language" that would overcome national language barriers by means of visual abstraction.[30]

Such constructivist collage, which operated in a visual register between architecture and film, is manifest also in another format that, rather than producing mere illusions, lent them physical reality in the form of an installation: the exhibition display system. In 1924, Frederick Kiesler designed a novel installation system for the *International Exhibition of New Theater Technique* held in Vienna, the "Leger- und Tragersystem," or "system of settings and supports" (fig. 7).[31] Freestanding wooden racks support horizontal and vertical panels made of wood slats that hold pictures and objects. The impression the visitors receive as they enter the room is that of a collage of orthogonal elements, a pictorial installation of overlapping structures. The beholders may interact with the installations thanks to flexible and height-adjustable parts that allow them to bring the pictures and objects to eye level. Kiesler also investigated the principles of human-object and human-environment interaction in further projects and elaborated a theory of "correalism," his term for the aspect of his work in architecture and design that addressed "the dynamics of continual interaction between man and his natural and technological environments."[32]

Iconic Criticism

The discussion of these three examples from the era of the avant-garde was not meant to construct a causal or genealogical relation between modernism and the conceptions of graphical interaction with the computer. Comparison across media as well as different moments in history of the various intentions behind structurally similar arrangements of images is a heuristic tool that enables us to arrive at a more precise analysis of the seemingly evanescent visual phenomena of digital world. On a more fundamental level, it also allows us to bring out the *iconic* nature of this digital phenomenon.

30 Hans Richter, "Universelle Sprache" (1920); for quotations from the lost original, see Goergen, ed., *Hans Richter: Film ist Rhythmus*, 17.

31 Friedrich Kiesler, *Internationale Ausstellung neuer Theatertechnik*, exhibition catalogue (Vienna: Würthle, 1924); Friedrich Kiesler, "Ausstellungssystem: Leger und Trager," *De Stijl* 6, nos. 10–11 (1924/1925): 138–41; cf. Christoph Grunenberg, "Espaces spectaculaires: l'art de l'installation selon Frederick Kiesler," in *Frederick Kiesler: Artiste-architecte*, exhibiton catalogue (Paris: Éditions du Centre Pompidou, 1996), 103–13; *Friedrich Kiesler: Art of This Century*, exhibition catalogue, ed. Österreichische Friedrich-und-Lilian-Kiesler-Privatstiftung (Ostfildern-Ruit: Hatje Cantz, 2003). On modernist exhibition conceptions, see Mary Anne Staniszewski, *The Power of Display: A History of Exhibition Installations at the Museum of Modern Art* (Cambridge, MA: MIT Press, 1998).

32 Frederick J. Kiesler, "On Correalism and Biotechnique: A Definition and Test of a New Approach to Building Design," *Architectural Record*, September 1939, 61.

The formal structure of digital window systems resembles that of modernist constructivist collages designed to disorient the beholder. The historic parallelism of form reveals how problematic and positively contradictory the methods of arrangement employed in user interfaces are when considered in light of the original intention to simplify: disorder and irritation are universal features of constructivist collages. It is part of the basic disposition of images and the interaction with them that they convey their peculiar intrinsic logic as well. With regard to the interface, it seems, Alan Kay was perfectly aware of the polysemy of images: "One of the most interesting puzzles in iconic programming (and iconic communication in general) is why there is such a disparity between how understandable images are while they are actively being constructed and how obscure even one's own constructions can appear even the next day."[33] He even emphasizes that "the second factor is a property of images in general—their unsortedness. In other words, unlike paragraphs and lists of words, images have no a priori order in which they should be understood. This means that someone coming onto an image from the outside has no strategy for solving it."[34] Although Kay affirms the "unsortedness" of images—they do not in themselves contain instructions to guide the attempt to understand them—his refusal to acknowledge an a priori iconic quality seems to miss an important point.

This a priori quality of any image, however, is rooted in its diachronic and historically evolved intrinsic logic and its adaptation to the beholder's "structural intuition," which Martin Kemp describes as an anthropological constant whose particular cultural manifestation is determined by factors such as stored visual experience and, more importantly, by deep structures and their realization in sensual experience and familiarity.[35] Today, we may respond to widespread criticisms of graphical user interfaces, their utility and visual style—they are seen as impenetrable and difficult to use—by drawing heuristic comparisons and building knowledge of historical examples of similar graphical procedures in order to arrive at a better understanding of the problems that beset the interaction with (digital) imagery.[36] Issues of the "usability" of software are more and more dependent on visual components of their user interfaces, as is evident in calls for "aesthetic computing" or software "design."[37] Beyond the mere drawing of analogies, placing the form of graphical user interfaces in the historical context of modernism's structural imagery not only points up their genuine aesthetic; it also allows for a more nuanced understanding of their (dys)functional quality, between imagery designed for instrumental orderly purposes and the general disorderliness of images.

33 Kay, "User Interface," 202.

34 Ibid.

35 Martin Kemp, introduction to *Visualizations: The Nature Book of Art and Science* (Oxford: Oxford University Press, 2000), 1–5; cf. Horst Bredekamp, Angela Fischel, Birgit Schneider, and Gabriele Werner, "Bildwelten des Wissens," in *Bilder in Prozessen*, ed. Horst Bredekamp and Gabriele Werner, *Bildwelten des Wissens: Kunsthistorisches Jahrbuch für Bildkritik*, vol. 1, no. 1 (Berlin: Akademie, 2003), 19; Horst Bredekamp, "Drehmomente: Merkmale und Ansprüche des Iconic Turn," in *Iconic Turn*, ed. Maar and Burda, 23–24; Horst Bredekamp, "Kunsthistorische Erfahrungen und Ansprüche," in *Bild und Medium: Kunstgeschichtliche und philosophische Grundlagen der interdisziplinären Bildwissenschaft*, ed. Klaus Sachs-Hombach (Cologne: Halem, 2006), 11–26.

36 Margarete Pratschke, "Why History Matters: Visual Innovation and the Role of Image Theory in HCI," in *Design, User Experience, and Usability: Theory, Methods, Tools and Practice*, vol. 1, ed. Aaron Marcus (Heidelberg: Springer, 2011), 277–84.

37 Terry Winograd, ed., *Bringing Design to Software* (New York: ACM Press, 1996); Paul Fishwick, ed., *Aesthetic Computing* (Cambridge, MA: MIT Press, 2006).

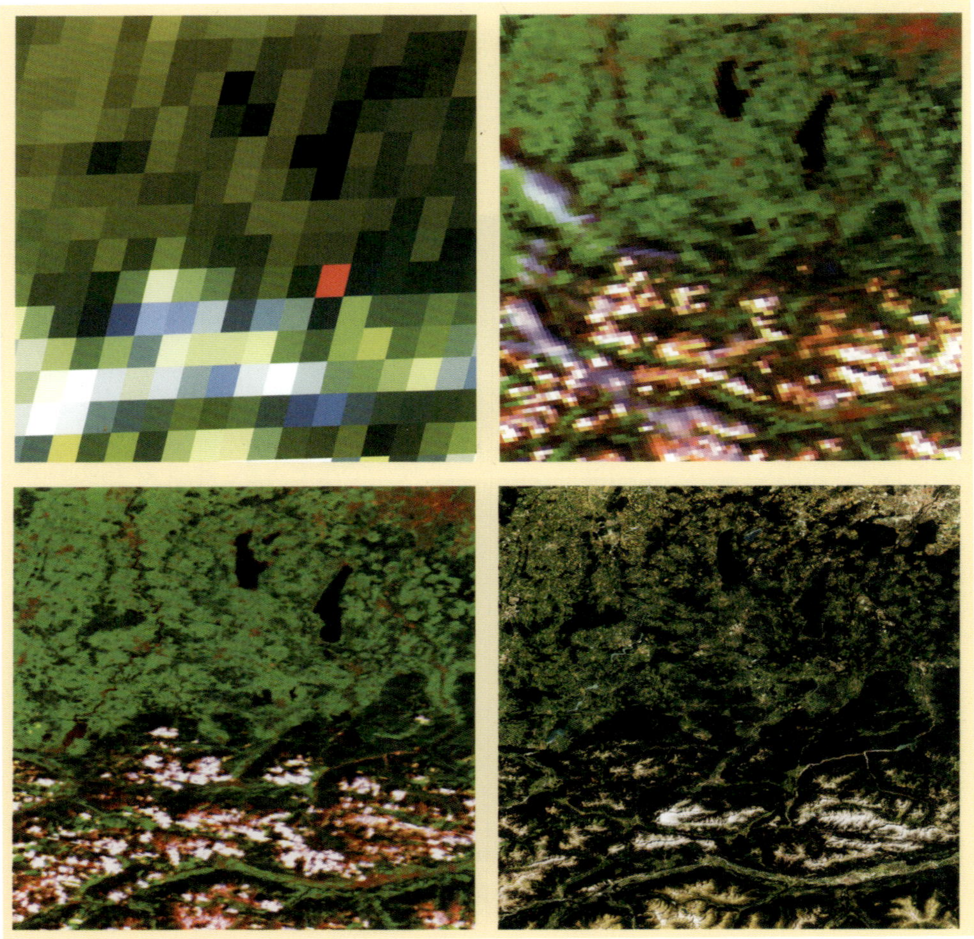

DIGITAL IMAGES

Several definitions have been offered for the term *digital*; authors have drawn on information theory, semiotics, or the conceptual history of the term, often contrasting it with *analog* (e.g., Goodman 1968; Seitter 2002; Schröter 2004). In the context of information theory, the digital is defined by segmentation, discretization, and coding, while the analog is regarded as continuous and dense. This division grew out of the history of media and technology and includes a history of technical images. It assumes that digital processing involves a semiotic process entirely different from that of analog processing. Today, however, scholars rarely postulate such a stark opposition in the sense of a historical era of the analog subsequently superseded by digitalization. Various works on media history have instead shown that this kind of succession from analog to digital does not apply in particular cases. Digital processes were already widespread before the advent of the computer, and mixed forms in many ways thwart the attempt to draw neat theoretical distinctions (Andriopoulus and Dotzler 2002; Zielinski 2002; Schneider 2007).

Defining the digital in the context of scientific imagery raises the question of the specific significance of computer-aided digital image production and ties together approaches from various disciplines. The account of what is specifically digital is subject to ongoing revision as the forms and functions of digital imagery vary—given the diversity of applications, ranging from Google Earth to digital satellite photography (fig. 1) and from computer tomography to climate simulation models, it is impossible to define the digital image once and for all, just as there is no single answer to the question "What is an image?" Dynamic and interactive visual products and simulated spaces further challenge the fundamental concept of the image.

FIG. 1 (opposite): Images from four different satellites showing the same view (an area south of Munich). Different resolution capabilities of satellite cameras are suitable to different applications, such as weather forecasting (lower resolution) or land use (higher resolution). From left to right: 5 × 5 km / 1 × 1 km / 250 × 250 m / 30 × 30 m per pixel. The higher the resolution, the longer the computing time. The current maximum resolution is 5 cm. (Images: DLR 2002.) Walter Hauser, ed. *Klima: Das Experiment mit dem Planeten Erde.* Exhibition catalogue (Darmstadt: Wissenschaftliche Buchgesellschaft, 2002): 79, fig. 2.1.4 a–d. a) © DLR / EUMETSAT. b) © DLR / NOAA. c) © DLR / NASA. d) © DLR / USGS.

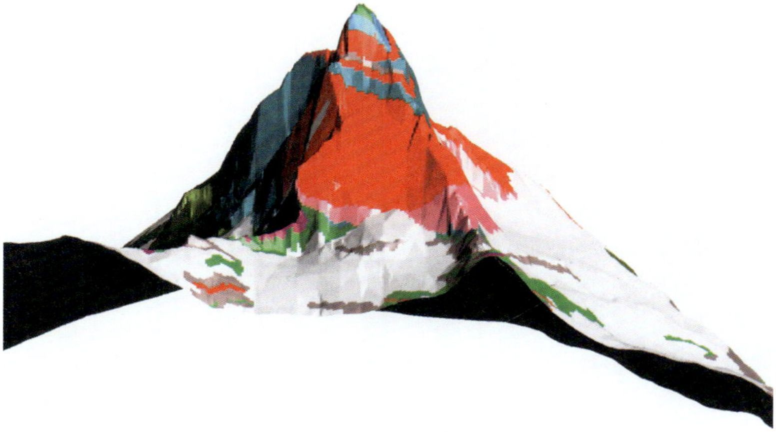

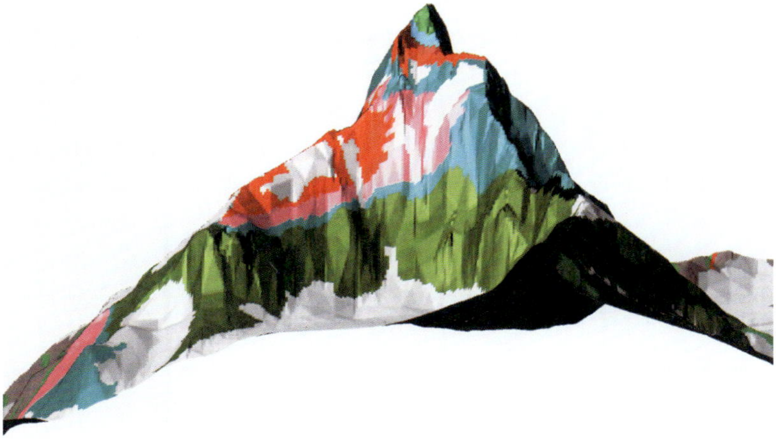

FIG. 2 (right): The geology of the Matterhorn in a digital elevation model made in 2003 at the Swiss Federal Institute of Technology, Zurich. The combination of a three-dimensional elevation model with geological data has enabled the production of a representation that, unlike printed maps, offers various options in terms of viewing angles. The top image is a view of the Matterhorn's geology from the east, while the bottom image shows it from the west. © ETH Zürich, 1995; Digitales Höhenmodell DHM25: Bundesamt für Landestopografie swisstopo, Schweiz.

So how can the focus on the digital be rendered productive for an analysis of its epistemic effects on scientific images? Beyond the many attempts to evaluate digital images in ontological terms, which range between the poles of deception and virtuality—Jean Baudrillard's work is paradigmatic of the genre—the study of scientific imagery should analyze the significance of digitalization techniques with respect to their referentiality, generation, and uses and with a view to the levels of phenomena (fig. 2), equipment, operational steps, and algorithms. If scientific images are instrumental in nature, this aspect must be studied in the case of digital images as well; the operative aspects of these images need to be considered. So we must not only subject the form in which images are presented to visual analysis and evaluate it, for example, with the tools of stylistic history (Pratschke 2005). Images must also be related to each step along the long chain of translations of measurements and raw data and to their application contexts and utility. Concepts such as the "computed" or the "algorithmic image" (Kittler 2001) have revealed the deeper programming dimension of digital images. Consideration of the resulting images alone—to the extent that the results even are images in the narrow sense—would fall victim to the obfuscating tendency of (digital) images: the manipulation and production processes involved are not necessarily discernible in the finished product. Yet this tendency should not be lamented as the fundamental loss of a putative truth of images; it must be soberly regarded as a property of digital image processing. Focusing on algorithms, operations, decisions concerning parameters and threshold values, and the interactive uses of images in diagnostics, research, or evaluation (figs. 3 and 4), the study of digital imagery may penetrate to the core

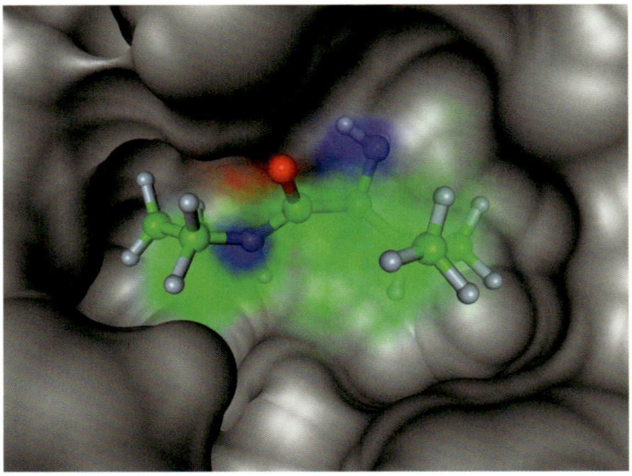

FIG. 3 (left): This image shows the result of a simulation of the so-called invariant density of a diabetes mellitus type 2 agent in the binding pocket of a protein, dipeptidyl peptidase 4 (DPP4). DPP4 deactivates certain hormones in the gastrointestinal tract that are desirable in patients with diabetes mellitus type 2 because they stabilize blood sugar levels. Computer-aided agent design is used to identify a molecule that will inhibit the effect of the protein and may be used as medication in therapy. The protein is shown as a surface that approximately represents its external form. The agent is symbolized by its atoms, which are color-coded by element, and the bonds between them. The colored cloud around the agent visualizes its probability of presence within the protein's binding pocket. Alexander Bujotzek and Marcus Weber, "Efficient Simulation of Ligand-Receptor Binding Processes Using the Conformation Dynamics Approach," *J. Bioinform. Comput. Biol.*, no. 5 (2009): 811–31.

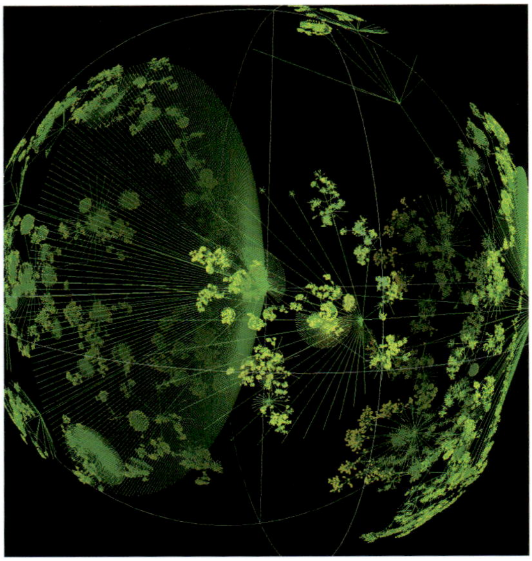

FIG. 4 (right): A network consisting of 535,102 nodes and 601,678 links is visualized using the graph visualization program Walrus. Unlike programs such as Chart Tools, which translate spreadsheet calculations into conventional diagram forms, programs like Walrus are designed to provide three-dimensional forms to represent especially large and complex data networks and trees that may be interactively turned and rotated. The image renders the structure of the network, not its content. The software was developed by the Cooperative Association for Internet Data Analysis (CAIDA) in San Diego, California, in 2008. © 2003 The Regents of the University of California. All Rights Reserved.

of scientific visualization. Such analysis of generative processes and uses highlights the crucial significance of the ways in which digital imaging practices have changed scientific work, of the point at which data is visualized, and of the ways in which digital processes are re-shaping knowledge as well as the value that is added by digitalizing already conventionalized processes such as diagrammatics (fig. 5) and cartography. —BS

LITERATURE

Adelmann, Ralf, Jan Frercks, Martina Hessler, and Jochen Hennig, eds. *Datenbilder. Zur digitalen Bildpraxis in den Naturwissenschaften*. Bielefeld: Transcript, 2009.

Andriopoulus, Stefan, and Berhard Dotzler, eds. *1929: Beiträge zur Archäologie der Medien*. Frankfurt am Main: Suhrkamp, 2002.

Böhnke, Alexander, and Jens Schröter, eds. *Analog/Digital—Opposition oder Kontinuum? Zur Theorie und Geschichte einer Unterscheidung*. Bielefeld: Transcript, 2004.

Dencker, Klaus Peter, ed. *Weltbilder, Bildwelten: Computergestützte Visionen*, Interface 2. Hamburg: Verlag Hans-Bredow-Institut für Rundfunk und Fernsehen, 1995.

Flusser, Vilém. *Into the Universe of Technical Images*. Minneapolis: University of Minnesota Press, 2011.

Foley, James D., et al. *Computer Graphics: Principles and Practice*. 3rd ed. Reading, MA: Addison-Wesley, 2009.

Franke, Herbert W. *Computer Graphics: Computer Art*. London: Phaidon, 1971.

Friedberg, Anne. *The Virtual Window: From Alberti to Microsoft*. Cambridge, MA: MIT Press, 2006.

Goodman, Nelson. *Languages of Art: An Approach to a Theory of Symbols*. Indianapolis: Bobbs-Merrill, 1968.

Helmerdig, Silke, and Martin Scholz. *Ein Pixel, zwei Korn: Grundlagen analoger und digitaler Fotografien und ihre Gestaltung*. Frankfurt am Main: Anabas, 2006.

Heßler, Martina. "Von der doppelten Unsichtbarkeit digitaler Bilder." *Zeitenblicke* 5, no. 3 (2006). www.zeitenblicke.de/2006/3/Hessler (January 2012).

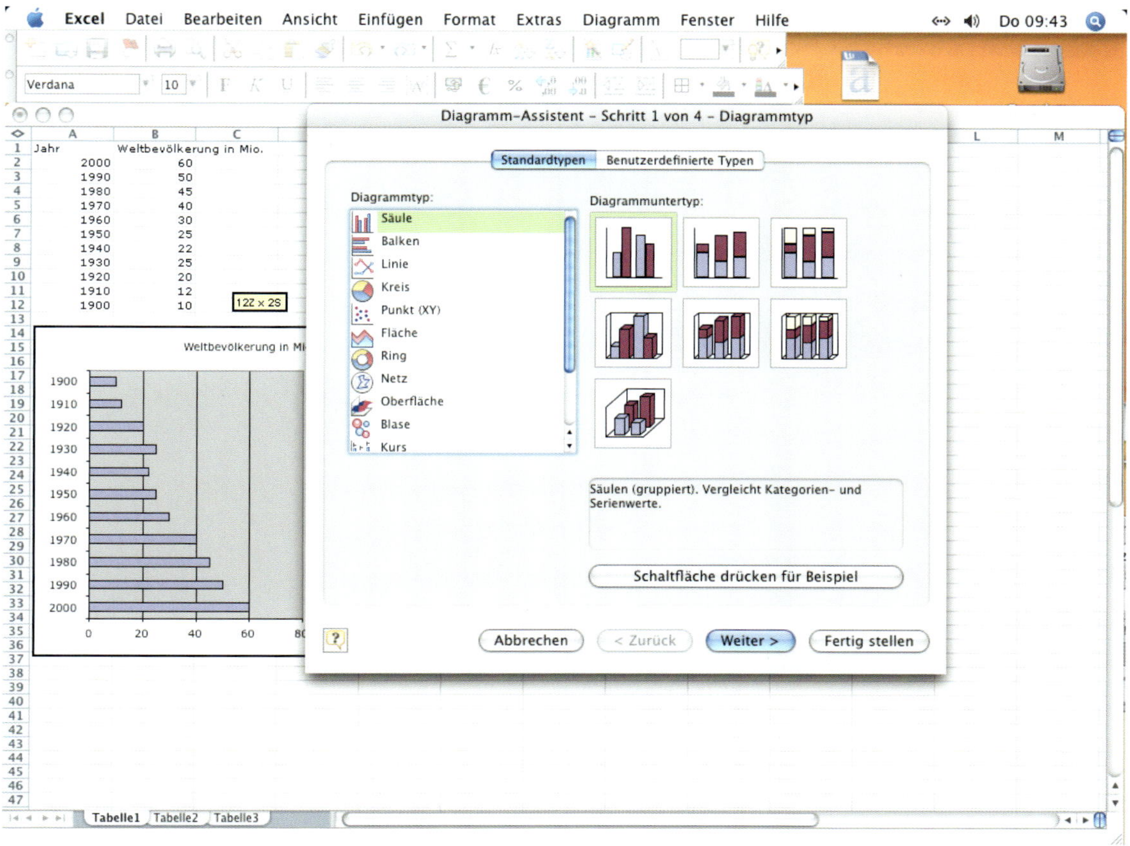

FIG. 5: ChartTools, the "diagram assistant" of the Microsoft Excel spreadsheet calculation program. Excel offers fourteen basic types of diagram that can be used to visualize the values of a table. Series of figures can thus be quickly shown as graph, bar, bubble, or scatter charts. The application was primarily designed to meet the needs of economists and statisticians. Screenshot of the program interface, archive DTB. Used with permission from Microsoft.

Hui Kyong Chun, Wendy, and Thomas Keenan, eds. *New Media, Old Media: A History and Theory Reader.* London: Routledge, 2006.

Kane, Carolyn L. "Digital Art and Experimental Color Systems at Bell Laboratories, 1965–1984: Restoring Interdisciplinary Innovations to Media History." *Leonardo* 43, no.1 (February 2010): 53–58.

Kelty, Christopher, and Hannah Landecker. "A Theory of Animation: Cells, L-Systems, and Film." *Grey Room* 1, no.17 (2004): 30–63.

Kittler, Friedrich A. "Computer Graphics: A Semi-Technical Introduction." *Grey Room*, no. 2 (2001): 30–45.

Kittler, Friedrich A. *Literature, Media, Information Systems: Essays.* Amsterdam: Gordon & Breach, 1997.

Manovich, Lev. *The Language of New Media.* Cambridge, MA: MIT Press, 2001.

Mindell, David A. *Between Human and Machine: Feedback, Control, and Computing before Cybernetics.* Baltimore: Johns Hopkins University Press, 2002.

Mitchell, William J. *The Reconfigured Eye: Visual Truth in the Post-Photographic Era.* Cambridge, MA: MIT Press, 1994.

Mumford, Lewis. *Technics and Civilization.* New York: Harcourt, Brace and Co., 1934.

Pias, Claus. "Das digitale Bild gibt es nicht: Über das (Nicht-)Wissen der Bilder und die informatische Illusion." *Zeitenblicke* 2, no. 1 (2003). www.zeitenblicke.de/2003/01/pias/ (January 2012).

Pratschke, Margarete, ed. *Digitale Form. Bildwelten des Wissens: Kunsthistorisches Jahrbuch für Bildkritik*, ed. Horst Bredekamp, Matthias Bruhn and Gabriele Werner, vol. 3, no. 2. Berlin: Akademie Verlag, 2005.

Schneider, Birgit. *Textiles Prozessieren: Eine Mediengeschichte der Lochkartenweberei.* Zurich: Diaphanes, 2007.

Seitter, Walter. *Physik der Medien: Materialien, Apparate, Präsentierungen.* Weimar: Verlag und Datenbank für Geisteswissenschaften, 2002.

Thacker, Eugene. "Data Made Flesh: Biotechnology and the Discourse of the Posthuman." *Cultural Critique* 53, no. 1 (2003): 72–97.

Warnke, Martin, Wolfgang Coy, and Georg Christoph Tholen, eds. *HyperKult II. Zur Ortsbestimmung analoger und digitaler Medien.* Bielefeld: Transcript, 2005.

Zielinski, Siegfried. *Deep Time of the Media: Toward an Archaeology of Hearing and Seeing by Technical Means.* Cambridge, MA: MIT Press, 2006.

Pictorial Tradition and Difference—Visual Knowledge Acquisition in Science: The Case of Scanning Tunneling Microscopy

Jochen Hennig

Taking scanning tunneling microscopy, an imaging technique developed by physicists in the 1980s, as a case study, the following chapter describes a fundamental challenge in analyzing scientific and technical images and visual knowledge acquisition. On the one hand, scientific images constantly produce new und unexpected results; on the other hand, it can be shown that even these images create meaning through references to previous images. Differences from and similarities to other, preceding images are among the essential characteristics of scientific images and shape the creation and structuring of knowledge, influencing their epistemic status. In this case study, taking similarities as well as differences into account in the analysis of scientific images is regarded not as a contradiction but as providing mutually complementary perspectives. In subjecting the images to a formal image analysis, they are initially briefly detached from the context of their provenance before being embedded in the concerns and material cultures of scientific practice.

Instrumental-experimental records provide experimenters with images they would not have been able to access without the instruments they use. X-ray images provide insights into the human body in medicine and into crystalline structures in the area of materials science, insights that were not available before this technology was developed. PET scans detect an injected radioactive contrast agent, making it possible to see metabolic processes in the brain that would otherwise have remained inaccessible to human senses. Such imaging techniques open up representational spaces by creating the parameters and measurements used to create images.[1] While scientific techniques of this kind are fundamentally designed to generate new knowledge, laypersons encounter PET images of the brain in everyday culture as familiar representations that fit in with their habits of seeing. PET images show the brain as a wide public audience would expect it to look. The complexity of this measuring technique, which detects the positrons emitted from an inhaled radioactive isotope of oxygen, is not perceptible in the images. A similar dynamic is at work with popular images of nanotechnology, in which apparently familiar landscapes are sometimes seen that are in fact the result of complex experiments. One such example from nanotechnology is presented below: images produced by a scanning tunneling microscope. Examining the early stages of this research field at the University of Basel, the case study offers an opportunity to observe how imaging strategies develop through the exertion of personal influence, before these strategies go on to become part of the canon of pictorial representation.[2]

1 On the concept of the representational space, see Hans-Jörg Rheinberger, "Alles, was überhaupt zu einer Inskription führen kann," in *Wissensbilder: Strategien der Überlieferung*, ed. Ulrich Raulff and Gary Smith (Berlin: Akademie, 1999), 265–77.

2 The following observations draw on structured, open-ended interviews conducted with the Basel professor of physics Hans-Joachim Güntherodt, the former doctoral candidate Markus Ringger, and the electronics engineer Hans-Rudolf Hidber. They provided previously unpublished imagery, which was used to focus discussions. Their support made it possible to develop this case study.

The Atom as a Pictorial Object

A scanning tunneling microscope—also referred to below as STM—basically operates by scanning a surface using an atom-sharp needle held extremely close to it. Voltage is applied between the tip and the sample to create a weak current between them. The tip's trajectory is adjusted to keep the current invariant, and the signals recorded identify the positions of constant current. These positions are not properties of an atomic surface but exist only during the experiment as a result of the interaction between the sample and the tip. The measurements that are then translated into images have the status of a visualization created by the interaction between the instrument and the sample.

Research into scanning tunneling microscopy was launched in Basel on April 23, 1982, in a lecture given by Gerd Binnig, who had developed the instrument together with Heinrich Rohrer in the preceding years in the IBM laboratory at Rüschlikon with support from the workshops there. In 1986, Binnig and Rohrer were awarded the Nobel Prize for their invention.[3] Inspired by Binnig's lecture, the head of the Basel Institute of Physics electronics laboratory, Hans-Rudolf Hidber, after consulting with physics professor Hans-Joachim Güntherodt, began building a scanning tunneling microscope. Güntherodt and Hidber believed that building an STM would be a demanding but manageable task and that they had the required know-how. Güntherodt hoped to be able to use this new instrument to study metallic glasses, on which his research at the time was focused, so he had a concrete application in mind. He could not have foreseen that his research would concentrate on scanning probe microscopy for the next twenty years and that his group would help shape the progress of research in this area.[4]

Shortly after Güntherodt and Hidber decided to go into this field, Markus Ringger, a graduate student and later a doctoral candidate, showed an interest in working together with the laboratory to build and operate a scanning tunneling microscope. In the resulting distribution of tasks, it was Hidber's job to develop an electronic system to move the tip across

3 On Binnig and Rohrer's view of the development of STM, see also their Nobel Prize acceptance speech: Gerd Binnig and Heinrich Rohrer, "Scanning Tunneling Microscopy: From Birth to Adolescence," *Reviews of Modern Physics* 59, no. 3 (1987): 615–25. On the history of instruments in the perspective of the sociology of science, see Cyrus Mody, *Instrumental Community: Probe Microscopy and the Path to Nanotechnology* (Cambridge, MA: MIT Press, 2011).

4 The importance of the group working in Basel around Prof. Güntherodt was evident, for example, during the International Conference on Nanoscience and Technology, held from July 30 to August 4, 2006, at which the twenty-fifth anniversary of the scanning tunneling microscope and twentieth anniversary of the atomic force microscope were celebrated in the presence of the pioneers of the method. Many of Güntherodt's former students are now professors in the field of scanning probe microscopy.

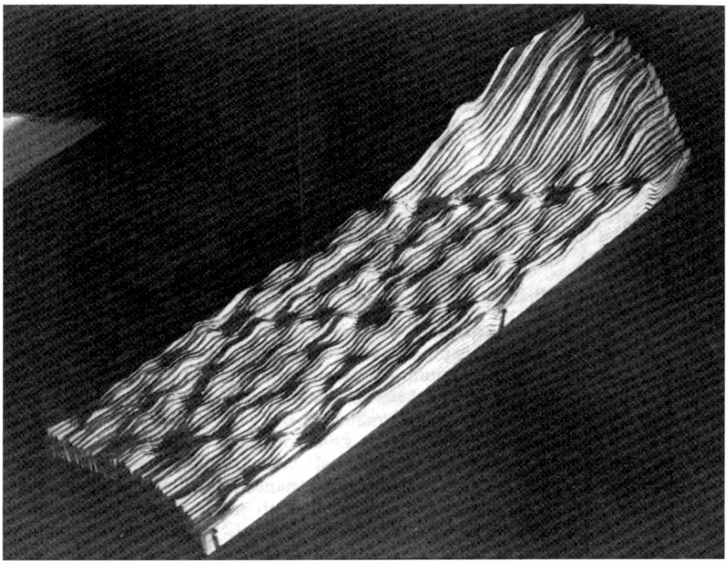

FIG. 1: Photograph of a paper model based on a scanning tunneling microscopic examination of silicon. The image has been widely published. Gerd Binnig and Heinrich Rohrer, "Scanning Tunneling Microscopy," *Helvetica Physica Acta* 55 (1982): 731.

FIG. 2: Copy of the original unpublished printout of October 18, 1982, on which the paper model (see fig. 1) was based. Archive of the IBM research laboratory Rüschlikon.

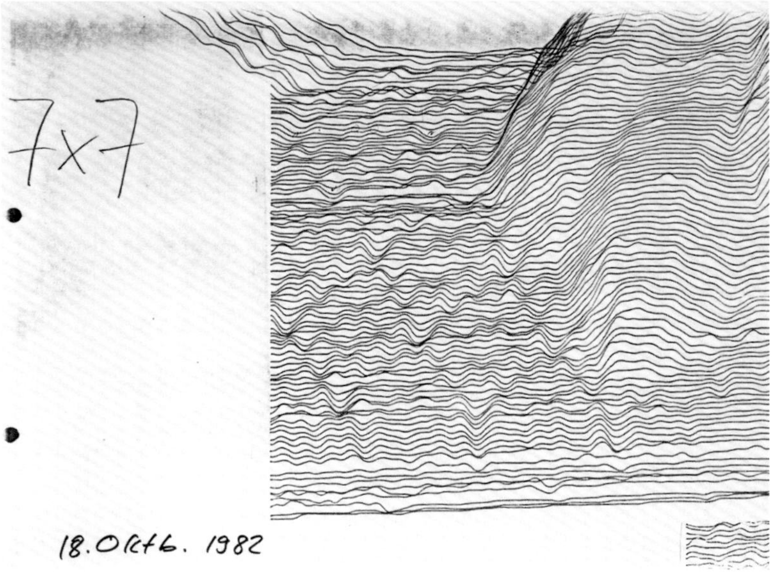

the sample surface and to record signals using an x-y recorder. Hidber's goal was to turn the recorded data into pseudo-3D images. As the model for this form of image, Hidber explicitly used a scanning tunneling microscopic image of Binnig and Rohrer's (fig. 1),[5] the publication of which had aroused great interest. This was a photograph of a paper model that Binnig had made based on an image produced by a scanning tunneling microscopic survey of silicon. Binnig had copied the original, unpublished image (fig. 2) many times, cut out individual strips of paper and stuck them together.[6] He illuminated the paper model to strongly highlight the shadows of the undulating surface; part of the lamp is visible near the top left corner of the photograph of the model (fig. 1).

In contrast to the original image with its individual, unconnected lines, the photo of the model gives the impression of being a relief, so that it shows a pictorial object.[7] Binnig and Rohrer could correlate each of the now-visible mounds, apparently features of a real object, with the position of a single atom in their interpretation of the measurements.[8] Yet the apparent depiction of individual atoms in the form of mounds was a product of intentional pictorial representation by Binnig and Rohrer.

Remodeling the Model

The publications on silicon turned out to be a breakthrough for scanning tunneling microscopy.[9] The photo of the paper model, moreover, subsequently became the paradigmatic pictorial representation of STM, as its reproduction by Hans-Rudolf Hidber shows. He developed an electronic system designed to render the recorded data measurements of a scanning tunneling microscope in a pseudo-3D form, for which Hidber initially created images of simulations (fig. 3), trying out isometric forms of representation.

Hidber used the electronic system he developed and tested in simulations to represent the data generated by the first Basel scanning tunneling microscope, and Markus Ringger produced scanning tunneling micro-

5 In the interview, Hidber referred to Binnig and Rohrer's paper as a model for his representations.

6 I would like to thank Hartwig Thomas, formerly of IBM's Rüschlikon laboratory, for the copy of this original print.

7 On the terminology of the "pictorial object showing itself," see the interview with Lambert Wiesing, "Ornament, Diagramm, Computerbild: Phänomene des Übergangs. Ein Gespräch der Bildwelten des Wissens mit Lambert Wiesing," in *Diagramme und Bildtextile Ordnungen*, ed. Birgit Schneider, *Bildwelten des Wissens: Kunsthistorisches Jahrbuch für Bildkritik*, vol. 3, no. 1 (Berlin: Akademie, 2005): 115, 124–25. Martina Heßler has pointed out that communication contexts determine whether images in a scientific context are read as signs or regarded as pictorial objects. Martina Heßler, "Einleitung: Annäherung an Wissenschaftsbilder," in *Konstruierte Sichtbarkeiten: Wissenschafts- und Technikbilder seit der Frühen Neuzeit*, ed. Martina Heßler (Munich: Fink, 2006), 34–36.

8 Gerd Binnig, Heinrich Rohrer, Christoph Gerber, and Edmund Weibel, "7x7 Reconstruction on Si(111) Resolved in Real Space," in *Physical Review Letters* 50, no. 2 (1983): 121.

9 See Mody, *Instrumental Community*, ch. 3.

scopic recordings in accordance with the parameters of this electronic system (fig. 4). Each recording of a measurement Ringger made was transformed into an isometric projection, which a viewer invariably regards as a representation in perspective. The experimenter was merely free to vary the angle between the axes using a potentiometer.[10] Binnig's paper model, created by the simplest means, became influential after Hidber used it as a guideline in designing the instrument's electronic system, so Markus Ringger could only produce images within these parameters.

While the scanning tunneling microscope defines its general representational space by virtue of its mode of operation and the representation of the positions of constant current, a more specific representational mode had thus been incorporated into the instrument that embedded all the recorded graphs in an object, suggesting an objective quality not evident in the measurements alone. Binnig and Rohrer chose a comprehensible visualization in publishing the photo of the paper model, which was designed to ensure "credibility"[11] and be persuasive. This persuasiveness resulted in the model also being adopted by the group in Basel. The pseudo-3D representation of an object with a smoothed relief surface used to illustrate scanning tunneling microscopic measurements not only was reproduced in Basel but soon was also circulated widely. In *Nanotechnology: Research and Perspectives*—published in 1992, it was probably the first book to have the word *nanotechnology* in its title—the first thing the reader finds among the color plates is an image of a three-dimensional object with an undulating surface (fig. 5).[12] The visual language no longer focuses on quantum-physical processes measured using a scanning tunneling microscope, instead suggesting the reification and attendant

10 Interview with Markus Ringger, April 21, 2005.

11 Binnig and Rohrer, "Scanning Tunneling Microscopy," 620.

12 B. C. Crandall and James Lewis, *Nanotechnology: Research and Perspectives* (Cambridge, MA: MIT Press, 1992).

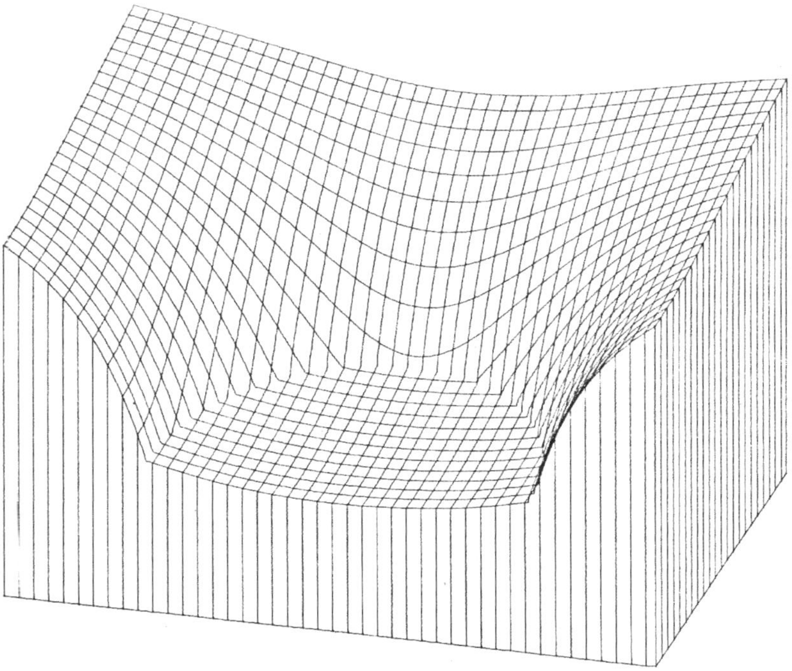

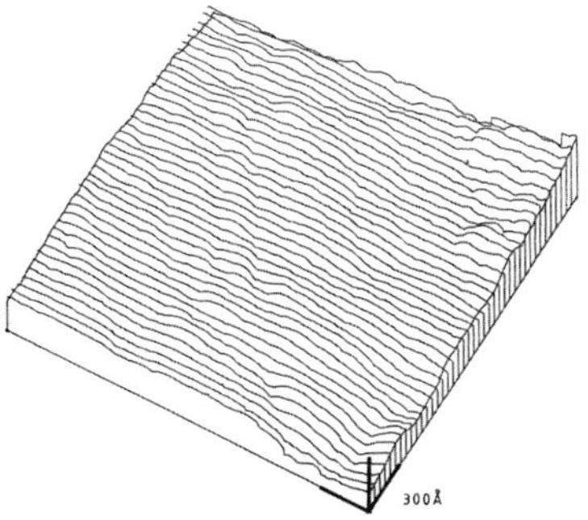

FIG. 3 (left): Visualization of values simulated using analog equipment from Hans-Rudolf Hidber's experimentation with possible visualizations of scanning tunneling microscopic research, ca. 1983. Private archive of Hans-Rudolf Hidber.

FIG. 4 (above): Image of a scanning tunneling microscopic measurement of amorphous Pd81Si19 by Markus Ringger, based on the visualization framework implemented in the technology, 1985. Markus Wilhelm Ringger, "Tunnelmikroskopie in Basel" (PhD diss., University of Basel, 1986). Courtesy Markus Ringger.

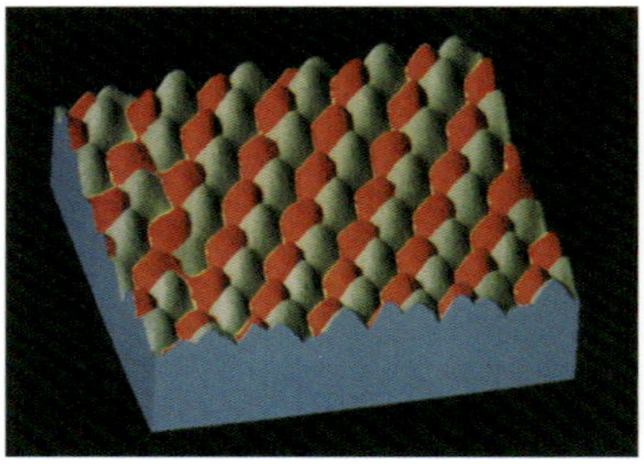

FIG. 5 (above): Scanning tunneling microscopic image from B. C. Crandall and James Lewis, eds., *Nanotechnology: Research and Perspective: Papers from the First Foresight Conference on Nanotechnology* (Cambridge, MA: MIT Press, 1992), fig.1. Courtesy, Jun Nogami.

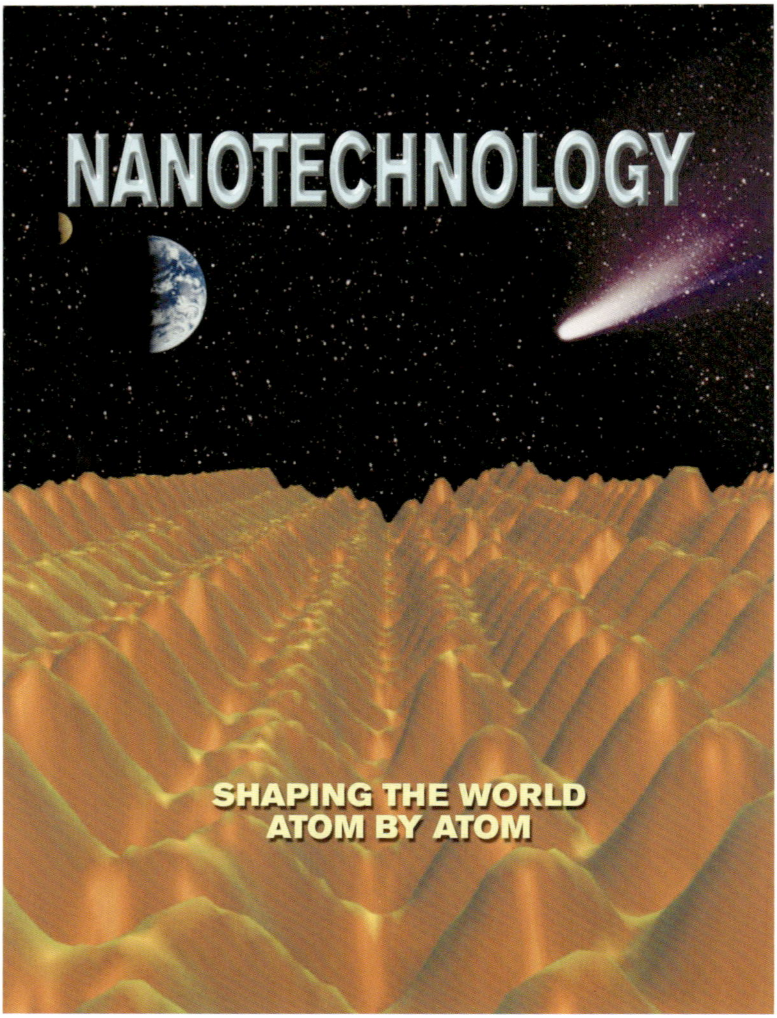

FIG. 6 (right): Cover picture of an information brochure of a government-funded American nanotechnology initiative, 1999. The image is a digitally edited composite created by Lloyd J. Whitman at the U.S. Naval Research Laboratory (NRL), currently at the U.S. National Institute of Standards and Technology, using the following images: (1) the atomic-scale topography of a silicon single crystal surface with a (115) orientation as revealed by scanning tunneling microscopy (from work by A. A. Baski and L. J. Whitman, NRL); (2) the earth and moon as photographed by NASA's Galileo spacecraft (NASA Photo Number P-41508C); and (3) Comet P/Halley as photographed by W. Liller (NASA Photo LSPN-1725). http://www.wtec.org/loyola/nano/IWGN.Public.Brochure/IWGN.Nanotechnology.Brochure.pdf (accessed November 2013). © Lloyd J. Whitman, US Naval Research Laboratory (NRL), 2012.

utilization and controllability of atomic structures. It is not the possibilities of the tunneling microscope per se but the pictorial representation used that creates these nano-worlds and atomic landscapes; it already constitutes an interpretation of the data.[13] Nanotechnology's claim to control the atomic dimension is commensurate with this visual language. The American Nanotechnology Initiative's motto, "Shaping the world atom by atom," expresses nanoscience's engineering and technological mission in physics, foregrounding the utilization and control of the atomic scale. The philosopher of science Alfred Nordmann incisively pointed out that the title and design of the title page of the brochure, published in 1999[14] and showing a controllable atomic landscape (fig. 6), seem to render the claims of technology almost unlimited by suggesting that designing technology coincides with designing nature.[15]

Producing and Interpreting Differences

A design paradigm like the one used to create nano-landscapes was actively produced by scientists and technicians and became influential

13 On the discrepancy between the original intentions of scanning tunneling microscopic experiments in the laboratory and their subsequent reception and embedding in nanotechnological visions, see Jochen Hennig, "Vom Experiment zur Utopie: Bilder in der Nanotechnologie," in *Instrumente des Sehens*, ed. Angela Fischel, *Bildwelten des Wissens: Kunsthistorisches Jahrbuch für Bildkritik*, vol. 2, no. 2 (Berlin: Akademie, 2005): 9–18.

14 Available at www.wtec.org/loyola/nano/IWGN.Public.Brochure/ (accessed March 2012).

15 Alfred Nordmann, "Shaping the World Atom by Atom: Eine nanowissenschaftliche WeltBildanalyse," in *Technikgestaltung zwischen Wunsch und Wirklichkeit*, ed. Armin Grunwald (Berlin: Springer, 2003): 191–99.

because of nanotechnology's immense claims. Yet each of these images was also incorporated into research and interpreted in the context of complex practices. The following discussion will address an exemplary interpretation of an image by the Basel group for the purposes of physics.

Markus Ringger, the first graduate student and later a doctoral candidate in the field of scanning tunneling microscopy in Basel, initially studied metallic glasses, as Güntherodt had planned (fig. 7). He visualized the expected granular structure and traced over it again for clarification.[16] He then expanded his research to include STM samples such as graphite, which was popular in the mid-1980s.[17] The focus of Ringger's research moved away from that of the Basel group, and he increasingly concentrated on materials that seemed to be promising for the scanning tunneling microscopic method.[18] In his research into graphite, Ringger sought to increase his instrument's performance up to atomic resolutions. He was not looking for new findings on the properties of graphite, which was his sample at that time, but trying to stabilize the instrument, which thus became the object of research.[19]

In his studies of graphite, Ringger did not get resolutions high enough to show a surface relief. Instead he was confronted with shaky individual lines, which demonstrated certain regularities (fig. 8). What is striking in the lower image of figure 8 is the divergence of a single line, which can be explained by changes to the tip during measurement. No visual tradition exists for this sort of "outlier" because the form of the individual lines is due to the contingency of interference.[20] The latter was

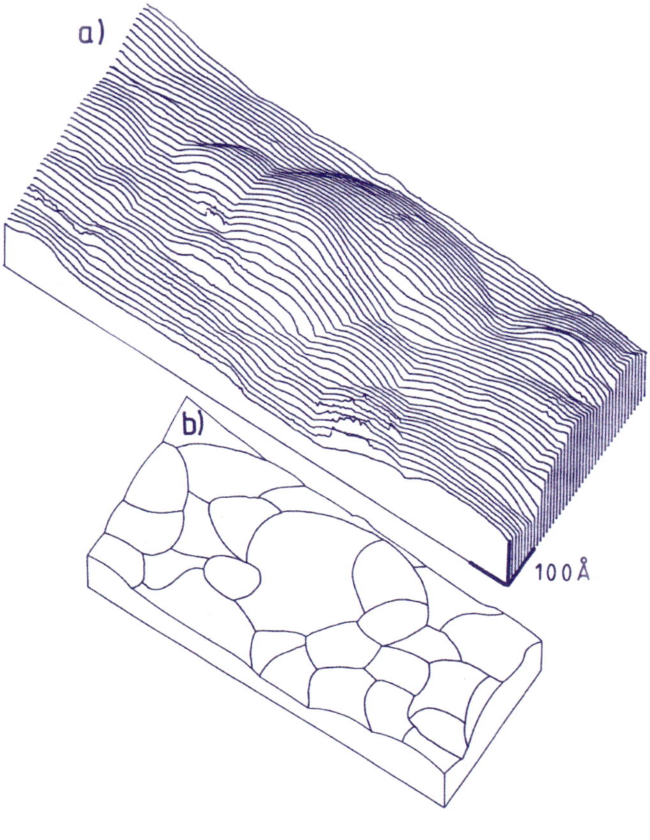

FIG. 7 (left): Markus Ringger's first scanning tunneling microscopic studies at the University of Basel, 1985. Markus Wilhelm Ringger, "Tunnelmikroskopie in Basel" (PhD diss., University of Basel, 1986). Courtesy, Markus Ringger.

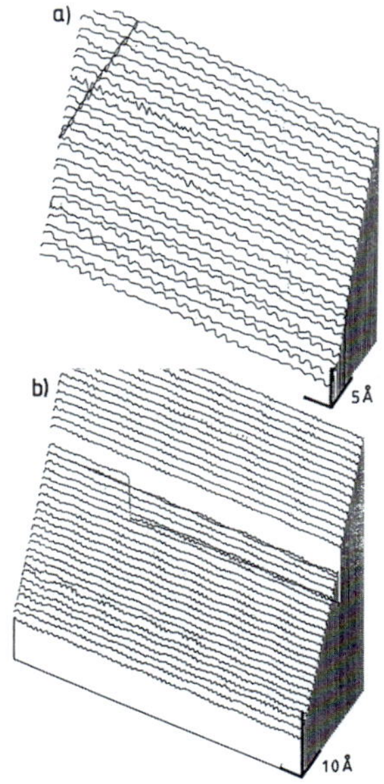

FIG. 8 (below): Scanning tunneling microscopic study of graphite carried out by Markus Ringger, 1985. Markus Wilhelm Ringger, "Tunnelmikroskopie in Basel" (PhD diss., University of Basel, 1986). Courtesy, Markus Ringger.

16 Markus Ringger, "Tunnelmikroskopie in Basel" (PhD diss., Philosophisch-Naturwissenschaftliche Fakultät der Universität Basel, 1986), 92–93.

17 Ibid., 99–104.

18 On the importance of graphite research in developing and stabilizing scanning tunneling microscopy, see also Mody, *Instrumental Community*, ch. 3.

19 Hans-Jörg Rheinberger describes the switch between object of research (epistemic things) and stabilized technical component, which can occur at any time, with his concept of the experimental system. Hans-Jörg Rheinberger, *Toward a History of Epistemic Things: Synthesizing Proteins in the Test Tube* (Palo Alto, CA: Stanford University Press, 1997), 74.

20 On the observation that images showing such interference are printed only in certain phases of a method's development—figure 4 comes from Ringger's

doctoral thesis—and then disappear from publications in favor of images free of interference, see Jochen Hennig, "The Instrument in the Image," in *Instruments in Art and Science: On the Architectonics of Cultural Boundaries in the 17th Century*, ed. Helmar Schramm, Ludger Schwarte, and Jan Lazardzig (Berlin: Walter de Gruyter, 2008): 348–61.

21 Interview with Markus Ringger.

22 Rheinberger, *Toward a History of Epistemic Things*, 74–79.

23 Interview with Markus Ringger.

24 Rheinberger, *Toward a History of Epistemic Things*, 28–31.

immediately recognizable to the experimenter and not interpreted as such only after long investigations; it did not contribute to the progress of the scientific process.

But the image diverged from Ringger's expectations in another way, which would unexpectedly determine his next course of action. The distances between the maxima, the small spikes in the recordings, were not as he had expected. Ringger had derived his expectations from values in the literature, which were in turn based on experimental research.[21] Such unpredictable discrepancies are at the core of all research practice. They correspond to those "differences" identified by historian of science Hans-Jörg Rheinberger that essentially determine the progress of research.[22]

For Ringger, the surprising distances of the maxima created a new problem, which led him to question the calibration of the piezo motors that guide the tip over the sample's surface and measure the distances between the maxima. Originally just technical components, the piezo motors themselves became an object of research, and Ringger investigated them with help and infrastructure obtained from other scientists.[23] In the terms of Rheinberger's description of the dynamic in experimental systems, the motors had become "epistemic things." According to Rheinberger, such an "epistemic thing" is not an object of investigation at the outset but becomes a focus of attention only during the course of the experimental process.[24] Interpretation of the graphite recordings was ambiguous because the image showed unexpected elements that were without precedent in the tradition. When Markus Ringger recalibrated the piezo motors, he found that their properties diverged from the originally hypothesized values because soldering connections had changed their physical properties. Ringger was able to use this knowledge in further research and pass it on to subsequent doctoral candidates in the Basel working group.

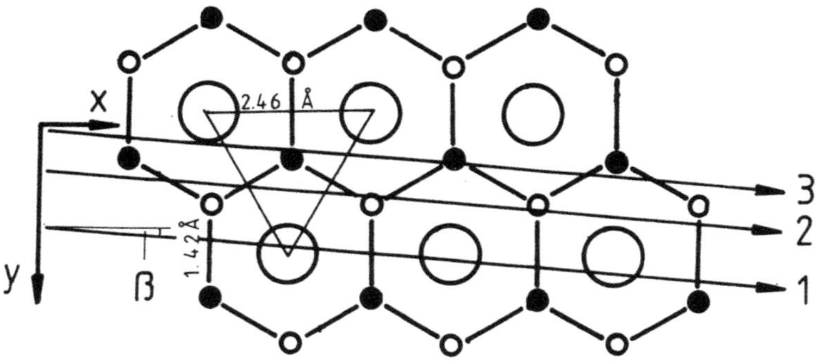

FIG. 9: Graphical analysis of how scan lines relate to the positions of individual atoms in a survey of graphite, 1985. Markus Wilhelm Ringger, "Tunnelmikroskopie in Basel" (PhD diss., University of Basel, 1986). Courtesy, Markus Ringger.

The disposition of the maxima raised further research questions after the piezo motors were recalibrated. Ringger's most pressing technical problem was now not being able to bring together individual scans at greater densities and measure distances between single atoms, a problem he tried to solve by geometric means, showing how individual scans behaved in terms of the atomic direction on the surface (fig. 9). Using a sketch, he plausibly showed that the scanning process did not record the positions of all of the atoms and that a divergence of the scanning direction from the primary direction of the graphite had caused the pattern and the distances in the tunneling microscope image. Quantification and geometric considerations led to the interpretation of the original image (fig. 8) with whose forms and disposition of the maxima Ringger had been confronted. These did not confirm previous models, instead producing unexpected results that required explanation and shaped Ringger's subsequent research. The disposition of the maxima was a crucial challenge for the experimenter. In the perspective of the historical analysis of imagery, it defies inclusion in a framework of pictorial references, as in the development of a design paradigm for the representation of solid bodies. But in interpreting his research into graphite, Markus Ringger looked not primarily at the isometric representation of an object as implemented by Hidber but beyond this predefined framework, at the recordings of individual lines.

The Simultaneity of Reproduction and the Production of the New

Scanning tunneling microscopic images are shaped by instrumental preconditions. The instrument opens up a representational space, because the positions of constant tunneling current do not exist outside the instrument, and Hidber, with his electronic system, had used an isometric projection in the first Basel instrument, which established the representation of bodies with a relief surface. Yet instrumental preconditions are not codified in scientific processes, so technical components in experimental systems can themselves become objects of research, as happened in Ringger's experiments with the piezo motors.

A comprehensive analysis of scientific image practices, then, will have to address matters of form and visual design, because only through such analyses can the representation of atomic landscapes and the immense claims of nanotechnological visions to be able to control and shape the atomic dimension be understood as the outcome of a visual strategy. Formal analysis would seem necessary and meaningful as long as there is an awareness of its limited applicability to the scientific image—a limitation that is due to the generation of differences that are unexplained and unknown at the moment of their emergence and ipso facto cannot derive from a formal tradition or stand in relation to the familiar. But that certain features—the interferences and differences—of a scientific image necessarily evade inclusion in the traditions and conventions of form does not render comparative image analysis obsolete, because scientific images are no less the products of a history.[25] The production of a visual surplus is a constructive part of scientific imaging practice, so pictorial representation contributes to configuring the promises of a research area such as nanotechnology.

25 Whereas art-historical discussions and reflections on scientific and technical images usually focus on analyzing image traditions, analyses in the history of science often start with the dynamics of experimental processes and the production of differences. These positions have been explicitly taken by Horst Bredekamp and Gabriele Werner on the one hand and Michael Hagner on the other. See the interview with Michael Hagner, "Bildunterschätzung—Bildüberschätzung," in *Bilder in Prozessen*, ed. Horst Bredekamp and Gabriele Werner, *Bildwelten des Wissens: Kunsthistorisches Jahrbuch für Bildkritik*, vol. 1, no. 1 (2003): 103–11. While these two positions seemed contradictory in the interview, they have been combined in this essay so as to do justice to the diverse facets of scientific imaging practice.

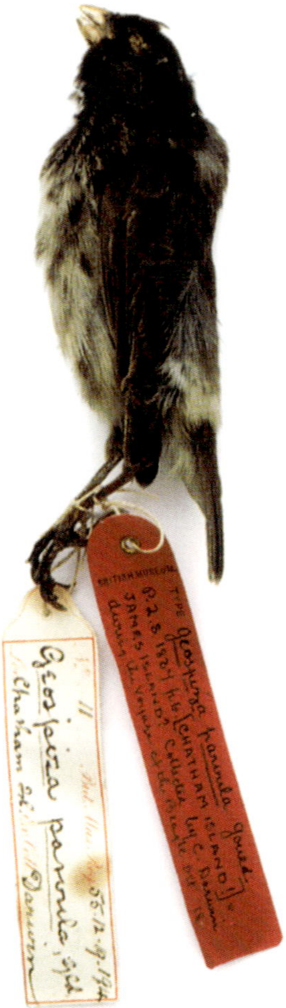

1a

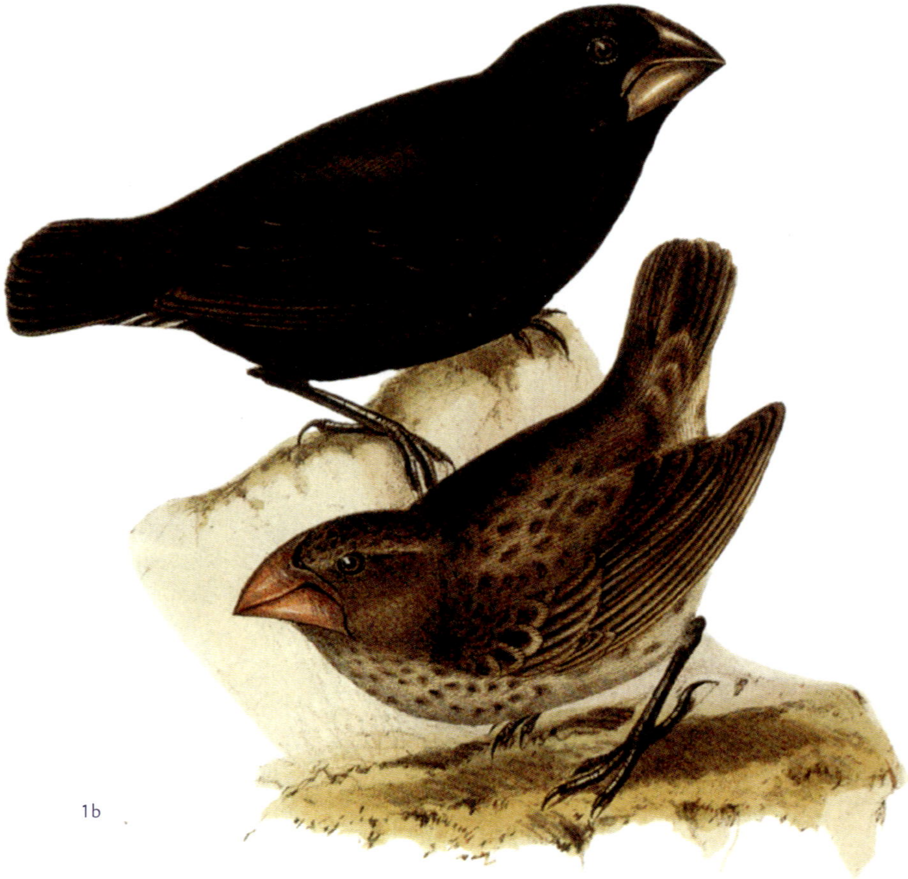

1b

CHAINS OF REPRESENTATIONS

According to a traditional understanding, scientific images attain a purpose outside themselves by referring to phenomena and data. Since the 1980s, however, scholars in the field of science studies have increasingly pointed out that the traditional notion of a correspondence between a scientific image and a discrete object is untenable because it is impossible to clearly correlate the representation with what is represented. Consequently, the question of the referent in science cannot be posed in the sense of a simple illustration of, and references to, reality. The philosopher of science Ian Hacking contributed a vital impetus to the discussion of these issues in the early 1980s with his book *Representing and Intervening*. He emphasized that representation is only possible through the experimenter's intervention, so the represented can only emerge out of the intervening construction of a representation (Hacking 1983). According to this view, representation cannot be understood in terms of references to reality or a correspondence with a discrete object; it must be conceived in terms of constructive processes (Lynch 1990 and1994; Hagner 1997). For this reason, the philosopher of science Hans-Jörg Rheinberger has proposed dispensing with the word *representation* in the context of the experimental-instrumental acquisition of findings and speaking instead of "visualizations" (Rheinberger 2001).

Furthermore, in studying experimental practices, the history of science has highlighted the reference of representations to other representations. These references are shaped by complex referential practices. Many authors writing on the history and philosophy of science therefore use the term *chains of representations* (Pickering 1995, Latour 1999) to metaphorically describe the process of one representation transforming into

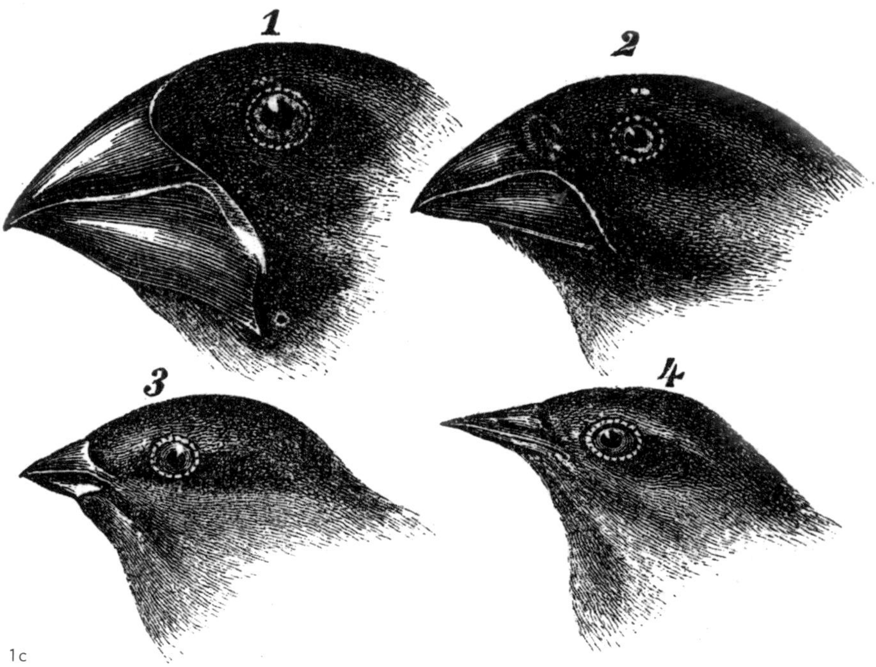

1c

FIGS. 1a–1c: In her book *Darwins Bilder: Ansichten der Evolutionstheorie 1837–1874* (2007, 80–81), Julia Voss reconstructs a chain of representations in Latour's sense in the example of Charles Darwin's Galápagos finches. Shot by Darwin and his companions on the Galápagos Islands in the Pacific in 1835, the birds were preserved as skins (fig. 1a: Galápagos Island finch with a museum label in Darwin's own hand at the Natural History Museum, London). After they were sent to London in 1837, John Gould, curator of the Zoological Society, identified them as a genus of finch and sketched them for the first time. Based on these initial sketches, Gould's wife Elizabeth then made drawings, which she transferred to a stone that was subsequently used as the template in the lithographic printing process. The resulting lithographs (fig. 1b: *Geospiza strenua*; Elizabeth Gould's lithographic plates from *The Zoology of the Voyage of H.M.S. Beagle*, 1841) presented the finches as species of a new genus in color and with a section of landscape. Darwin examined, trimmed, and arranged the plates anew, creating a comparative sequence of images of four of the finch species collected on the Galápagos Islands (fig. 1c: The Galápagos finches in the German edition of Darwin's *A Naturalist's Voyage round the World*, 1899). Only in this overall view did the gradual changes in the finches become apparent, helping Darwin to pave the way for his theory of evolutionary speciation. Fig. 1a: Julia Voss, *Darwins Bilder. Ansichten der Evolutionstheorie 1837–1874* (Frankfurt a.M.: Fischer Taschenbuch Verlag, 2007), fig. 13. Fig. 1b: Charles Darwin, ed., *The Zoology of the Voyage of H.M.S. Beagle, Part III: Birds* (London: Smith, Elder & Co., 1841), plate 37. Fig 1c: Charles Darwin, *Gesammelte Werke: Reise eines Naturforschers um die Welt*, 2nd ed. (Stuttgart: E. Schweizerbart'sche Verlagshandlung, 1899), 413.

another. This approach has always been practiced in the field of art history, where works are related back to their preliminary studies. Sketches, designs, and models also raise the question of how steps in the transformation from preliminary stages to the final work are to be interpreted (Morgan et al. 1984).

The ethnologist and sociologist of science Bruno Latour has made a central contribution to the concept of a chain of representations with his field study of a survey of the Amazon rainforest, which describes in detail how the scientists initially gathered and sorted plant and soil samples so as to reduce them over several stages to sketches and graphics. Latour emphasizes that, with each of these transformations, there is discontinuity, while there must also be constancy and resemblance. Latour sees the reference here not in a mimetic correspondence between the chain's first and final links—in his example, between the jungle on the one hand and the diagrams of the jungle that were eventually published on the other—but in the properties of the chain. Each link in the chain must necessarily refer to a previous one, and the chain must be able to be traced back to its beginnings (Latour 1999). The universality of this approach lies in the description of the transition and transformation between different media, such as objects in collections, graphs, drawings, and diagrammatic images (Figs. 1a–1c and 2a–2g). In this connection, Latour emphasizes the existence of specific conventions for different forms of representation. This suggests the importance of the media characteristics of different forms of representation: knowledge, too, is shaped by the media it is presented in. What should be investigated, then, is how the knowledge represented changes with such transformations of the media (Mersch 2006).

This approach may be productive in a historical *Bild-*

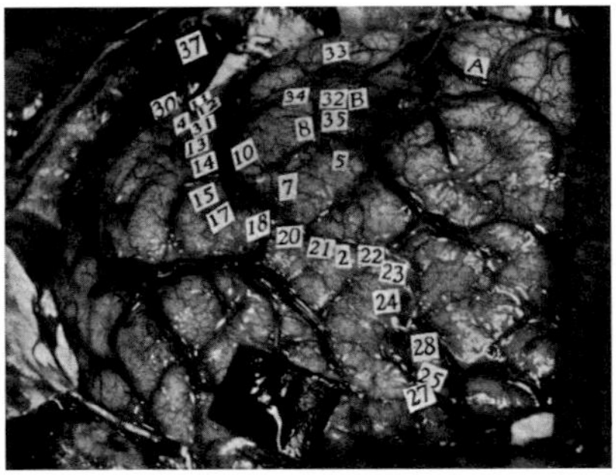

2a

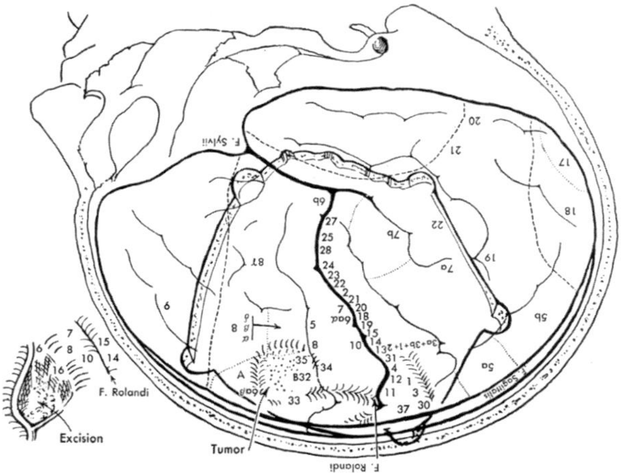

2b

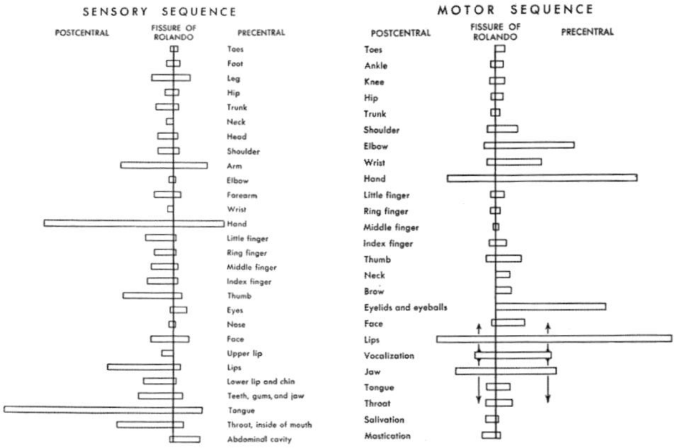

2c, 2d

FIGS. 2a–2g: "It is a cartoon of representation in which scientific accuracy is impossible." With these words, the neurosurgeons Wilder Penfield and Theodore Rasmussen commented on their renowned images of the sensory and motor homunculi (figs. 2e and 2f), which summed up the central results of their study *The Cerebral Cortex of Man: A Clinical Study of Localization of Function* (1950). Penfield and Rasmussen's "cartoon of representation" is in fact the preliminary end of a "chain of representations." Figures 2a and 2b demonstrate the methods Penfield and his colleagues used to collect the data the study was based on. They relate to an operation carried out under local anesthetic on the open brain of a patient suffering from epileptic fits caused by a tumor. In order not to damage any of the brain's sensory and motor functions during the removal of the tumor, these functions were localized by stimulation with an electrode. If the doctors received a positive reaction from the patient to a stimulus, such as a sensation in the tongue or a movement of the hand, they indexed it by attaching a small, consecutively numbered paper ticket to the place (fig. 2a). The doctor finally documented the positions of all the paper tickets by charting them in a brain map (fig. 2b). It could also be verified subsequently through photographs (fig. 2A). Penfield and his colleagues collected data on the localization of sensory and motor functions in the brain over a period of nineteen years from almost four hundred operations in the form of written reports, brain maps, and photographs. Figures 2c and 2d subsequently analyze the data in two bar graphs. The vertical axis represents the brain's fissure of Rolando, along which the stimulation fields of selected body parts are arrayed in order. The length of the bar is in proportion to the number of positive reactions in the area of the gyrus postcentralis and gyrus praecentralis. It thus is clear that sensory functions are primarily represented in the gyrus postcentralis of the brain (fig. 2c), while motor functions are located mainly in the gyrus praecentralis (fig. 2d). These results were then turned into figurative representations (figs. 2e and 2f). For the sensory and motor functions, separated from each other, individual parts of the body were positioned along a cross section of the hemisphere in accordance with their representation in the cortex. The size of each body part is proportional to the number of positive reactions. The resulting distorted representations of the body's anatomy are called the sensory (fig. 2e) and motor homunculi (fig. 2f).

The two figurative representations from 1950 in turn cross-reference Wilder Penfield and Edwin Boldrey's sensory-motor homunculus from 1937 (fig. 2g). This earlier version of the homunculus shows that the fundamental principle of proportionally distorted parts of the body projected onto the cortex was already developed in 1937. In the 1950 version (figs. 2e and 2f), the representations of sensory and motor functions were separated from each other, so Penfield and his colleagues arrived at an itemized result compared with the earlier version. Figs. 2a–f: Wilder Penfield and Theodore Rasmussen, *The Cerebral Cortex of Man: A Clinical Study of Localization of Function*, 3rd ed. (New York: Macmillan, 1955). Fig. 2g: Wilder Penfield and Edwin Boldrey, "Somatic Motor and Sensory Representation in the Cerebral Cortex of Man as Studied by Electrical Stimulation," *Brain: A Journal of Neurology* 60, no. 4 (1937): 432. 2a: Photo: Charles Hodge and H. S. Hayden. 2e–f: Drawing, H. P. Cantlie. 2g: Drawing: H. P. Cantlie, by permission of Oxford University Press.

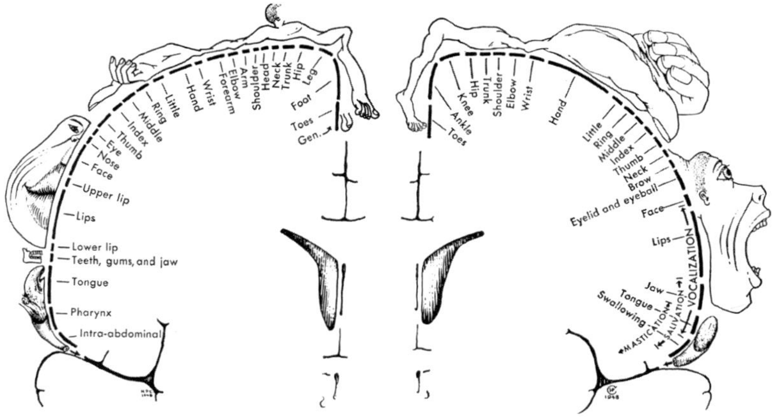

2e, 2f

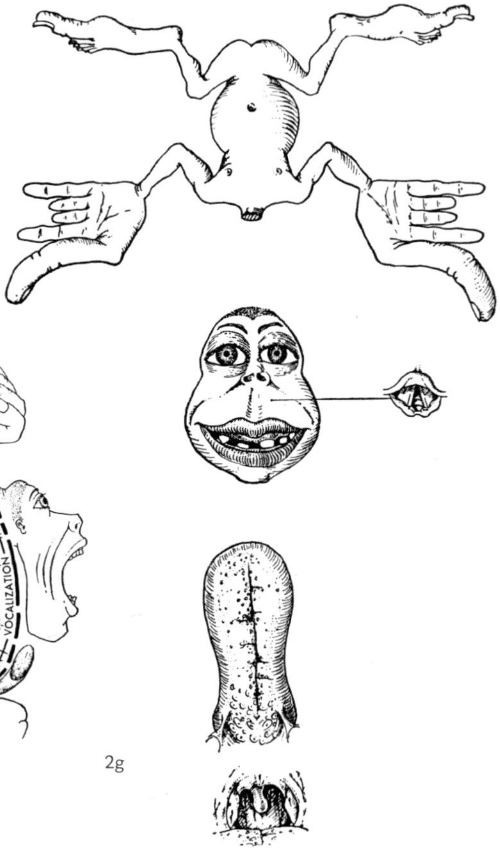

2g

wissenschaft if individual scientific images are considered as links in a chain of representations and their epistemic status is determined based on this reference. Such studies may then be linked to an investigation of individual images, determining their visual surplus through reference to images outside the chain of representations. —JA/JH

LITERATURE

Bredekamp, Horst, and Franziska Brons. "Fotografie als Medium der Wissenschaft: Kunstgeschichte, Biologie und das Elend der Illustration." In *Iconic Turn: Die Neue Macht der Bilder*, ed. Hubert Burda and Christa Maar, 365–81. Cologne: DuMont, 2004.

Hacking, Ian. *Representing and Intervening: Introductory Topics in the Philosophy of Natural Science*. Cambridge: Cambridge University Press, 1983.

Hagner, Michael. "Zwei Anmerkungen zur Repräsentation in der Wissenschaftsgeschichte." In *Räume des Wissens: Repräsentation, Codierung, Spur*, ed. Hans-Jörg Rheinberger, Michael Hagner, and Bettina Wahrig-Schmidt, 339–55. Berlin: Akademie Verlag, 1997.

Latour, Bruno. "Circulating Reference: Sampling the Soil in the Amazon Forest." In *Pandora's Hope: Essays on the Reality of Science Studies*, 24–79. Cambridge, MA: Harvard University Press, 1999.

Latour, Bruno. "Drawing Things Together." In *Representation in Scientific Practice*, ed. Michael Lynch and Steve Woolgar, 19–68. Cambridge, MA: MIT Press, 1990.

Lynch, Michael. "Representation Is Overrated: Some Critical Remarks about the Use of the Concept of Representation in Science Studies." *Configuration* 1, no. 2 (1994): 137–49.

Lynch, Michael, and John Law. "Lists, Field Guides, and the Descriptive Organization of Seeing: Bird-Watching as an Exemplary Observational Activity." In *Representation in Scientific Practice*, ed. Michael Lynch and Steven Woolgar, 267–99. Cambridge, MA: MIT Press, 1990.

McLuhan, Marshall. *Understanding Media: The Extensions of Man*. New York: McGraw-Hill, 1964.

Mersch, Dieter. "Naturwissenschaftliches Wissen und bildliche Logik." In *Konstruierte Sichtbarkeiten. Wissenschafts- und Technikbilder seit der frühen Neuzeit*, ed. Martina Heßler, 405–19. Munich: Fink, 2006.

Morgan, Mary S, and Margaret Morrison, eds. *Models as Mediators: Perspectives on Natural and Social Sciences*. Cambridge: Cambridge University Press, 1999.

Pickering, Andrew. *The Mangle of Practice: Time, Agency, and Science*. Chicago: University of Chicago Press, 1995.

Rheinberger, Hans-Jörg. "Objekt und Repräsentation." In *Mit dem Auge denken: Strategien der Sichtbarmachung in wissenschaftlichen und virtuellen Welten*, ed. Bettina Heintz and Jörg Huber, 55–61. Vienna: Springer, 2001.

Rheinberger, Hans-Jörg. *Towards a History of Epistemic Things: Synthesizing Proteins in the Test Tube*. Palo Alto, CA: Stanford University Press, 1997.

Wallis, Brian, ed. *Art after Modernism: Rethinking Representation*. New York: New Museum of Contemporary Art, 1984.

Thinking with Models: On the Genesis of James Watson's *Molecular Biology of the Gene*

Reinhard Wendler

The following pages discuss the graphic design of the first editions of James Watson's textbook Molecular Biology of the Gene.[1] *They describe the novel design concepts that were implemented in the book and trace the considerable influence it has had on subsequent textbooks of molecular biology, which continues to this day. The discussion sets the innovations in the layout and the role of illustrations in relation to Watson and Crick's speculative model building in their investigation of the structure of DNA, advancing the hypothesis that the layout and illustrations of* Molecular Biology of the Gene *reflect and carry forward the playful form of thinking with material models that was of fundamental significance to establishing the structure of DNA. Watson's textbook represents the emergence of a self-conception of molecular biology as a discipline that relies on visual and tactile models and leaves the paradigms of representation and exactitude behind.*

1 James D. Watson, *Molecular Biology of the Gene* (New York: W. A. Benjamin, 1965).

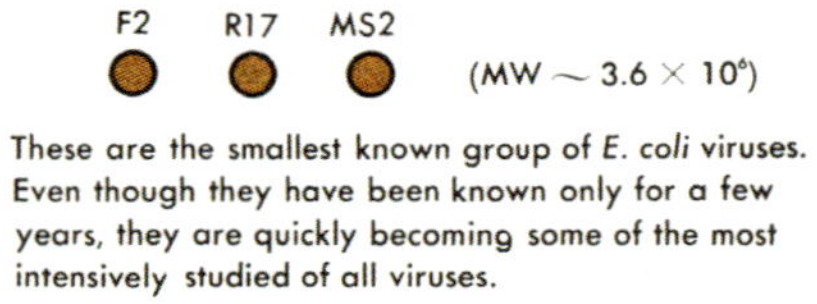

FIG. 1: Schematic illustration of viruses F2, R17, and MS2, from J. D. Watson, *Molecular Biology of the Gene*, 3rd ed. (Menlo Park, CA: Benjamin/Cummings, 1977). © 1965, p. 202. Adapted by permission of Pearson Education, Inc., Upper Saddle River, NJ.

James Watson's *Molecular Biology of the Gene* (1965) occupies a special place in the history of textbooks of molecular biology: it pioneered the use of several forms of visual argumentation in teaching molecular biology and defined the style for the entire genre. A single small schematic illustration (fig. 1) encapsulates how the book harnessed visual means to present its argument. It consists of three simple bronze-colored circles with black borders that bear designations from molecular biology; a specification of their size appears in parentheses. The three dots belong in the context of a number of additional graphical illustrations, which they serve, among other functions, as a standard of scale (fig. 2). The caption notes: "These are the smallest known group of E[scherichia] coli viruses," a proposition that is not as plain as it may seem at first glance—in fact, taken literally, it is false, since what we see are not viruses but three abstract color compositions on paper.

202 MOLECULAR BIOLOGY OF THE GENE

combat bacterial diseases. Phages were unfortunately never medically useful, because their bacterial hosts mutate readily to forms resistant to viral growth. Thus almost everyone lost interest in the phage, and almost no one studied how it multiplied until the late 1930s. Then a small group of biologists and physicists, intrigued by the problem of gene replication, began to investigate the reproduction of several phages which multiplied in *E. coli*. They chose to work with phages

FIGURE 7–8 *Some bacterial viruses that have been important in the study of the chemistry of genetics.*

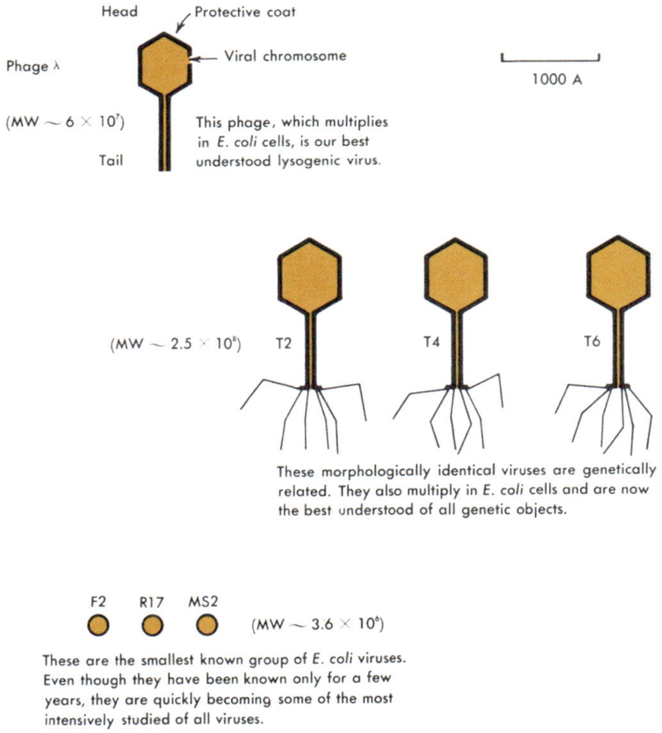

FIG. 2: The graphical context of fig. 1.

The illustration and the caption are each incomplete in their own ways: regarded in isolation, the dots do not refer to any specific object, and nothing about them suggests that they should be taken to represent molecular structures. What becomes concrete in them, Ludmilla Jordanova writes, is an “interpretive gap for viewers to fill in, the ‘beholder’s share’ in Gombrich’s words.”[2] The caption, it is true, bridges that gap, but for its part covers up the complexity of the iconic reference by suggesting that the viruses themselves rest on the page. The caption—that is the crucial point—is not a description of the illustration but an instruction on how to read it: the dots are *not* viruses, they are *meant to be construed as though they were* viruses. The beholder’s share consists in filling the

2 Ludmilla Jordanova, “Material Models as Visual Culture,” in *Models: The Third Dimension of Science*, ed. Soraya de Chadarevian and Nick Hopwood (Palo Alto, CA: Stanford University Press, 2004), 447.

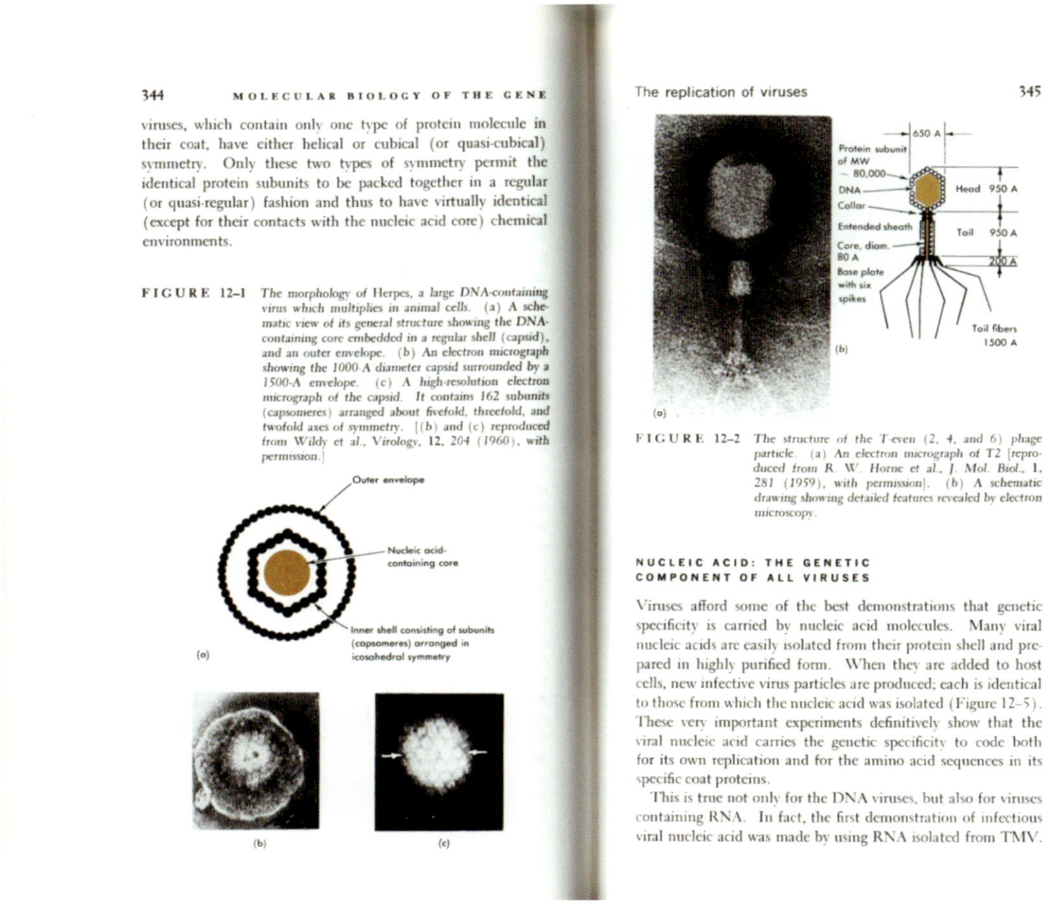
344 MOLECULAR BIOLOGY OF THE GENE

viruses, which contain only one type of protein molecule in their coat, have either helical or cubical (or quasi-cubical) symmetry. Only these two types of symmetry permit the identical protein subunits to be packed together in a regular (or quasi-regular) fashion and thus to have virtually identical (except for their contacts with the nucleic acid core) chemical environments.

FIGURE 12-1 The morphology of Herpes, a large DNA-containing virus which multiplies in animal cells. (a) A schematic view of its general structure showing the DNA-containing core embedded in a regular shell (capsid), and an outer envelope. (b) An electron micrograph showing the 1000-A diameter capsid surrounded by a 1500-A envelope. (c) A high-resolution electron micrograph of the capsid. It contains 162 subunits (capsomeres) arranged about fivefold, threefold, and twofold axes of symmetry. [(b) and (c) reproduced from Wildy et al., *Virology*, 12, 204 (1960), with permission.]

The replication of viruses 345

FIGURE 12-2 The structure of the T-even (2, 4, and 6) phage particle. (a) An electron micrograph of T2 [reproduced from R. W. Horne et al., *J. Mol. Biol.*, 1, 281 (1959), with permission]. (b) A schematic drawing showing detailed features revealed by electron microscopy.

NUCLEIC ACID: THE GENETIC COMPONENT OF ALL VIRUSES

Viruses afford some of the best demonstrations that genetic specificity is carried by nucleic acid molecules. Many viral nucleic acids are easily isolated from their protein shell and prepared in highly purified form. When they are added to host cells, new infective virus particles are produced; each is identical to those from which the nucleic acid was isolated (Figure 12-5). These very important experiments definitively show that the viral nucleic acid carries the genetic specificity to code both for its own replication and for the amino acid sequences in its specific coat proteins.

This is true not only for the DNA viruses, but also for viruses containing RNA. In fact, the first demonstration of infectious viral nucleic acid was made by using RNA isolated from TMV.

FIG. 3: Different forms of representation appear in juxtaposition, challenging the reader to make comparative observations and pushing the text into the intervening spaces. J. D. Watson, *Molecular Biology of the Gene,* 3rd ed. (Menlo Park, CA: Benjamin/Cummings, 1977). © 1965, p. 202. Adapted by permission of Pearson Education, Inc., Upper Saddle River, NJ.

interpretive gap with a sort of pragmatic fiction, to use Hans Vaihinger's term, reading the three dots as viruses.[3]

Watson's book tests this constellation of different indeterminacies and the beholder's involvement in forever-new variations. That is one of the ways in which it marks a paradigm shift away from earlier textbooks of molecular biology, in which language was the primary medium of argumentation. Its size mimics the formats of text-dominated books, but unlike the latter, it contains a wealth of images from a wide range of sources: diagrams, charts, structural formulae, schematic visualizations of molecular processes, pictures of ball-and-stick models, photographs, and illustrations generated using a variety of imaging techniques (fig. 3).[4] Yet what truly sets Watson's textbook apart from earlier representatives of the genre such as Haggis et al.'s *Introduction to Molecular Biology*[5] is not just the considerably larger number and variety of visual means of presentation but, more importantly, the fact that Watson uses pictures as the primary medium for conveying information: in many instances, the text merely serves to explain them.

Mainly responsible for this innovation were Cyrus Levinthal, Keith Roberts, Bill Prokos, and, not least, James Watson himself. Cyrus Levinthal, the editor of the Biology Teaching Monograph Series, in which Watson's *Molecular Biology of the Gene* was published, is considered one

3 See Hans Vaihinger, *Die Philosophie des Als Ob: System der theoretischen, praktischen und religiösen Fiktionen der Menschheit auf Grund eines idealistischen Positivismus* (Leipzig: Meiner, 1927; repr., Aalen: Scientia, 1986).

4 Including autoradiographs, electron micrographs, photomicrographs, X-ray crystallographs, etc.

5 Geoffrey Harvey Haggis et al., *Introduction to Molecular Biology* (London: Longman, 1964).

of the founders of what is called molecular graphics.[6] In 1966, he wrote an essay for *Scientific American* in which he described attempts to unlock the potential of computer-aided visual thinking for research into proteins: "We realized that our best hope of gaining insight into unexpected structural relations—relations that had not been anticipated—lay in getting the computers to present a three-dimensional picture of the molecule."[7] Keith Roberts designed the illustrations in *Molecular Biology of the Gene*. A student at King's College, Cambridge, Roberts had come to Watson's attention because of his skills as a draftsman. The biologist and historian of science Errol Friedberg believes that Roberts was the author of crucial design principles that underlie the illustrations in *Molecular Biology of the Gene*: "Roberts, aware of the prosaic quality of figures in most high school and college texts, developed fundamental rules for his illustrations: no extraneous labels or information in a figure, a single figure had to tell a complete story, and if a figure was too complicated to understand in ten seconds, it was split in two."[8] To explain what that means, Friedberg refers to an illustration created for the textbook's first edition (fig. 4). He reconstructs three drafting stages the image went through: a conceptual sketch by Roberts; a drawing for the use of the executing designer, Bill Prokos; and the final result as it was printed. Friedberg remarks: "The most important stage was the first, which requires both scientific knowledge and understanding of how to present ideas visually. Watson recognized that Keith Roberts had that rare combination of attributes."[9] Watson himself thanked Roberts in the preface to the third edition of his textbook: "The illustrations again are largely the work of Dr. Keith Roberts. [...] Very few individuals are highly talented in both science and art and I was most fortunate in again obtaining his most unique assistance."[10] The final graphical implementation was the work of the New York–based painter Bill Prokos.

With Levinthal, Roberts, and Prokos, people surrounded Watson who each embodied a different combination of scientific and design practices. Another major influence on Watson's understanding of visual communication was Francis Crick's wife, the artist Odile Crick, who created the diagrammatic illustration of the double helix for Watson and Crick's paper in the April 25, 1953, issue of *Nature*.[11] Yet Watson himself was also profoundly aware of the significance of visual and tactile thinking, and we may assume that the visual strategy of *Molecular Biology of*

6 See Éric Francoeur, "Cyrus Levinthal, the Kluge and the Origins of Interactive Molecular Graphics," *Endeavour* 26, no. 4 (December 2002): 127–31.

7 Cyrus Levinthal, "Molecular Model Building by Computer," *Scientific American* 214, no. 6 (June 1966): 50.

8 Errol C. Friedberg, *The Writing Life of James D. Watson* (Cold Spring Harbor, NY: Cold Spring Harbor Laboratory Press, 2005), 142.

9 Friedberg, *Writing Life*, 143.

10 James Watson, *Molecular Biology of the Gene*, 3rd ed. (New York: W. A. Benjamin, 1976), viii. To this day, Roberts works at the interfaces between the sciences and art; he leads several art associations and gives out awards in the field called Sci-Art. See www.uea.ac.uk/ssf/cue-east/people; www.photo-id.org.uk/cms.php?categoryid=8.

11 See Horst Bredekamp, "Bild, Beschleunigung und das Gebot der Hermeneutik," in *Weltwissen: 300 Jahre Wissenschaften in Berlin*, ed. Jochen Hennig, Udo Andraschke, exhibition catalogue (Munich: Hirmer, 2010), 50–57.

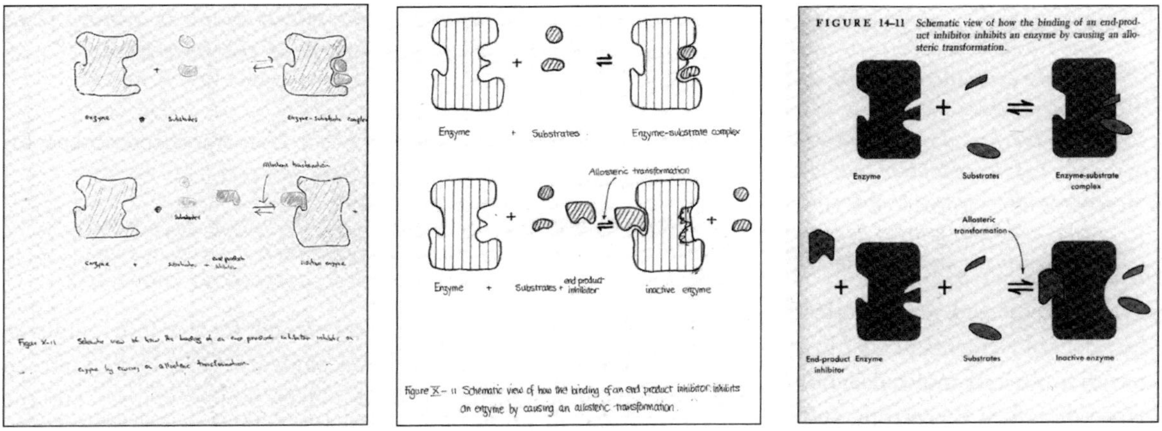

FIG. 4: Three phases of the genesis of an illustration for *Molecular Biology of the Gene*: from left to right, one of Keith Roberts's earliest sketches, a more definitive draft, and the ready-for-print design. Errol C. Friedberg, *The Writing Life of James D. Watson* (Cold Spring Harbor, NY: Cold Spring Harbor Laboratory Press, 2005), 143. Courtesy of the James D. Watson Collection, Cold Spring Harbor Laboratory Library and Archives.

the Gene was developed primarily at his instigation. As is well known, when Watson and Crick sought to determine the structure of DNA, one of the methods they relied on was a sort of "*bricolage* with macroscopic models,"[12] which is to say, a form of scientific thinking to which tactile and visual interaction with the material models were fundamental. As the following pages will show in the example of the genesis of the DNA model, Watson's work on this project led him to profound insights into the import of visual thinking for molecular biology.

Watson and Crick's programmatic work with scale model elements goes back directly to the chemist Linus Pauling and his use of models. Pauling's success in determining the structure of the α-helix, in particular, served them as a blueprint for their own undertaking. Pauling, according to his own recollections, drew a structural formula on a sheet of paper he then folded to produce a cylindrical shape. This allowed him to create a structure whose existence seemed at least compatible with the laws of chemistry.[13] His account was especially significant for Watson and Crick, since Pauling's colleagues Lawrence Bragg, John Kendrew, and Max Perutz had failed to solve the same problem, even though they, too, built models. The key to Watson and Crick's decision to adopt "solid fiddling with [...] molecular models"[14] as their own method may accordingly be found in what sets it apart from the approach Pauling's competitors had taken; Bragg, Kendrew, and Perutz, Crick argued, had built and used their models in a way that could not have led them to a solution: "Unfortunately they did not let the models take up their most favorable configurations."[15] What Crick means is that the other researchers sought to build a model that would passively represent the available data. By contrast, Pauling's success, Crick believed, was made possible by his allowing his models to fold of their own accord, as it were, breaking free of the requirement to represent something: "Pauling had not attempted to make the structure with an integer screw but had let the models fold naturally into any screw they were comfortable with."[16] Crick's choice of words suggests that he ascribed a sort of volition or self-will to the models, which Pauling's *bricolage* rendered fruitful, whereas Bragg, Kendrew, and Perutz had sought to suppress it.

Pauling's model as conceived by Crick thus emerges as what the historian of science Marx W. Wartofsky would have described as an action or a mode of action. In his 1968 essay "Telos and Technique: Models as Modes of Action,"[17] Wartofsky sets models in the context of Kant's categorical imperative, describing them as embodied actions that, like an action performed in accordance with the categorical imperative, we are called upon to emulate. The model in this sense, Wartofsky writes, is a "mode of action" and moreover a "call to action," inviting a particular action determined in characteristic fashion by the model in its situative performance. The model communicates that "this is how it ought to be done; this is what needs to be understood; this is how one ought to operate."[18] This notion of a "call to action" issued by a model is also at the root of Crick's observation that models seem to prefer certain configurations and that it may be fruitful to let them have their way.

Watson and Crick's conviction that models held the key to uncovering the structure of DNA led them to take scale model elements and assemble them in tentative fashion. The most famous insight gained

12 Hans-Jörg Rheinberger, "Kurze Geschichte der Molekularbiologie," in *Geschichte der Biologie: Theorien, Methoden, Institutionen, Kurzbiographien*, ed. Ilse Jahn (Heidelberg: Spektrum Akademischer Verlag, 2000), 651. Rheinberger's term translated here as "bricolage," *Basteln*, alludes to the concept in Lévi-Strauss; see Claude Lévi-Strauss, *The Savage Mind* (Chicago: University of Chicago Press, 1966), 16–34.

13 Robert Olby, *The Path to the Double Helix: The Discovery of DNA* (New York: Dover, 1994), 280–81; Horace Judson, *The Eighth Day of Creation: Makers of the Revolution in Biology* (Cold Spring Harbor, NY: Cold Spring Harbor Laboratory Press, 1996), 62; Rheinberger, "Kurze Geschichte der Molekularbiologie," 651.

14 James Watson, *The Double Helix: A Personal Account of the Discovery of the Structure of DNA* (New York: Atheneum, 1968), 77.

15 Francis Crick, *What Mad Pursuit: A Personal View of Scientific Discovery* (New York: Basic Books, 1988), 57.

16 Ibid., 58.

17 Marx W. Wartofsky, "Telos and Technique: Models as Modes of Action," in *Models: Representation and the Scientific Understanding* (Dordrecht: D. Reidel, 1979), 140–53.

18 Ibid., 143; Michael Erlhoff similarly remarks: "'That is how it should be,' the model says. Or at least, 'that is how it might be if everything went well.'" Michael Erlhoff, "Modelliert: Model the Model," *form: Zeitschrift für Gestaltung*, no. 198 (September–October 2004): 49.

using these elements concerns base pairing, which enabled them to create the presentation model that would later be well known and widely popular. In his book *The Double Helix*, published in 1968, Watson recalls the event as follows: "I spent the rest of the afternoon cutting accurate representations of the bases out of stiff cardboard. [...] Suddenly I became aware that an adenine-thymine pair held together by two hydrogen bonds was identical in shape to a guanine-cytosine pair held together by at least two hydrogen bonds. All the hydrogen bonds seemed to form naturally; no fudging was required to make the two types of base pairs identical in shape."[19] If this story is to be believed, physical—which is to say, tactile and visual—aspects played a central role in the discovery of base pairing. It is a replay, as it were, of Pauling's discovery of the structure of the α-helix, and Watson's "seemed to form naturally" echoes Crick's observation that Pauling allowed his models to fold as they pleased. When the physicist Maurice Wilkins later noted that "model building is not a mere illustration of thought, but enables the mind to explore and find new structures that may otherwise not appear out of imaginative processes,"[20] Watson's discovery was an important example.

The visual strategy of *Molecular Biology of the Gene* directly reflects these lessons. The example discussed above demonstrates how the productive nature of the models might be rendered by didactic images. Just as Pauling construed his diagram on the folded sheet of paper *as* the molecular structure of the α-helix, just as Watson took the four pieces of cardboard he had cut *to be* bases, the student of molecular biology is to read the three bronze-colored dots as viruses, using them as visual elements in a playful process of understanding. These tactile and visual principles of speculative model building were a crucial factor in the great success of *Molecular Biology of the Gene* as well as of the style it pioneered, which, as Errol Friedberg notes, molecular biology textbooks emulate even today: "The textbook *Molecular Biology of the Gene* [...] comprehensively documented the discipline of molecular biology for the first time. Its design, structure, and clarity would unalterably change the landscape of textbook publishing in biology."[21]

Yet the book's popularity is explained not only by the fertility of its approach but also by the fact that the underlying principles of tentative concretization ran diametrically counter to the conceptions of the research process and especially of the role illustrations and models played in it that prevailed from the 1940s to the 1960s. The theoretical frameworks of, say, Kenneth Craik,[22] Arturo Rosenblueth and Norbert Wiener,[23] or Alfred Tarski[24] and Patrick Suppes,[25] all of which grew directly or indirectly out of Heinrich Hertz's work,[26] regarded the existence of an isomorphism between a model and the object it referred to as the foundation for any permissible application of observations in the model to the actual object of interest. These approaches treat models more or less strictly as final products, passive illustrations of insight that has already been gained, and not as objects an exploratory research process engages with in a tentative concretization whose failures may be no less significant than their successes. In the perspective of these theories, it must remain incomprehensible that speculative model building as practiced by Linus Pauling or Watson and Crick might generate scientific insight.

19 Watson, *Double Helix*, 194–95; see also Judson, *Eighth Day of Creation*, 149; Olby, *Path to the Double Helix*, 412.

20 Maurice Wilkins, *The Third Man of the Double Helix: The Autobiography of Maurice Wilkins* (Oxford: Oxford University Press, 2003), 231.

21 Friedberg, *Writing Life*, 16.

22 Kenneth Craik, *The Nature of Explanation* (Cambridge: Cambridge University Press, 1952), 51–52 and passim.

23 Arturo Rosenblueth and Norbert Wiener, "The Role of Models in Science," *Philosophy of Science* 12, no. 4 (October 1945): 316–21.

24 Alfred Tarski, "Über den Begriff der logischen Folgerung," *Actes du congrès international de philosophie scientifique*, vol. 7: *Logique* (Paris: Hermann, 1936), 1–11.

25 Patrick Suppes, "A Comparison of the Meaning and Uses of Models in Mathematics and the Empirical Sciences," in *The Concept and the Role of the Model in Mathematics and Natural and Social Sciences*, ed. Hans Freudenthal (Dordrecht: D. Reidel, 1961), 165.

26 Heinrich Hertz, *Die Prinzipien der Mechanik in neuem Zusammenhange dargestellt* (Leipzig: Geest & Portig, 1894), 197–99, and see also ibid., 1–5; cf. Lopes R. Coelho, "Der Begriff des Bildes bei Hertz," *Logos*, Neue Folge, 3, no. 4 (1996): 271–92.

So James Watson faced an interesting situation when, in the 1960s, he began to draft his textbook of molecular biology: Pauling's discovery of the structure of the α-helix and Crick's and his own discovery of the structure of DNA had proven exceptionally consequential, but the way these discoveries were made contrasted sharply with the era's prevailing theoretical definitions of what a model was. At this juncture, the models were not just productive tools: they also became the hallmarks of a new thought style that had left the paradigm of representation behind and instead turned to a play with models and a fruitful dialogue with media and materials, with the self-will of things. From now on, the visibility of models not only served a playful investigation of unknown molecular constellations and the didactic presentation of findings in molecular biology; it also became the distinguishing feature of a new visual discipline that broke free of an obsolescent traditional understanding of research in the theory of science.

In light of this conjunction between the tools of thinking and the medium of popularization and given the outright rejection of these tools by the classical approaches in the theory of science, the wide variety of pictorial representations in Watson's *Molecular Biology of the Gene* may thus also be read as the visual expression of the disciplinary autonomy the young field of molecular biology had attained. The illustrations remind the reader of the most important events in the field's history, rendering the history of molecular biology in condensed form and indicating the ascendancy of a visual thought style. As the three bronze-colored dots show, the images themselves function as "calls to action," as visual models that, by virtue of their "interpretive gaps,"[27] challenge the reader not to wait passively for insight to arrive but to play with them, as Pauling and Watson and Crick had played with their models.

27 Jordanova, "Material Models as Visual Culture," 447.

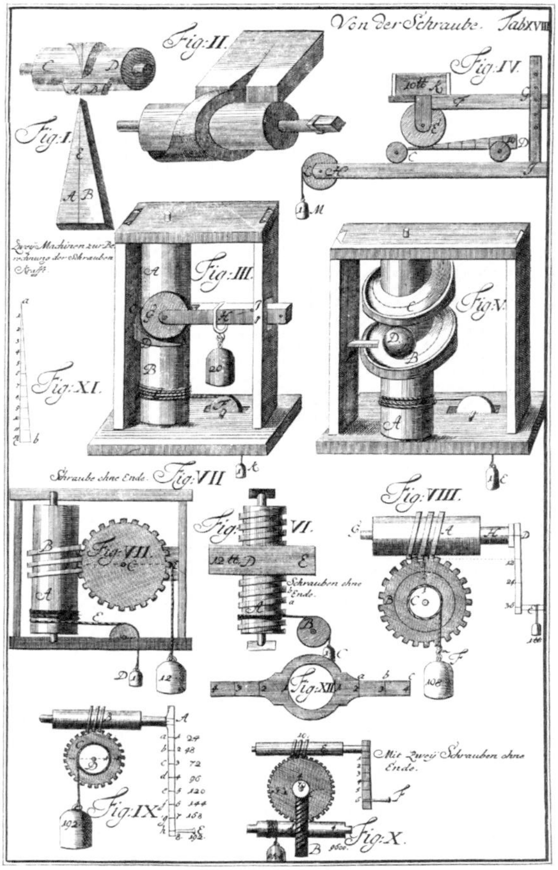

FIG. 1: "Von der Schraube" (On the screw), from Jacob Leupold's *Theatrum machinarum Generale* (Leipzig, 1724), plate 18. A screw is here defined as a wedge wound around an axle. In further illustrations, the plate shows various machines for calculating the strength of screws in different views, with scales and precise details. Jacob Leupold, *Theatrum Machinarum Generale. Schau-Platz Des Grundes Mechanischer Wissenschafften* (Leipzig: Zunkel, 1724), plate 18.

ARRANGING IMAGES AS TABLEAUX

The formal principle of graphically arranging several pictorial elements in a plate or image area so as to put them into a certain relationship with each other has so far only been studied in isolated cases with respect to scientific and technical images. Analyzing the way images are formally arranged to provide an overall view reveals iconic argumentation structures, which, in the form of image tableaux, define a specific tradition of scientific-technical image formats originating in the early modern period (fig. 1).

The term *tableau* has long been used in art history for a panel or painting (Stoichita 1998). Since the nineteenth century, however, a second meaning has come to include overviews in a scientific context in which individual elements of an image are arranged next to each other in an illustrated plate. Such overviews may show a zoological subject in various stages of development, views, proportions, etc., with individual elements positioned next to each other across the image (fig. 2). This allows for a simultaneous portrayal of chronologically or spatially separate phenomena for the purposes of comparative seeing. Tableaux provide an overview, convey information concisely, and have been used as charts in teaching and as illustrated plates in scientific texts, often in educational contexts (Te Heesen 1997). They thus often contribute to popularizing scientific knowledge (fig. 3).

In a broader meaning, the term *tableau* refers to the arrangement of images and juxtaposition of illustrated plates or image fields for the purpose of visual argumentation and comparison (Tufte 1997). Art historians have become interested in this form of image arrangement, especially in connection with Aby Warburg's *Mnemosyne Atlas* of images (fig. 4). Warburg's visual

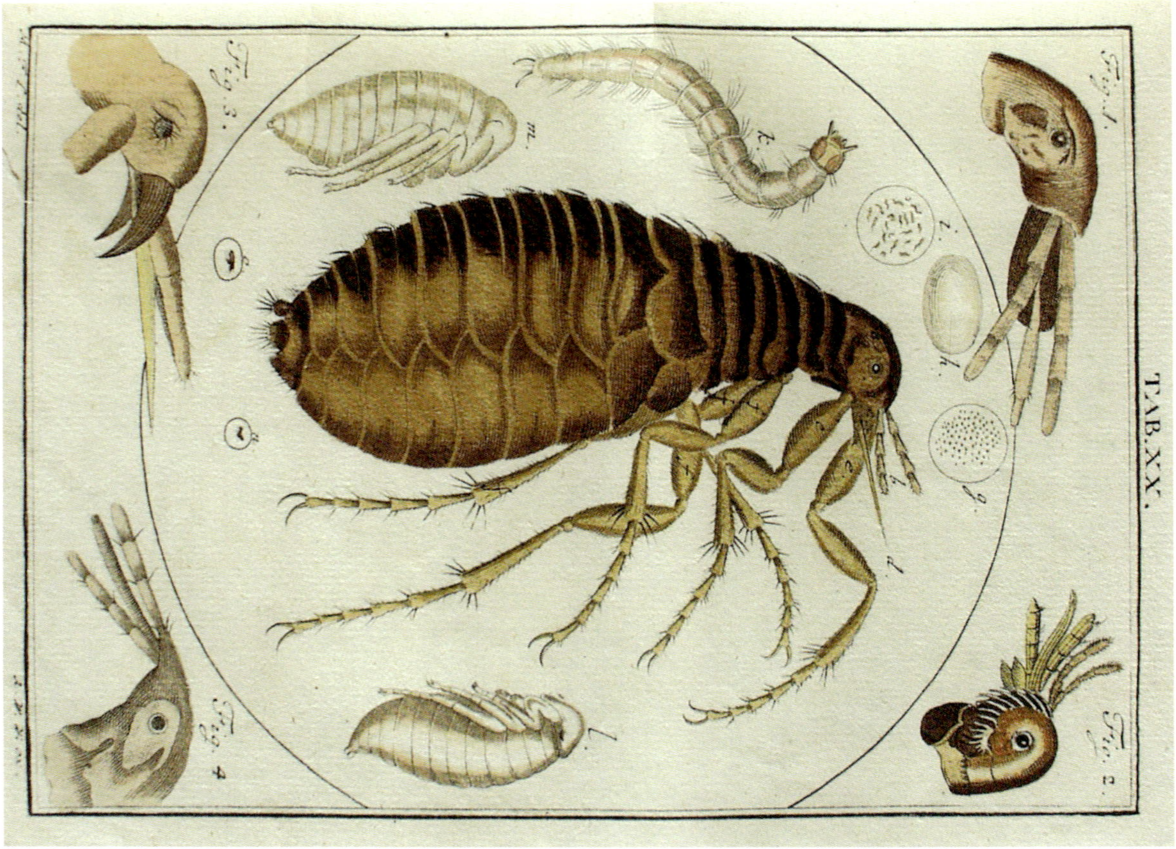

FIG. 2: Colored copperplate engraving from Martin Frobenius Ledermüller's *Mikroskopische Gemüths- und Augen-Ergötzung* (Microscopic delights for the eye and the mind), 1763. It portrays the various developmental stages of a flea. Two smaller circles depict the specimen in its original size and are juxtaposed for comparison in a central, circular detail, which simulates the gaze through the microscope. The image's corners show four views of the head of the flea. Martin Frobenius Ledermüller, *Mikroskopische Gemüths- und Augen-Ergötzung. Bestehend in ein Hundert nach d. Natur gez. u. mit Farbe erleuchteten Kupfertafeln samt deren Erklärung* (Nuremberg: Winterschmidt, 1763), plate 20.

organization of individual images covering a range of subjects from different disciplines and periods in the form of photographs is the intensely discussed nucleus of *Bildwissenschaft,* which focuses on form (Warburg 2000; Michaud 2004). Visual argumentation through an arrangement of images can be made by means of a grid structure (fig. 5), as in Lavater's engraved plates of physiognomy, or in the form of a successive series, as in catalogues of images (e.g., Comenius's *Orbis sensualium pictus*) and pictorial encyclopedias (e.g., Diderot's *Encyclopédie*). In their most abstract form, such arrangements serve to order data visually into individual fields or sectors and graphically collate and structure them into tables or grids (Campbell-Kelly et al. 2003). Ordering images is therefore also the starting point for a symbolic ordering in the form of collecting and classifying images and objects, as expressed in the act of ordering the materials of field research or in archives and collections to create an epistemic overview (Latour 1999).

The various principles of visual argumentation conveyed by different arrangements of images have not yet been adequately investigated. Issues such as the argumentative coherence of the object represented and the leading of the viewer's gaze, presenting various details, aspects, fragments, cross sections, different scales, and perspective views, as well as issues of the constancy and variance of the object shown, planar or spatial representation principles, composition, etc., can yield information about the suggestive argumentation and communicative strategies being pursued in the technical-scientific field through images. To investigate these strategies further, art historical research into visual

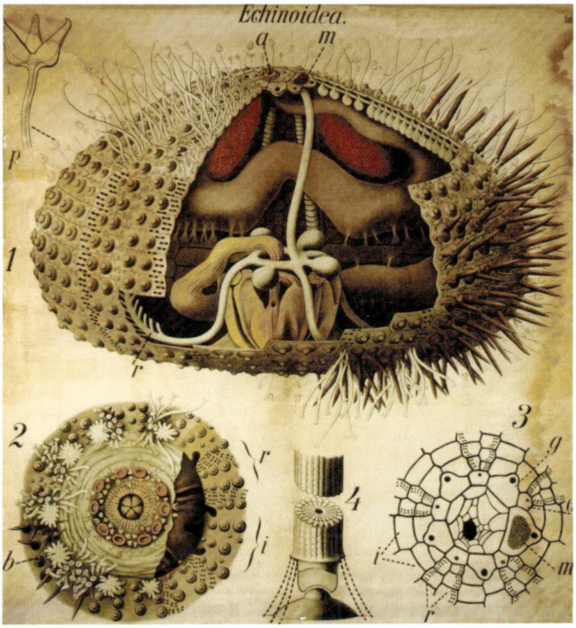

FIG. 3: This wall chart of "Echinoidea" by Paul Pfurtscheller (1890) shows the structure of a common sea urchin in four different views and representational modes. Pfurtscheller, a Viennese high school teacher, originally drew the charts for his classes, so they were designed for use in a school, but their outstanding scientific and artistic quality meant that they were soon used in zoological departments at universities. Pfurtscheller made a total of thirty-eight such charts. Collection of the Department Vergleichende Zoologie, Humboldt University of Berlin, Germany.

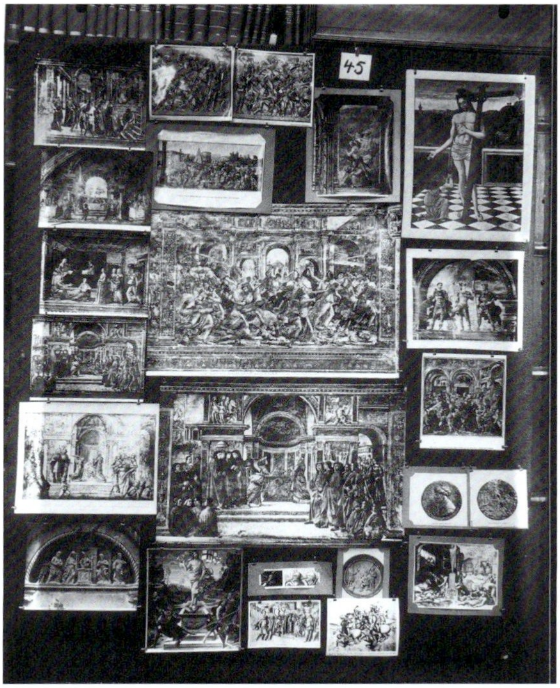

FIG. 4: Plate 45 from Aby Warburg's *Mnemosyne Atlas*. This plate from Warburg's unfinished Atlas project, which he began working on in 1929, unites "Superlative der Gebärdensprache" (superlatives of the language of gesture) in reproductions of paintings, frescoes, sculptures, and medallions by Italian artists working mainly around 1500. The two large reproductions in the middle show frescoes by Domenico Ghirlandaio, the "Massacre of the Innocents" and the "Annunciation to Zachariah," from the Capella Tornabuoni in S. Maria Novella in Florence, created around 1485–1490. The latter scene is reiterated in a smaller version in the left register in the context of further frescoes. Archive of The Warburg Institute, United Kingdom. © The Warburg Institute.

argumentation, the narrative structures and arrangement of images, e.g., in stained-glass windows or comic strips, and the pictorial principle of images within images should be taken into account (McCloud 1993; Kemp 1997; Bann 2001; Bogen 2005; Bender 2010; Harmon 2012).

Specific arrangements of scientific and technological images in the form of overviews also emerge from chronological and successive imaging techniques and the data measured. The resulting images are presented in series, as in chronophotography (fig. 6) or the sectioned images of computer tomography. Here individual images are compiled into tableaux that impose a certain linear direction for their reading. Apart from imaging techniques used in laboratories or experimental scientific contexts, new means of visual argumentation are emerging in research communication with the use of digital processes and software like PowerPoint that make arrangements of images more dynamic and interactive. —MP

LITERATURE

Bann, Stephen. *Parallel Lines: Printmakers, Painters, and Photographers in Nineteenth-Century France*. New Haven, CT: Yale University Press, 2001.

Bender, John B., and Michael Marrinan. *The Culture of Diagram*. Stanford, CA: Stanford University Press, 2010.

Bogen, Steffen. "Verbundene Materie, geordnete Bilder: Reflexion diagrammatischen Schauens in den Fenstern von Chartres." In *Diagramme und bildtextile Ordnungen*, ed. Birgit Schneider, 72–84. *Bildwelten des Wissens: Kunsthistorisches Jahrbuch für Bildkritik*, ed. Horst Bredekamp, Gabriele Werner, and Birgit Schneider, vol. 3, no. 1. Berlin: Akademie Verlag, 2005.

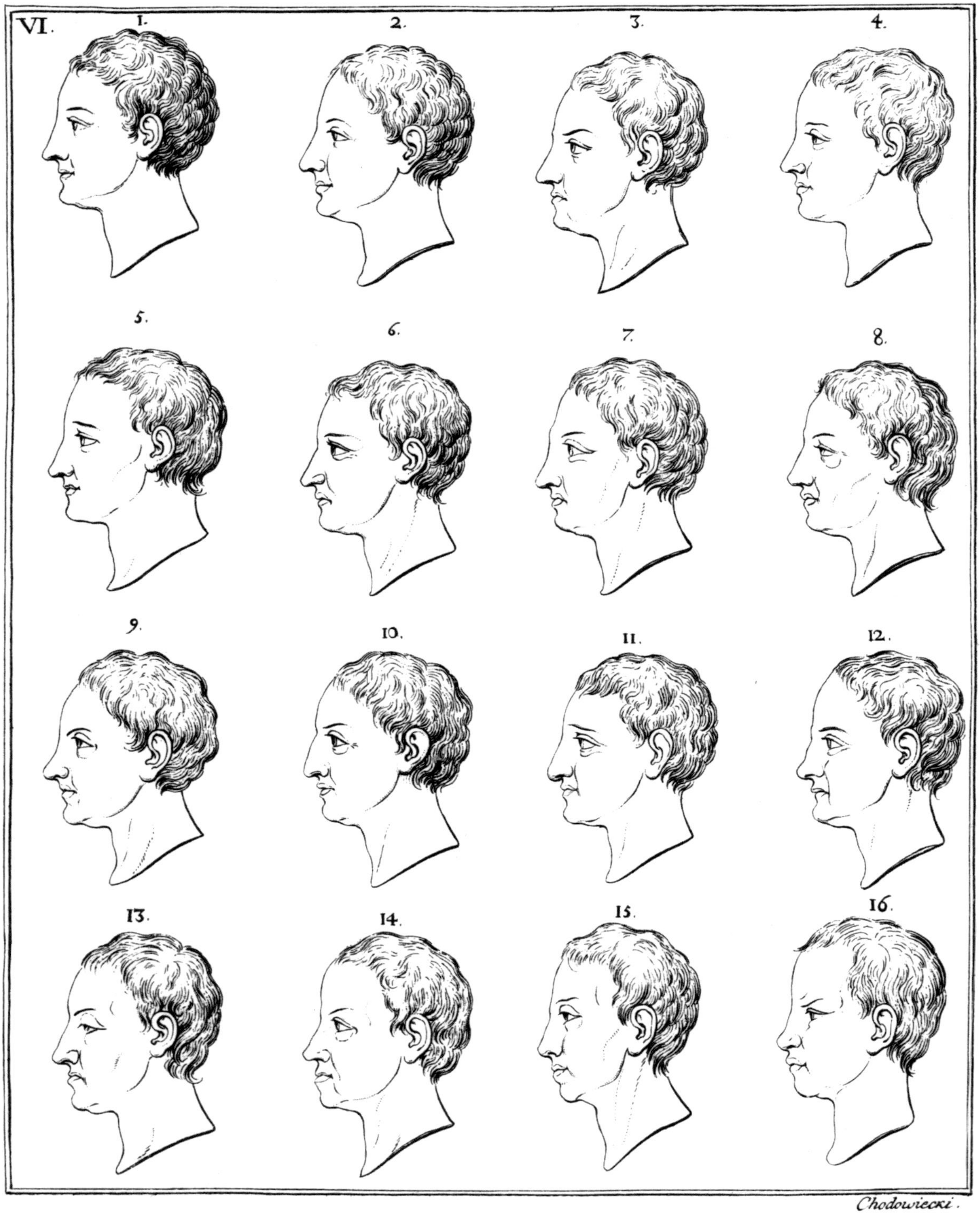

FIG. 5: Sixteen idealized heads in profile based on Chodowiecki and reproduced in Lavater's *Physiognomische Fragmente* (Physiognomical fragments), 1776, under the heading "schwache, thörichte Menschen" (weak, foolish people). Lavater identified people's characters based on their facial features, which he varied here in a series from figs. 1 to 16 and arranged in sequence so that the characteristics of the "thörichte Menschen" successively intensify. Johann Caspar Lavater, *Physiognomische Fragmente, zur Beförderung der Menschenkenntnis und Menschenliebe* (Leipzig: Bey Weidmanns Erben und Reich, und Heinrich Steiner und Compagnie, 1776): sixteenth fragment, plate 6.

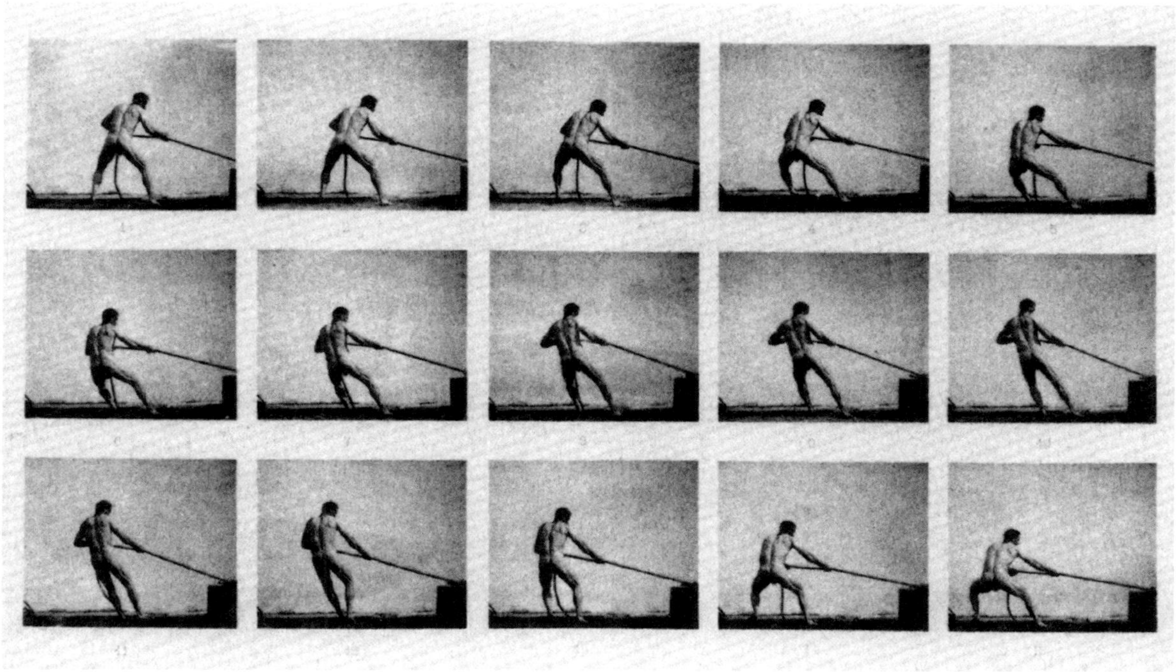

FIG. 6: Étienne-Jules Marey, Chronophotography of a man pulling on a rope from *Études de physiologie artistique faites au moyen de la chronophotographie,* 1893. Serially ordering photographs taken at brief chronological intervals made it possible to analyze sequences of motion. They are arranged here into a tableau that is designed to be read in a specific direction. Bodo v. Dewitz, ed., *Ich sehe was, was du nicht siehst! Sehmaschinen und Bilderwelten. Die Sammlung Werner Nekes.* Exh. cat. (Göttingen: Steidl, 2002), 372, fig. 7.

Campbell-Kelly, Martin, Mary Croarken, Raymond Flood, and Eleanor Robson, eds. *The History of Mathematical Tables: From Sumer to Spreadsheets.* Oxford: Oxford University Press, 2003.

d'Alembert, Jean le Rond, and Denis Diderot. *Encyclopédie ou Dictionnaire raisonné des sciences, des arts et des métiers.* 28 vols. Paris: Briasson; David l'aîné; Le Breton; Durand, 1751–1772.

Hamann, Byron Ellsworth. "Drawing Glyphs Together." In *Past Presented: Archaeological Illustration and the Ancient Americas,* ed. Joanne Pillsbury, 230–81. Washington, DC: Dumbarton Oaks Research Library and Collection, 2012.

Hunter, Matthew C. "Cascade, Copper, Collection: Constellations of Images in 1670s Experimental Philosophy." In *Wicked Intelligence: Visual Art and the Science of Experiment in Restoration London,* 125–58. Chicago: University of Chicago Press, 2013.

Kemp, Wolfgang. *The Narratives of Gothic Stained Glass.* Cambridge: Cambridge University Press, 1997.

Kunzle, David. "The Early Comic Strip: Narrative Strips and Picture Stories in the European Broadsheet from c. 1450 to 1825." In *History of the Comic Strip,* vol. 1. Berkeley: University of California Press, 1973.

Latour, Bruno. "Circulating Reference: Sampling the Soil in the Amazon Forest." In *Pandora's Hope: Essays on the Reality of Science Studies,* 24–79. Cambridge, MA: Harvard University Press, 1999.

McCloud, Scott. *Understanding Comics: The Invisible Art.* Northampton, MA: Kitchen Sink Press, 1993.

Michaud, Philippe-Alain. "Zwischenreich: Mnemosyne, or Expressivity without a Subject." In *Aby Warburg and the Image in Motion,* 251–76. New York: Zone Books, 2004.

Middleton, Robin, and M. N. Baudouin-Matuszek. *Jean Rondelet: The Architect as Technician.* New Haven, CT: Yale University Press, 2007.

Schoell-Glass, Charlotte. "Serious Issues: The Last Plates of Warburg's Picture Atlas 'Mnemosyne.'" In *Art History as Cultural History: Warburg's Projects,* ed. Richard Woodfield, 183–208. Amsterdam: G+B Arts International, 2001.

Stoichita, Victor I. *The Self-Aware Image: An Insight into Early Modern Meta Painting.* Cambridge: Cambridge University Press, 1997.

Te Heesen, Anke. "Verbundene Bilder: Das Tableau in den Erziehungswissenschaften des 18. Jahrhunderts." In *Bilder als Quellen der Erziehungsgeschichte,* ed. Hanno Schmitt, Jörg W. Link, and Frank Tosch, 7–90. Bad Heilbrunn: Klinkhardt, 1997.

Te Heesen, Anke. *The World in a Box: The Story of an Eighteenth-Century Picture Encyclopedia.* Chicago: University of Chicago Press, 2002.

Tufte, Edward R. *Visual Explanations: Images and Quantities, Evidence and Narrative.* Cheshire, CT: Graphics Press, 1997.

Warburg, Aby. *Der Bilderatlas Mnemosyne.* Ed. Martin Warnke. Berlin: Akademie Verlag, 2000.

Technological Image Series: The Project "Technik im Bild" at the Deutsches Museum, Munich

Heike Weber

A series of images—be it a comic strip, a picture book, or a film—can tell a story through the interrelations between the individual images in their sequential arrangement. The "Technik im Bild" compilation of images at the Deutsches Museum, Munich, produced in the 1930s, relies on this principle of pictorial narration: historical and contemporary visual documents from a wide variety of fields of technology were assembled and published in the familiar format of the photographic album; the collection eventually encompassed around a hundred and fifty different titles. The chronologically arranged image series in these albums represent the evolution of a given technology as a history of successive advances that culminates in the highly differentiated procedures of modern industrial and mass production; series of photographs taken in contemporary manufacturing facilities, meanwhile, take the beholder on virtual tours of industrial production processes, from raw materials to the finished products. The medium of the image served to present technology in an easily comprehensible fashion: the image series break up technical workflows into understandable stages; individual images offer impressions of equipment and its operation or give schematic representations of the principles of technology. The image held the promise of drawing the interest of the layperson, introducing broad popular audiences to technological knowledge and guiding them through technical processes. Photography, for its part, was regarded as a medium of precise documentation as well as cheap reproduction.

A photograph of an engraving showing two overflowing vats over a fire (fig. 1) opens a photo series compiled for a portfolio that bears the title "Glas" ("Glass"). As the beholder turns the volume's pages, sheets of black cardboard with photographs mounted on both sides alternate with pages of paper bearing brief typescript remarks to complement the pictures. After a while, the reader reaches image no. 18 (the handwritten number appears to the top left of the picture): a company photograph of furnace batteries in a factory building devoid of workers (fig. 2). Considered in isolation, these two photographs have little more in common than the fact that they illustrate some sort of technological matter; they are, moreover, instances of different pictorial genres, and even the dimensions of the prints vary. They make coherent sense only as well-ordered links in a long chain of images that unfolds as the beholder pages through the portfolio. In around a hundred and twenty photographs, it illustrates the history of glass production and its present-day forms. Compiled by an instructor at a trade school of glass technology who drew on a wide

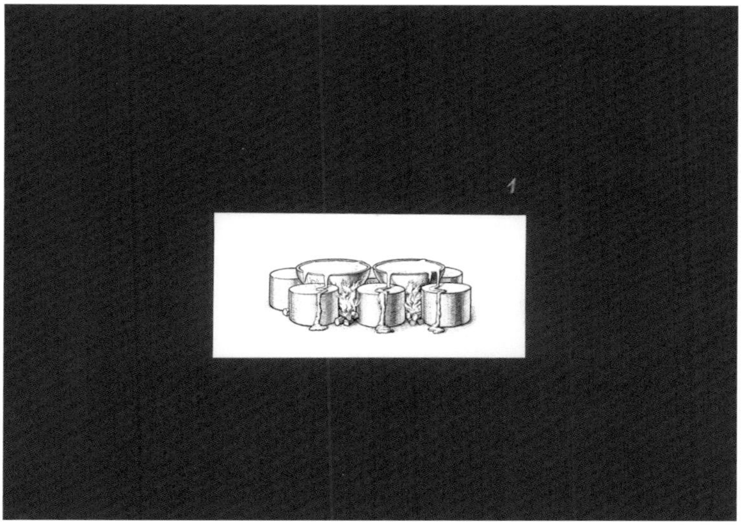

FIG. 1: Picture no. 1 from the "Glass" portfolio. Archives of the Deutsches Museum, Munich, Germany.

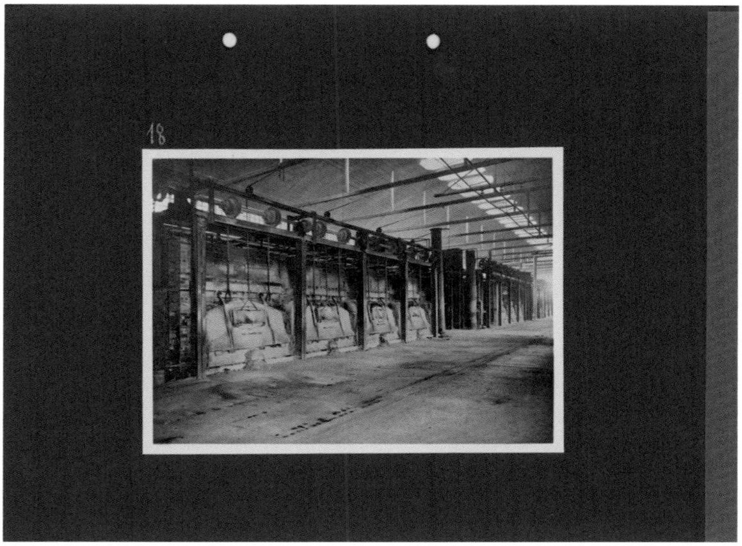

FIG. 2: Picture no. 18 from the "Glass" portfolio: front view of a pot furnace used to produce plate glass. Archives of the Deutsches Museum, Munich, Germany.

variety of visual sources, the portfolio itself is a unique copy. At the same time, it is part of a continuously growing body of images entitled "Technik im Bild" ("Technology in the Image") that was created at the Deutsches Museum starting in 1933; the albums were laid out in the museum's library, where the public was invited to inspect them.[1] The following pages will discuss the composition, effects, and aims of these technological image series.

Using Images to Popularize Technology: On the Idea behind "Technik im Bild"

"Technik im Bild" was initiated in 1932 by Oskar von Miller (1855–1934). He had founded the Deutsches Museum in 1903 as an institution of popular education in the natural sciences and technology; complementing the museum's established avenues of the dissemination of knowledge—the exhibition and the library—"Technik im Bild" was meant to provide

1 I would like to thank Dr. Wilhelm Füßl, the director of the Deutsches Museum's archives, who not only brought the "Technik im Bild" project to my attention but also assisted me in my search for additional sources. The following pages discuss two of the "Technik im Bild" portfolios contained in the museum's image archive: "Glass" (no portfolio no.; edited by Prof. Dr. Ludwig Springer, head of the department of chemical engineering at the Staatliche Fachschule für Glasindustrie, Zwiesel, Bavaria), Deutsches Museum, archive, BA 0651 and BA 0652; and "Brewing" (portfolio no. 88; edited by Dr. F. Eckhardt, a retired brewing chemist), Deutsches Museum, archive, BA 0782.

a decidedly image-based presentation platform. The "image show"[2] would treat distinct bodies of knowledge, primarily in technology and only secondarily in the natural sciences, in "portfolios" of between fifty and a hundred annotated photographs each. The volumes would find their place in the museum's library, one of the best technological and scientific libraries of the time. Oskar von Miller envisioned a set of no less than three hundred—perhaps even five hundred—such portfolios or, in other words, a collection of perhaps up to thirty-five thousand photographs;[3] even so, "Technik im Bild" would not cover all fields of technology. Rather, the aim was to explore those areas of knowledge where "the visual material, without the admixture of a particularly scientific element and without any pretension to exhaustive coverage, may, as real 'picture-books for adults,' prove effective entertainment for wide audiences," as the minutes of a meeting on September 26, 1932, note.[4] The project's encyclopedic aspiration was set aside in favor of the diverting example.

How many such portfolios, with what titles, ultimately graced the library's shelves cannot be said with certainty. When Oskar von Miller resigned from his office as the museum's honorary director in May 1933, he was able to inaugurate "Technik im Bild," his last major project at the museum, which then encompassed around a hundred portfolios; additional titles were under preparation. Most of the image shows were assembled by external specialists from industry, trade associations, and universities, whom the museum assisted with the technical aspects of actually making the portfolios. Members of the staff at the Deutsches Museum also conceived a few titles. By the mid-1930s, the collection of portfolios had probably grown to around a hundred and fifty volumes. After von Miller's retirement, however, no focused effort was made to carry on the project; on the other hand, this relatively low profile also prevented its exploitation for the technology propaganda of the Nazis.[5]

When the library was reestablished after the war, "Technik im Bild" was not reconstituted. Today, the museum's archive contains fifty-one image series, some of them in multiple parts; only one is in the original condition.[6] The remaining series were transferred into ring binders. To this end, the lateral folds, the cord binding, and the hard covers of the picture portfolios were removed; the papers and cardboards, originally wide formats, which had an integrated fold and staggered tabs for easier page turning, were cropped. Only the remnants of a fabric adhesive strip on the edges of many of the cardboard backings suggest the lavish craftsmanship that went into the original design. To render the pages compatible with standardized DIN A4 binders, holes were also punched in all of them—but along the top edge rather than the sides, where the original binding ran. Although everyone involved with the project spoke of *Mappen* ("portfolios"), that is to say, the volumes were actually photo albums with images permanently attached to the pages.

The Image Series as a Pictorial Program of Education

Few documents concerning the specifics of how the idea of "Technology in the Image" was to be implemented are extant today. Yet the content and configuration of the surviving portfolios or albums illustrate the goals and rhetoric of the erstwhile image show. Although some photographs have been torn from the cardboard backings, most of the albums present

2 The term *image show* came up during the committee meeting in May 1932 at which Miller presented his idea; see Oskar von Miller, "Bericht über die Museums-Bibliothek," in *Verwaltungs-Bericht über das 28. Geschäftsjahr Mai 1931 bis Mai 1932 und Bericht über die 21. Ausschuss-Sitzung des Deutschen Museums*, 30.

3 These figures are drawn from various extant administrative files concerning the "Technik im Bild" project; see Deutsches Museum, archive, VA 5043.

4 See the transcript "Besprechung mit Herrn Dr. von Miller am 26. 9. 32 nachmittags 4.30." The meeting was convened to discuss the image show. Deutsches Museum, archive, VA 5043.

5 For a more extensive discussion, see Heike Weber, "Technik im Fotoalbum: Die Bilderschau 'Technik im Bild' am Deutschen Museum," in *Konstruieren, Inszenieren, Präsentieren: Bilder von Wissenschaft und Technik*, ed. Alexander Gall (Göttingen: Wallstein, 2007), 397–434.

6 "Der Bau des Deutschen Museums" ("The Construction of the Deutsches Museum"), portfolio no. 59, edited by the architect Karl Bäßler.

the original sequence of pictures on cardboards and text on paper, which follows a fairly consistent formal pattern.

Each portfolio opens with a printed title page of uniform design. Beneath the identifiers "Deutsches Museum—Bibliothek—Technik im Bild," a blank, not always filled, provides space for the portfolio number; another field is reserved for the name of the editor of the individual volume. This title page is often, though not always, followed by a "table of contents" giving brief titles of all pictures or else a list of "picture credits" specifying the provider of each photograph or reproduction, but without precise information about the original source. In some portfolios, such indices appear near the end; that is also where some contain thematically specific bibliographies.

The sequential arrangement materialized in the album constitutes the basic formal principle of the image show proper. The bound album invites the beholder to turn its pages, look at the individual images, and read them together as suggested by the sequence. By flipping back and forth ad libitum, the individual reader may draw additional connections guided by her own interests and adapt the reception of the images to her own pace and comprehension. The arrangements of images and text differ considerably among portfolios and sometimes within one and the same portfolio: the photographs are of different formats; some cardboards hold one large picture, others, up to four smaller photographs; the pictures often, but not always, come with captions on strips of paper; depending on the editor of a given portfolio, the texts vary widely in terms of style, length, and their relation to the picture. Other features, however, are sustained throughout the collection: the page-by-page alternation of images and text and the consistent arrangement of the photographs on the cardboards, which identifies the album as a vehicle of information. In keeping with the format of the photo album, the images are arranged chronologically as well as grouped by subtopic and often also numbered.

In order to get an image series to tell a story, the individual pictures must evince a sequential order that is meaningful to the eye and the understanding.[7] The editors of the portfolios thus faced the challenge of arranging the imagery required to treat their particular subject so as to trace the contours of an argument. To be convincing, an image series needs elements that connect each picture to the preceding and subsequent images, tying them together like the links of a chain. This structural requirement made the format particularly suitable to two narrative patterns in the popularization of technology that are widely current even today: the description of the history of technology as one of linear progress, with each new invention building on the one that preceded it, and the description of the genesis of a commodity from raw material to finished product.

7 See Detlef Hoffmann, ed., *Erzählende Bilder* (Oldenburg: Isensee, 1998).

The History of Technology as a Developmental Series in Pictorial Stages

Almost all portfolios begin with a series of images illustrating the historical development of the respective technology. The opening picture of the "Glass" portfolio (fig. 1), for instance, visualizes what we know about glassmaking in ancient Egypt from a written source, the oldest such document (ca. 1600 BCE). Sand and soda were melted together over an open fire. After reheating, the molten mass could be shaped into glass objects. Since there are no visual documents of this process, the folio's editor adapted a schematic illustration from a widely available book on glassmaking but listed only the Photography Department of the Deutsches Museum in the picture credit. That department produced numerous reproductions for this as well as the other portfolios; glass manufacturers, industrial associations, and technical colleges supplied additional photographs.

The photographically reproduced drawing reduces the ancient Egyptian technique to the basic elements of open wood fire and flat melting pans; only the accompanying note on the facing page informs the reader of the contents of the pans and the various processing stages. The following pages present photographs of ancient glass objects, including transparent—i.e., decolorized—containers resembling bottles. Next are around ten images showing the production of glass in a brick-built furnace, which allowed for higher melting temperatures. The "painting" reproduced as image no. 7a (fig. 3, left) is a miniature from a 1023 CE manuscript of Hrabanus Maurus's ninth-century treatise *De universo*. It is the earliest known depiction of a brick furnace. Mounted next to it and numbered 7b is the reproduction (fig. 3, right) of a late-medieval illustration of a glassworks from Jean de Bourgogne's accounts of his travels. Again, the image has probably been sourced from the available specialized literature; that is also suggested by the inaccurate dating—like some of the literature, the annotation puts the manuscript in the fourteenth rather than the early fifteenth century.[8] Placing them on the same cardboard treats the two pictures as representations of the same stage of technological development. The beholder looks for, and finds, similar tools and procedural steps, like the brick furnace installations, the use of the blowing iron, and the temporary storage of glass objects in segregated parts of the furnace after shaping. Precisely because the lay reader cannot make sense of many details in the pictures such as the manual operations the man on the left in the miniature (fig. 3, left) is performing, her attention is drawn to salient features and changes of the technology. The succession of images from cardboards no. 1 to no. 7 thus generates the argument that glassmaking developed in a steady series of innovations, with each later technology appearing as a consistent improvement on the earlier one. This narrative suggests not only

8 For example, the image appears in print in Hans Schulz, *Die Geschichte der Glaserzeugung* (Leipzig: Akademische Verlagsgesellschaft, 1928), 31, where it is dated as "ca. 1340." That this date is inaccurate was already pointed out by F. Rademacher, *Die deutschen Gläser des Mittelalters* (Berlin: Verlag für Kunstwissenschaft, 1933), where plate A (ibid., p. 1) presents a detail from the painting with a complete citation of the original source, a manuscript in the British Museum.

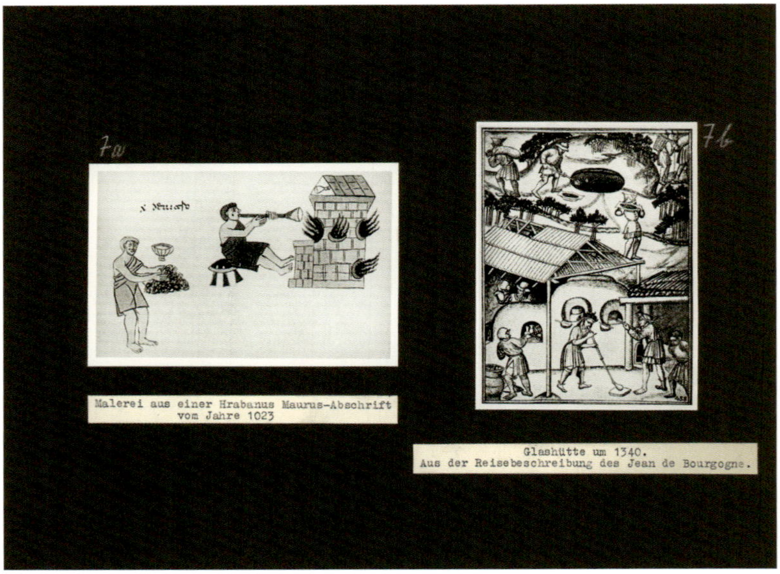

FIG. 3: Two reproductions of medieval manuscript illustrations visualizing the production of glass, arranged side by side on one cardboard backing. Archives of the Deutsches Museum, Munich, Germany.

a continuity of technological progress over time but also the existence of a space of knowledge that unites the geographical spaces of Egypt and Europe.

The visual experience of technological progress continues in the images that follow. Turning over cardboard no. 7, the beholder encounters, on the verso, two photographic reproductions of woodcuts from Georgius Agricola's handbook of mining and metallurgy, *De re metallica* (fig. 4). Glass is among the subjects Agricola addressed in the twelfth book of this richly illustrated treatise; all woodcuts he included to explain the furnaces, tools, and operations can be found in the "Glass" portfolio, where the captions identify them vaguely as showing methods "of Agricola's time." Image no. 8a (fig. 4, left) illustrates how the solidified mass obtained by cooling the initial melt is shattered. Standing next to the two-level furnace, a worker, raising a hammer, is about to strike. Mounted next to this image, picture no. 8b (fig. 4, right) represents additional furnace installations with vessels and containers set before them. Not least by virtue of their style, the woodcut sources for these reproductions create the impression that another milestone of technological progress has been passed: unlike the preceding paintings, they depict their objects in perspective; one of the furnaces is shown in a vertical section so its interior becomes visible; and even minor features such as the exterior arches supporting the melting furnaces are captured with precision. The "Glass" portfolio eschews textual explanations of such technological minutiae visible in the pictures; Agricola's book, by contrast, was designed as a closely interwoven and detail-rich ensemble of text and imagery. A legend, for example, identified the containers

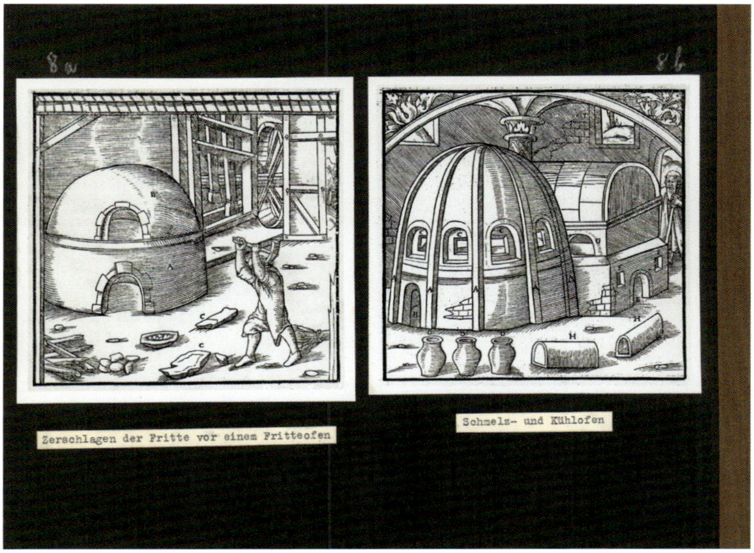

FIG. 4: Reproductions from Georgius Agricola, *De re metallica* (Basel: Froben und Bischoff, 1556), standardized and assembled on one cardboard support. Archives of the Deutsches Museum, Munich, Germany.

marked by the index *H* as earthenware vessels; Agricola's text recorded that the glass objects, after shaping, were set into these vessels and then placed in the openings *F* of the structure attached to the furnace for annealing. "Technik im Bild," however, is not designed to convey specialized knowledge about the images and technologies of the past; the goal is to highlight the different stages of a technological process as they appear crystallized in individual images.

The beholder has paged through a mere sixth of the "Glass" portfolio when a large number of engineering drawings and photographs are spread out before her to demonstrate the highly differentiated set of operations used to produce contemporary glass varieties. For instance, sectional drawings (fig. 5) that visualize the principle of the top-fired pot furnace are followed by image no. 18, discussed above without reference to its context (fig. 2). The company photograph shows the fronts of three furnace batteries based on this melting principle, which was used to make plate glass. The survey of contemporary manufacturing processes continues with photographs capturing, for example, the manual production of bottle glass as well as machines used in molded-glass fabrication, making glass bulbs for incandescent lamps, and automated bottle blowing. The sources for the reproductions include drawings from the training and workday reality of engineers as well as factory photographs but also illustrations in manufac-

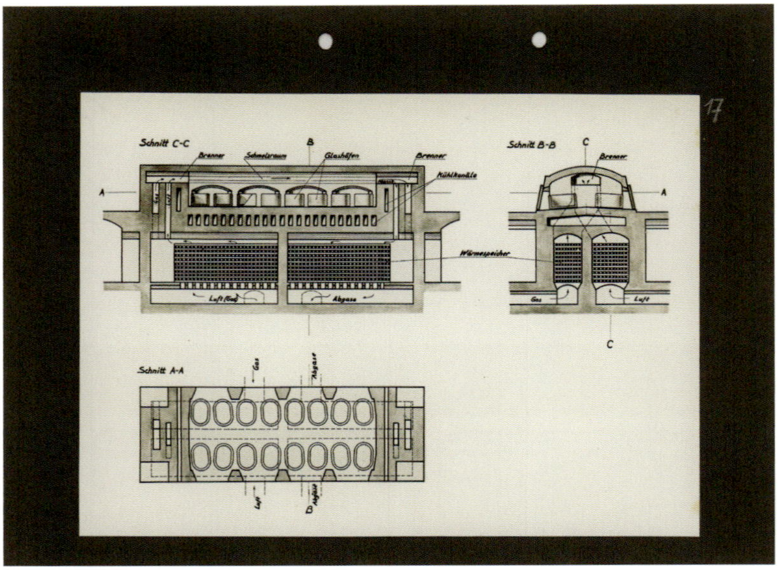

FIG. 5: Sectional drawings of a top-fired pot furnace. Archives of the Deutsches Museum, Munich, Germany.

turers' and machine-builders' catalogues. For instance, the European association of bottle manufacturers contributed the picture of the fifteen-arm Owens machine, a piece of automatic bottle-blowing equipment (fig. 6).

The profusion of contemporary images creates the impression of a rapid growth of technological knowledge and ability leading up to the present day. On the other hand, the focus on such specialization largely defeats the narrative function of the image series. The latter now serves to compile a wide variety of manufacturing variants, attempting to give a catalogue-like survey of a wide field with manifold subdivisions.

From Raw Material to Product: A Journey in Images through the Ideal-Typical Factory

In the section on modern glass production, the "Glass" portfolio places the main emphasis on the great diversity of contemporary manufacturing methods. Many other "Technik im Bild" portfolios, by contrast, visualize the ideal-typical manufacturing trajectory from raw material to finished product. In this structure of pictorial narration, which can be found as early as Diderot and d'Alembert's *Encyclopédie* (1751–1772), the com-

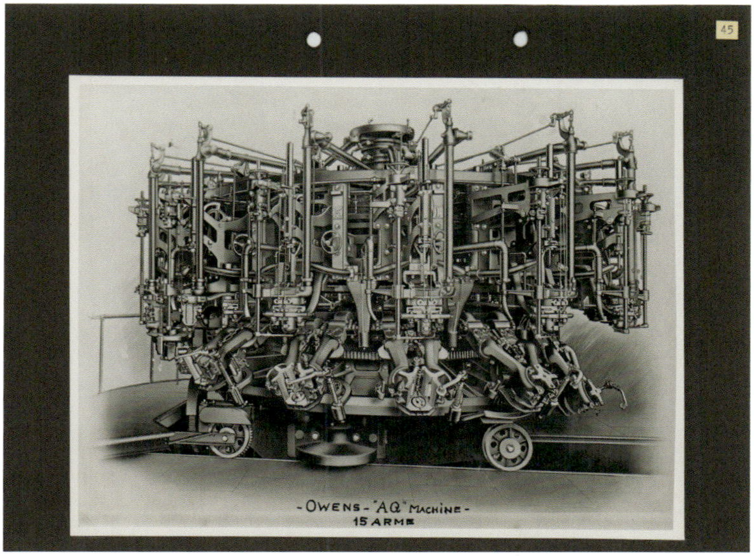

FIG. 6: Depiction of an automatic glass-blowing machine for bottle manufacturing. Archives of the Deutsches Museum, Munich, Germany.

modity, which acquires its definite shape through a successive series of transformations, is the connecting element holding the chain of images together. The corresponding organization in the case of mass manufacturing based on the division of labor is the virtual walk through the factory.

The portfolio "Brauerei" ("Brewery"), for example, takes the beholder on a visual tour of an ideal-typical large modern brewery. The editor of this volume, a retired brewing chemist, has eschewed all plate numbers and subtitles; there is no table of contents or index of plates. The portfolio begins with ten pairs of cardboard and text page illustrating the history of the brewing trade and various cultural practices of beer-making. The following image series on industrial brewing opens with a chart visualizing the significance of the industry for the German economy at large (fig. 7).

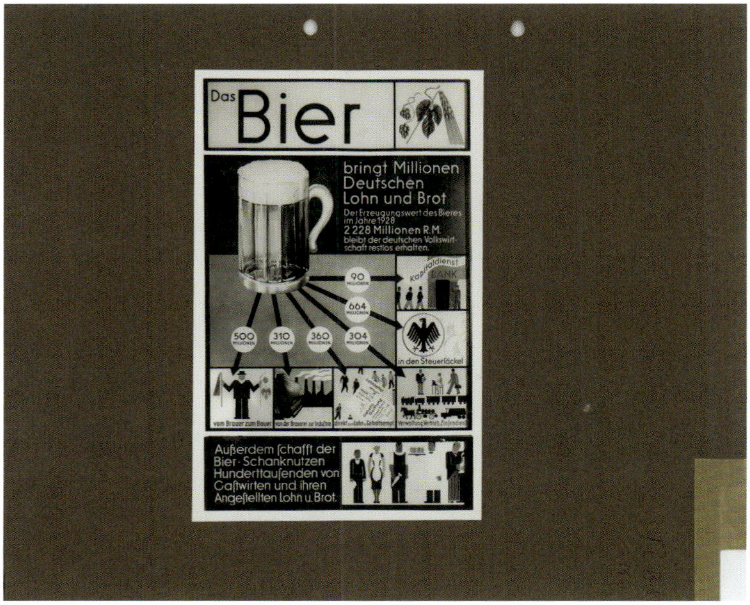

FIG. 7: Cardboard no. 11 from the "Brewing" portfolio. Archives of the Deutsches Museum, Munich, Germany.

The visual brewery tour starts in the facility where the barley is cleaned and concludes, around fifty image-stages later, with the loading of the finished beer into refrigerated railway cars (figs. 8–12). The pictures are recent industry photographs provided by different breweries, but they are assembled to suggest a walk through a single plant. They portray the equipment, usually in wide-angle shots designed so that the architecture of the production facilities frames the images as the beholder observes the workers operating the machinery from a remove, a vantage point that establishes distance; there are also several pictures that focus more closely on the workers or their actions.

A company photograph provided by the Schultheiss-Patzenhofer brewery (fig. 8) puts the beholder in the barley cleaning facility, where impurities of the barley such as dust and grains that are too small are

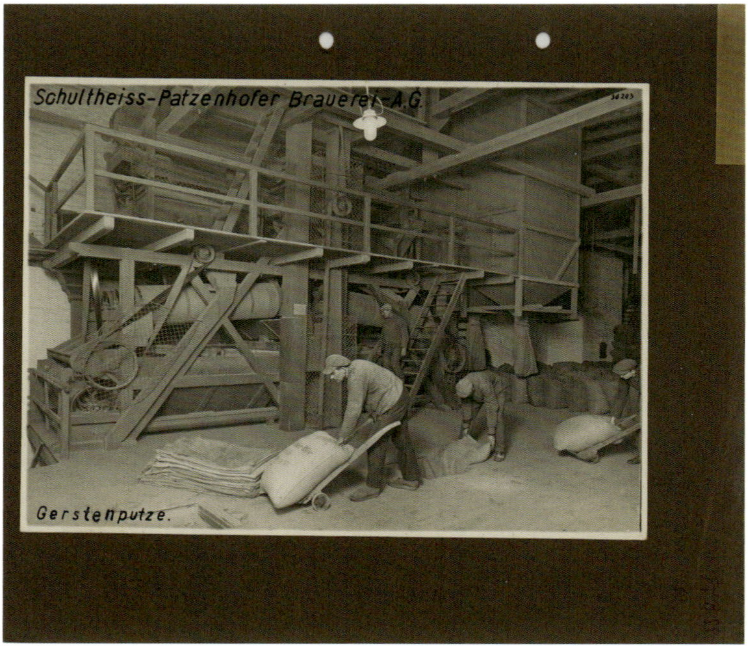

FIG. 8: In the barley-cleaning facilities at a Schultheiss-Patzenhofer brewing plant. Archives of the Deutsches Museum, Munich, Germany.

removed by mechanical means. In the background, we see the automated cleaning installation; in the foreground, three men are occupied with transporting the barley, which has by then been packaged in sacks, to the next stage. The following photograph, on the verso of the same cardboard, presents a glimpse of another division of the brewery: the steeping facility (fig. 9). Immersed in water, the barley is stored in a row of large vats to prepare it for the subsequent germination and malting. The next pictures illustrate different malting processes of varying degrees of automation, including a mechanical-pneumatic box malting installation (fig. 10). The photograph captures three men busily shoveling what must be finished green malt; behind them appears an installation of pipes that pneumatically feeds the malt into the next production stage, where it is dried in a kiln.

Like the cleaning, steeping, and malting of the barley, every subsequent stage of the beer-making process is shown in pictures, as are preparatory operations like the rinsing of empty return barrels before refilling (fig. 11). The individual operations captured in photographs are easy to

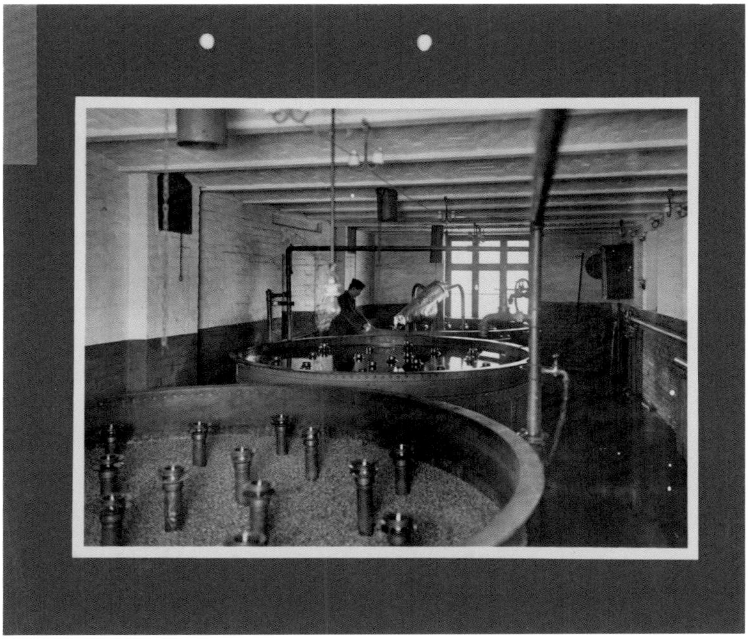

FIG. 9: A look at the barley-steeping stage of beer-making. Archives of the Deutsches Museum, Munich, Germany.

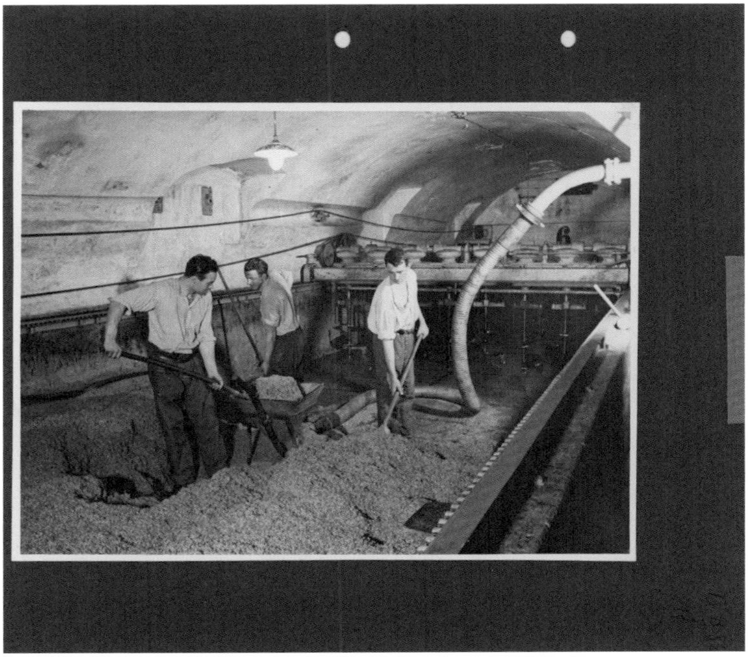

FIG. 10: View of a contemporary box malting installation. Archives of the Deutsches Museum, Munich, Germany.

understand because some element of the product-to-be in one picture is always recognizable in the next one. When the eye cannot discern such an immediate visual link, the connection may be supplied by the mind: the mass being steeped in the vats (fig. 9) must obviously be the same barley that appeared in the previous image as the content of grain sacks. Rarely does the beholder need to study the pages of text inserted between the picture cardboards to make sense of what she sees, although they always enable the curious reader to learn more.

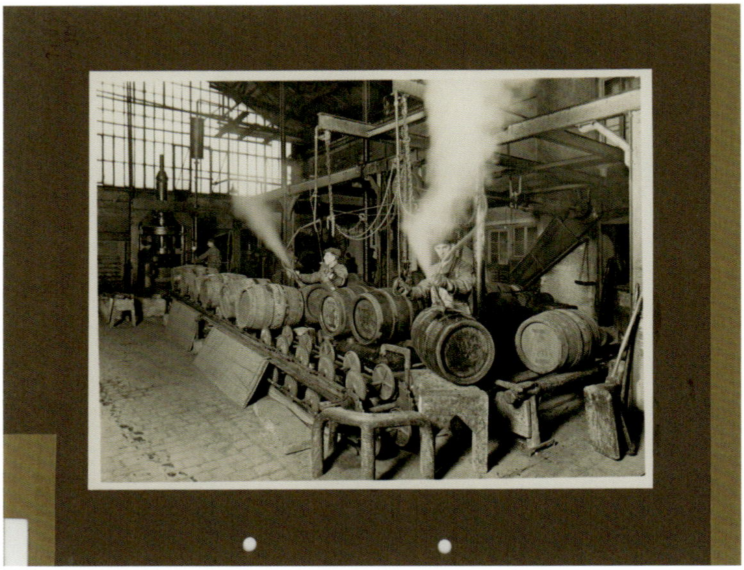

FIG. 11: View of empty beer barrels being rinsed. Archives of the Deutsches Museum, Munich, Germany.

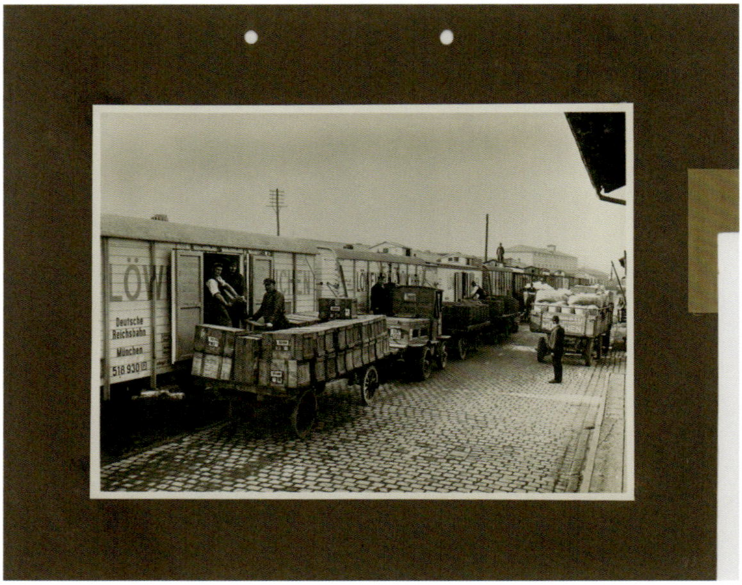

FIG. 12: Scene of beer being loaded into railway cars painted white to reflect the sunlight, Löwenbräu plant. Archives of the Deutsches Museum, Munich, Germany.

From Album to Archive

To the museum, the portfolio collection "Technik im Bild" not only represented another avenue of the popularization of technology; it also offered an efficient opportunity for a considerable expansion of the museum's own image library, to which its collection policies had heretofore given low priority. It moreover enabled the museum to present at least photographic documentation of technologies and implements that were not contained in the object collections and could not be purchased due to funding limitations. The album format chosen for the image show enabled the editors creating the portfolios to compile and arrange the materials to be presented in rapid fashion. Writing a textbook would have taken at least months; putting together an album, by contrast, was ideally a matter of no more than a few weeks. In a first step, the portfolio editors hunted for imagery, though they did not take the photographs themselves. Since their primary training was in technology and the sciences, they regarded photography as a method of objective reproduction and documentation that met scientific standards. Not only original photographs were used as "evidence in the historical trial";[9] so were photographic reproductions of paintings and drawings from the past. The image-hunters found suitable originals in the specialized collections of museums, businesses, and associations as well as the technical literature, from which they had "cutouts" made: the reprographic camera effectively served as a pair of scissors. No further editing to heighten the didactic value of the pictures was undertaken. A few diagrams were the only images the Deutsches Museum had created specifically for "Technik im Bild." Furnished with a definite selection and sequence of images and drafts of the texts, assistants at the museum probably then produced the portfolios by meticulous manual labor. After the hunt for images, the second step was handicraft work: texts and captions were typed on snippets of paper, pictures cropped; all elements were then arranged and mounted on the pages before the finished portfolio-collage was finally sent out to be bound.[10]

This one-of-a-kind quality of each portfolio may at the same time also have been the source of the problem that beset "Technik im Bild": by far not all of the image portfolios were produced with the same care as the titles on "Glass" and "Brewing." The collage-like nature of the albums invited an overly rapid and even slipshod assemblage that no longer allowed a narrative to emerge. The extant collection is highly heterogeneous, and some albums are impossible to understand without reading the texts. Moreover, because only one unique copy was made, "Technik im Bild" required interested parties to come to the library, restricting the project's potential audiences.

With the advent of the colorful and electronic media landscape of the postwar era at the latest, the handmade collage album of black-and-white photographs could no longer be regarded as a timely form for the dissemination of knowledge about technology. Moreover, the logic implicit in the arrangement of the albums seems to have become incompatible with the function that was now envisioned for the images the museum collected. Transferring the portfolios into ring binders—this was probably done in the 1960s—shattered the way the originals guided the beholder's eye: the pages are now held together not along the lateral fold but along the top edge; to see each page in its normal orientation, the reader must constantly rotate the binder. Many extant folders also show traces of cannibalization: photographs are often missing, whereas the text pages are virtually complete. The photographs torn from the cardboard backings were incorporated elsewhere in the image archive. The album was supplanted by the archive—and the unambiguous meaning of the album image, by the polysemy of the individual image. Today, the image archive is making efforts to regroup the "Technik im Bild" image stock by provenience and present it in its original context: the unique copy has itself become a singular item in the museum's collections.

9 Walter Benjamin, "The Work of Art in the Age of Its Technological Reproducibility," in *Selected Writings*, vol. 4: *1938–1940*, ed. Howard Eiland and Michael W. Jennings (Cambridge, MA: Harvard University Press, 2003), 258.

10 On such "cut-and-paste" practices in the early twentieth century, see Anke te Heesen, ed., *Cut and paste um 1900: Der Zeitungsausschnitt in den Wissenschaften* (Berlin: Vice Versa, 2002).

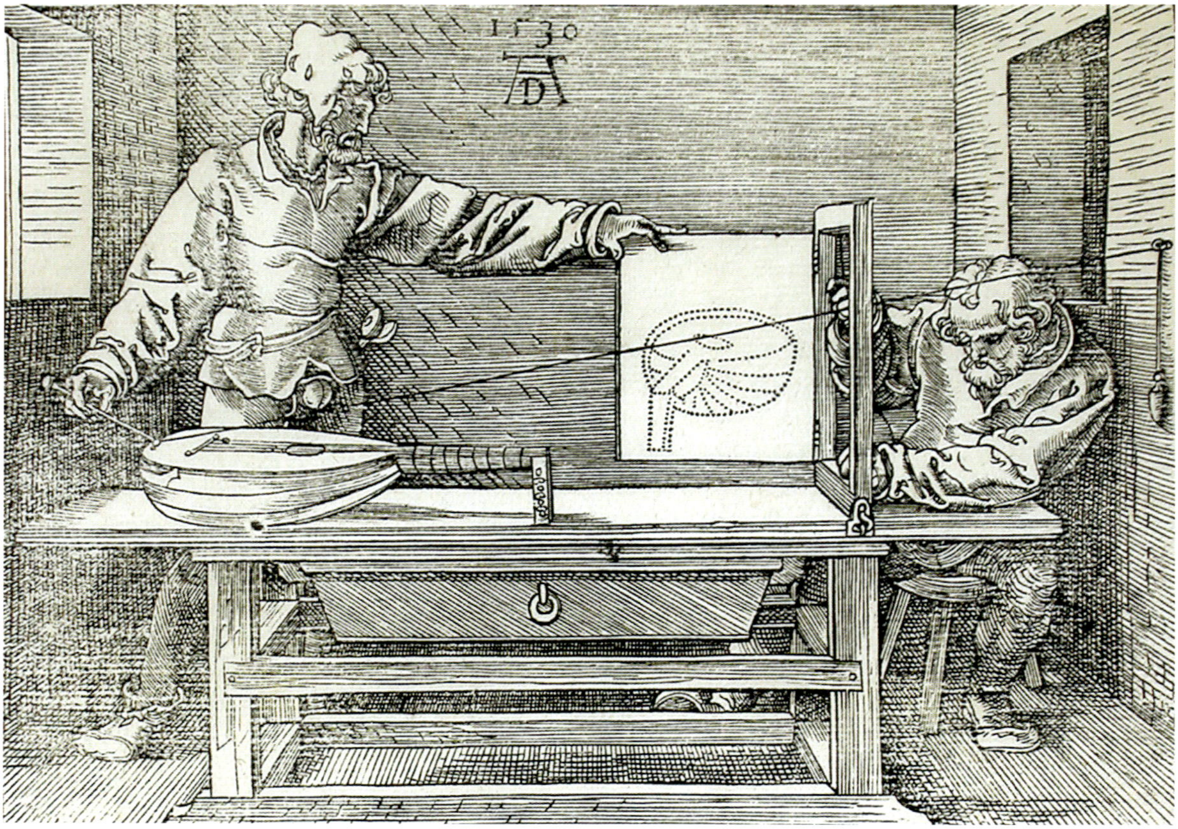

FIG. 1: The two men in this Dürer woodcut are taking part in exploring the practice of artistic perspective. With the help of a taut thread and a wooden frame they are able to produce a relatively accurate outline of the lute and its foreshortening effect as seen through the frame. Albrecht Dürer. *Two Men Drawing a Lute*, woodcut, 1523. Image courtesy History of Science Collections, University of Oklahoma Libraries.

OBSERVATION TECHNIQUES

Since the 1960s, the history of observation techniques has been a central topic in the history of science (Kuhn 1962; Feyerabend 1963; Hacking 1983; Shapin and Schaffer 1985). This continuing interest is remarkable given the prevailing consensus that even observation with the naked eye is a learned practice and therefore a technique. The use of observational instruments since the early modern era has been informed by a specific potential for conflict arising out of the tension between science's increasing orientation toward visual perception and the awareness that this perception is substantially shaped by instruments and other aids. Hans Blumenberg has described this as a "fundamental conflict of modernity" (Blumenberg 1979). He has also referred to the "theory-laden character" of scientific observation, which has been specifically investigated in the theory of science as well; see, for example, the work of Thomas Kuhn. Another discipline that has contributed to this issue is the philosophy of technology, which understands technology as an extension of man's organs (Mumford 1934; McLuhan 1964). This tradition originates in the writings of Ernst Kapp (Kapp 1877) and Ernst Cassirer (Cassirer 1985) and may be traced, for instance, in feminist theory (Haraway 1991) and critical reflections on the history of science (e.g., Latour 1999; Sarasin 2001).

Scientific observation, as Lorraine Daston and Peter Galison have pointed out, operates selectively (Daston and Galison 2007, 2011). The instrument predetermines which phenomena are the key to the essence of things. The classic objects of observation are themselves artifacts and are standardized in laboratory processes, slides, and atlas images. They help train the expert eye (Grasseni 2007). Against the claim that observational instruments falsify their objects, Joel Snyder has argued

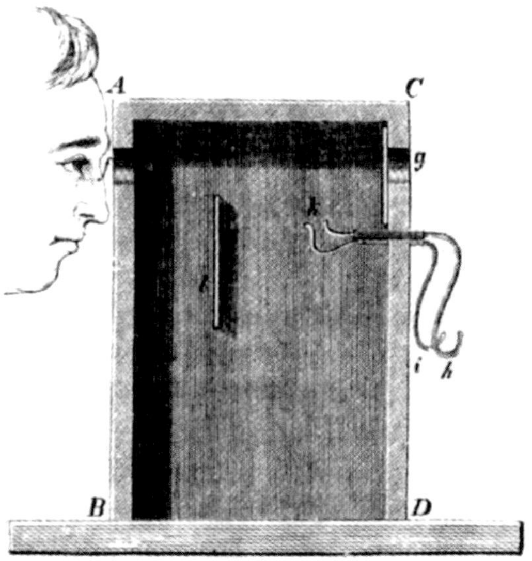

FIG. 2: Helmholtz's apparatus for investigating optical illusions, in which an image (*g*) on the back wall of the box is illuminated for a split second by the brief, bright reflected light of an electrical spark (from *k*), making it visible to the observer. Such flash illumination made it possible to exclude the effects of compensatory eye movements. Drawing from Helmholtz, *Handbuch der physiologischen Optik* (1867). Gerhard Kemner and Gelia Eisert, *Lebende Bilder: Eine Technikgeschichte des Films* (Berlin: Nicolai, 2000) (= Berliner Beiträge zur Technikgeschichte und Industriekultur, vol. 18), 80.

that the objects are neither falsified nor distorted because they would not exist without the instruments. Objects and areas of research are constituted solely by observational instruments and techniques. What is recorded is defined by the properties of the phenomenon as well as those of the observational instrument (Snyder 1998).

More recently, the issue of the reference of observations conditioned by technology has come to the forefront (Rheinberger 1997; Hoffmann 2006). A central topic was the fact that observational instruments rarely show visible phenomena but usually depict measurement data that result from the interaction between the object to be imaged and the instrument. So in addition to the properties of the sample, properties of the observational technique itself are incorporated into the image as well. This aspect is especially evident when technical equipment malfunctions, as in image errors and mishaps.

Since the mid-1980s, art historians have focused their attention on the observer and his technical and social constitution. Svetlana Alpers set standards in this field by choosing to analyze recording and observational techniques such as perspective machines (fig. 1). This allowed her to establish an alternative to the iconological analysis of early modern art (Alpers 1983). While Alpers focused on the artist as observer, a different position in the system of "art" was marked by the "viewer" or consumer. His status is the subject of the so-called aesthetics of reception. This approach begins with the hypothesis that the viewer's profile is already inscribed in the work. It analyzes the shaping of the observer by a work of art and the techniques of perception inherent in the work. Major impulses in this field have come from the history of photography, the first discipline to address the technical conditions of image production within art history (Kemp 1998). The art historian Barbara Maria Stafford also opened up many topics in the history of science and technology to art-historical research (Stafford 1991). With the publication of Jonathan Crary's *Techniques of the Observer* (Crary 1991), the concepts of "observation technique" and "instrumental observation" were finally established as art-historical terms. Crary linked theories of perception, such as Helmholtz's research on the physiology of the senses (fig. 2), to the history of observation techniques and especially of photographic technology (figs. 3–5).

The development of art-historical tools for the analysis of images from beyond the world of art has emerged as a rising branch of scholarship in recent years, with the issue of observational techniques playing an important role (Heintz and Huber 2001; Hoffmann and Berz 2001). Of special interest in this context is the graphical representation of scientific observations (Bredekamp 2001; Fiorentini 2004). The histories of microscopy and telescopy in particular are among the major fields of research in which the interplay of the pictorial tradition, the visualization of measured data, the question of the evidence and cogency of images, and the interpretation of images produced using instruments may be investigated (van Helden 1974; Wilson 1996; Bredekamp 2001; Ditzen 2008; Hennig 2011). —AF

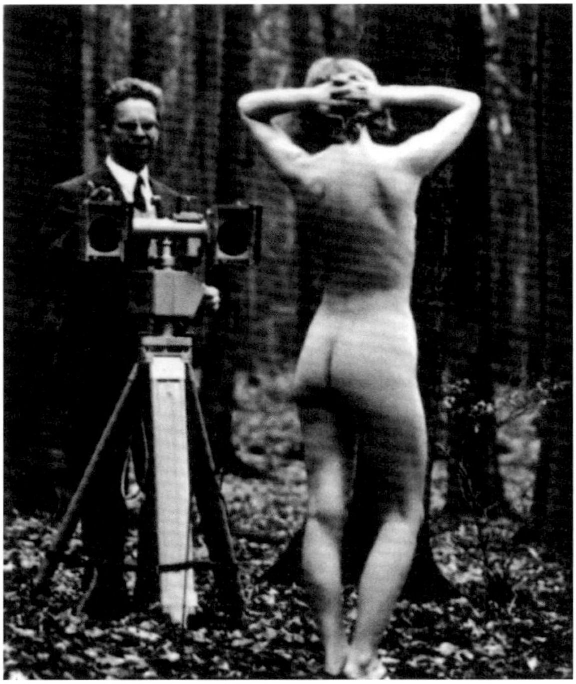

FIG. 3: A female nude in a forest in front of a stereo photogrammetric camera and a photographer. The stereo photogrammetric camera produced two photographic images of one object from different angles, creating a stereoscopic picture. Documentary picture of the photography setting, early 1970s. Dieter Lübeck, *Das Bild der Exakten — Objekt: Der Mensch; Zur Kultur der maschinellen Abbildungstechnik* (Munich: Moos, 1974), 58.

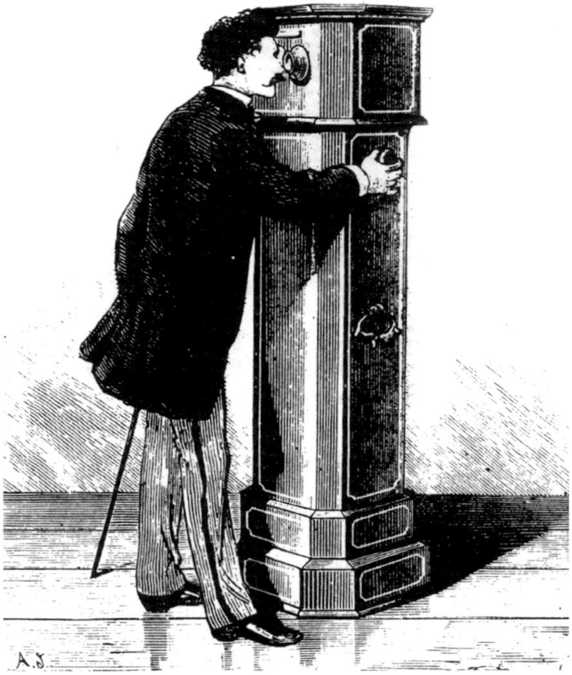

FIG. 4: Column stereoscope, around 1870. The stereoscope was the most popular way of viewing photographs in the nineteenth century. It presented two slightly different images of the same object separately to the viewer's two eyes in accordance with the eyes' angle of vision. The resulting impression of spatial depth made the object appear almost tangibly present. Jonathan Crary, *Techniken des Betrachters: Sehen und Moderne im 19. Jahrhundert* (Dresden and Basel: Verlag der Kunst, 1996), 138. Philo Fine Arts Stiftung & Co. KG.

LITERATURE

See also the bibliographical references on "Optical Instruments and Media of Seeing" in the bibliography, p. 188.

Alpers, Svetlana. *The Art of Describing*. Chicago: University of Chicago Press, 1983.

Blumenberg, Hans. *Die Lesbarkeit der Welt*. Frankfurt am Main: Suhrkamp, 1979.

Bredekamp, Horst. "Gazing Hands and Blind Spots: Galileo as Draftsman." In *Galileo in Context*, ed. Jürgen Renn, 153–92. Cambridge: Cambridge University Press, 2001.

Cassirer, Ernst. *Symbol, Technik, Sprache: Aufsätze aus den Jahren 1927–33*. Hamburg: Meiner, 1985.

Crary, Jonathan. *Techniques of the Observer: On Vision and Modernity in the Nineteenth Century*. Cambridge, MA: MIT Press, 1991.

Daston, Lorraine, and Peter Galison. *Objectivity*. New York: Zone Books, 2007.

Daston, Lorraine, and Elizabeth Lunbeck, eds. *Histories of Scientific Observation*. Chicago: University of Chicago Press, 2011.

Ditzen, Stefan. *Kunstformen instrumenteller Sicherheit. Etappen einer Bildgeschichte des Mikroskops*. Aachen: Shaker, 2008.

Feyerabend, Paul. "How to Be a Good Empiricist." In *Philosophy of Science: The Delaware Seminar*, ed. Bernard Baumrin, vol. 2, 3–39. New York: Interscience Publishers, 1963.

Fiorentini, Erna. "Subjective Objective: the Camera Lucida and Protomodern Observers." In *Instrumente des Sehens*, ed. Horst Bredekamp, Gabriele Werner, and Angela Fischel, 58–66. *Bildwelten des Wissens: Kunsthistorisches Jahrbuch für Bildkritik*, vol. 2, no. 2. Berlin: Akademie Verlag, 2004.

Grasseni, Cristina, ed. *Skilled Visions: Between Apprenticeship and Standards*. New York: Berghahn Books, 2007.

Hacking, Ian. *Representing and Intervening*. Cambridge: Cambridge University Press, 1983.

Haraway, Donna. *Simians, Cyborgs, and Women: The Reinvention of Nature*. New York: Routledge, 1991.

Heintz, Bettina, and Jörg Huber, eds. *Mit dem Auge denken: Strategien der Sichtbarmachung in wissenschaftlichen und virtuellen Welten*. Zurich: Edition Voldemeer, 2001.

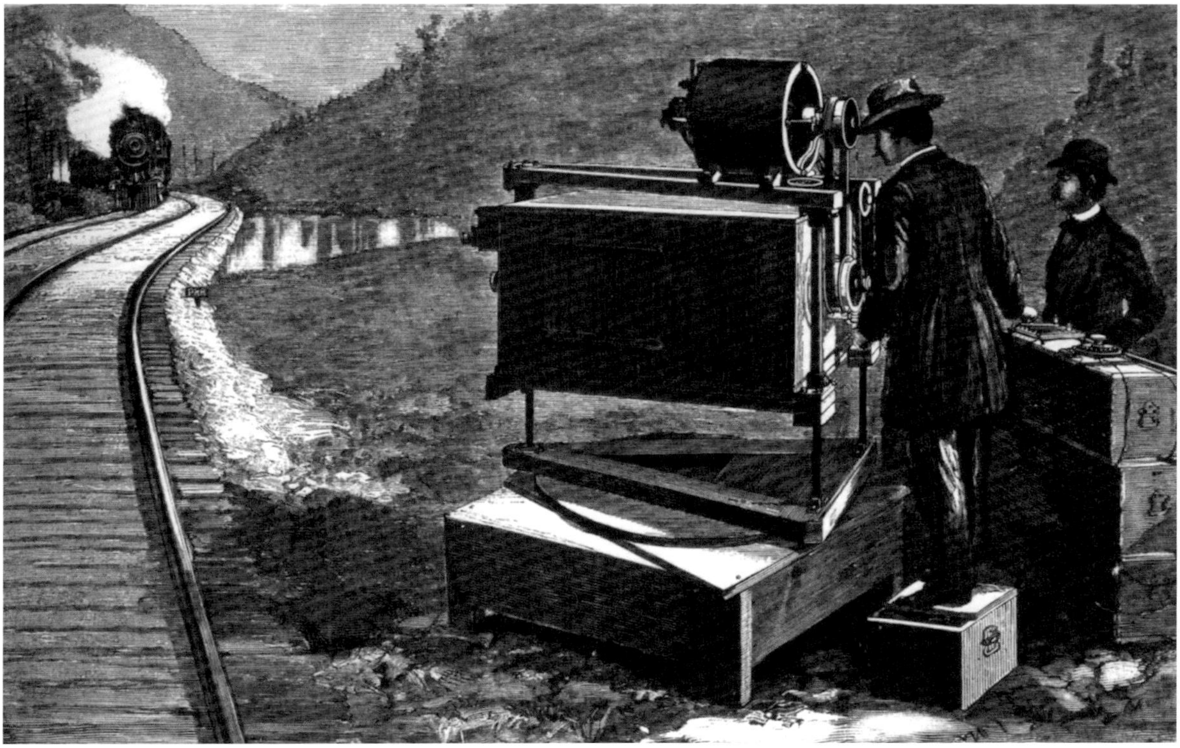

FIG. 5: Drawing showing the operation of the Mutograph and a popular Mutoscope motif, ca. 1897. The Mutograph, an electromagnetic camera, produced a series of images for the Mutoscope, a predecessor of the film projector, which worked like a flipbook. It was operated by means of a hand crank and had a roll of 800–1,000 photos. The device was specifically developed to compete with Edison's Kinetoscope. Dieter Lübeck, *Das Bild der Exakten — Objekt: Der Mensch; Zur Kultur der maschinellen Abbildungstechnik* (Munich: Moos, 1974), 45. Stiftung Deutsches Technikmuseum Berlin, Historisches Archiv.

Helden, Albert van. "The Telescope in the Seventeenth Century." *Isis* 65, no. 1 (March 1974): 38–58.

Hennig, Jochen. *Bildpraxis: Visuelle Strategien in der frühen Nanotechnologie*. Bielefeld: Transcript, 2011.

Hoffmann, Christoph, and Peter Berz, eds. *Über Schall: Ernst Machs und Peter Salchers Geschoss-fotografien*. Göttingen: Wallstein, 2001.

Hoffmann, Christoph. *Unter Beobachtung: Naturforschung in der Zeit der Sinnesapparate*. Göttingen: Wallstein, 2006.

Kapp, Ernst. *Grundlinien einer Philosophie der Technik: Zur Entstehungsgeschichte der Cultur aus neuen Gesichtspunkten*. Braunschweig: Westermann, 1877.

Kemp, Wolfgang. "The Work of Art and Its Beholder: The Methodology of the Aesthetic of Reception." In *The Subjects of Art History*, ed. Mark A. Cheetham, 180–96. Cambridge: Cambridge University Press, 1998.

Kuhn, Thomas. *The Structure of Scientific Revolutions*. Chicago: University of Chicago Press, 1962.

Latour, Bruno. *Pandora's Hope: Essays on the Reality of Science Studies*. Cambridge, MA: Harvard University Press, 1999.

McLuhan, Marshall. *Understanding Media: The Extensions of Man*. New York: McGraw-Hill, 1964.

Mumford, Lewis. *Technics and Civilization*. New York: Harcourt, Brace and Co., 1934.

Rheinberger, Hans-Jörg. *Towards a History of Epistemic Things: Synthesizing Proteins in the Test Tube*. Palo Alto, CA: Stanford University Press, 1997.

Sarasin, Philipp. *Reizbare Maschinen: Eine Geschichte des Körpers 1765–1914*. Frankfurt am Main: Suhrkamp, 2001.

Shapin, Steven, and Simon Schaffer. *Leviathan and the Air-Pump: Hobbes, Boyle, and the Experimental Life*. Princeton, NJ: Princeton University Press, 1985.

Snyder, Joel. "Visualization and Visibility." In *Picturing Science, Producing Art*, ed. Caroline A. Jones, 379–97. New York: Routledge, 1998.

Stafford, Barbara Maria. *Body Criticism: Imaging the Unseen in Enlightenment Art and Medicine*. Cambridge, MA: MIT Press, 1991.

Wilson, Catherine. *The Invisible World: Early Modern Philosophy and the Invention of the Microscope*. Princeton, NJ: Princeton University Press, 1996.

In the Eye of the Beholder: Emanuel Goldberg's Apparatuses at the *International Photographic Exhibition Dresden 1909*

Franziska Brons

The history of technical images, the aesthetic forces that shaped them, and their epistemic dimensions are inextricably linked to the question of how the very ideas of what seeing itself is undergo historical changes. Since the early modern era, optical media such as the microscope, the telescope, the camera obscura, and cameras of all sorts have been particularly fertile sources of metaphors and models of sense perception. As instruments of scientific insight, they influence the form and function of images. The invention of photography marked a turning point, adding new poignancy to the analogy between the human sense of vision and mechanical devices. The trope of the photographic plate as the researcher's new and improved retina frames a paragone *that, during the closing decades of the nineteenth century, pitted the eye against apparatuses. This contest itself was exhibited at the* International Photographic Exhibition Dresden 1909*: devices constructed by Emanuel Goldberg presented striking optical phenomena to let the beholder experience his own seeing as analogous to the laws of modern camera technology. But how can ephemeral phenomena such as afterimages, glares, and changes of focus be reconstructed at all? The absence of visual documentation poses a methodological challenge to art historians, who must analyze the surviving traces and verifiable contexts.*

Palace of Photography, Room 17

The catalogue accompanying the *International Photographic Exhibition Dresden 1909* contains a floor plan of the municipal exhibition palace, where this epoch-making exposition of the medium was headquartered (fig. 1).[1] The plan—flanked by a scale and a sketch of the layout of the upper floor, it takes up a full spread—bears various inscriptions that offer information on the location of individual departments. A seemingly endless arrow meanders through this diagrammatic representation of the building, guiding the visitor through sections such as "Geography and Ethnography," "Professional Photography," "Amateur Photography," and "Science." That last department is reached via rooms 1 and 17; it is accessible, that is to say, through the galleries devoted to the history of photography and to optics (in the catalogue, the latter is labeled "Education and Entertainment"). An interior view of room 1 gives an impression of the "Didactic Museum of Photography" that the doyen of the Dresden photography scene, Hermann Krone, presented in a series of tableaux (fig. 2). By contrast, no picture of the installation inside room 17 of the Internationale Photographische Ausstellung, or Ipha, survives to complement the exhibition plan. Still, reports, documents,

1 As Sigfried Giedion has noted, the term *exposition* with its French etymology has a much wider range of meanings than the German *Ausstellung* ("exhibition"), signifying "overview, juxtaposition, comparison, situation, or, in a figurative sense, even the presentation of a doctrine." See Sigfried Giedion, *Bauen in Frankreich: Eisen; Eisenbeton*, 2nd ed. (Leipzig: Klinkhardt and Biermann, 1928), 36.

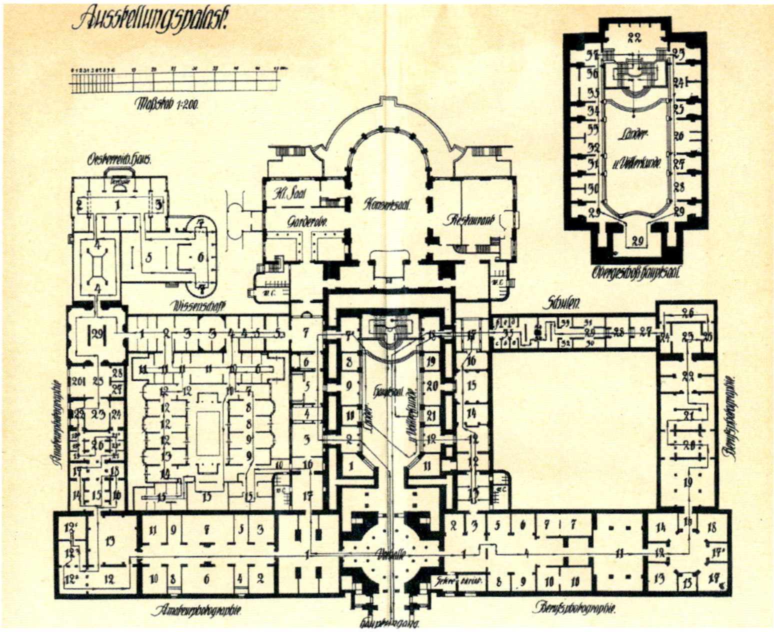

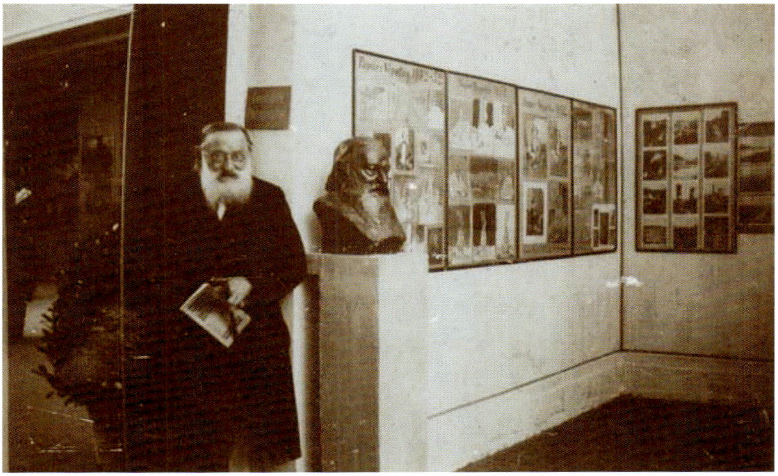

FIG. 1: Floor plan of the exhibition palace in the official catalogue of the *International Photographic Exhibition Dresden 1909. Offizieller Katalog der Internationalen Photographischen Ausstellung Dresden 1909*, 2nd ed. (Dresden: Baensch, 1909).

FIG. 2: A view of the section "History of Photography," room 1 of the division "Scientific Photography," showing Hermann Krone next to his bust and his Historisches Lehrmuseum für Photographie; picture postcard, mailed on June 18, 1909, recto. Rheinisches Bildarchiv Köln. © Rheinisches Bildarchiv Köln, rba_c021060.

and descriptions indicate that its creator had set up a veritable school of vision, presenting lessons that aimed to build a wider understanding of the equipment generating technical images and the aesthetic experiences they make possible. Optical devices activated the beholder; rather than inviting him to contemplate, this leg of the course through the exhibition prompted him to experience perception itself with even greater intensity. By the same token, the static medium of photography seemed to be animated, as the fundamental laws of optics appeared in the guise of evanescent phenomena such as optical illusions, blurs, and shifts of focus that no frame could capture. The focus of attention moved to the interplay between mechanically generated visual phenomena and subjective perception. The eye and the apparatus were on display as correlative entities. The room elevated the sense of sight, with explicit reference to instruments constructed specifically for this purpose, to the status of the exposition's thematic object. And yet not a single picture has been found of this part of the Dresden exhibition and the collection of apparatuses set up in it.

Besides specialist periodicals, handbooks, and the popular press, the grand exhibitions of the late nineteenth and early twentieth centuries were the most important medium for illustrations of the technological foundations and diverse applications of photographic images. In a contemporary report from the opening day of the Dresden show, the correspondent conveys his impression that "one must indeed call this a world exhibition, a general survey of photography that spans the globe, comprehending and presenting all interests and all goals of photography in unprecedented completeness."[2] The Ipha was an inventory of photography. The current stage of the medium's development as well as the spectrum of its applications in research, technology, and art appeared concentrated in one place, as the official poster also suggests (fig. 3). The world exhibition—the photography historian Ulrich Pohlmann called it "the preeminent photography exhibition event in Europe before World War I"[3]—represented a thorough stocktaking. At the same time, the exposition, encyclopedic in design, may be regarded as a catalyst that induced the manufacturing of artifacts and apparatuses to be exhibited. Beyond capturing the status quo, these exhibits prompted a revision of the generally accepted knowledge of the medium and neighboring fields. The eminent significance exhibitions such as the Ipha had for the history of modern art, of images, media, and the sciences, contrasts with the challenges they pose to the historian, who cannot hope to achieve more than an approximate reconstruction. On the one hand, the extant traces and splinters of room 17, which the following pages will seek to arrange and analyze, may stand as exemplary of the ephemeral constellations between exhibits and beholders that expositions establish, demonstrating how "ideas about perception and attention were transformed [...] alongside the emergence of new technological forms of spectacle, display, projection, attraction, and recording."[4] On the other hand, they confront the art-historical examination of technical images with the question: Must its objects necessarily all be "visible, discerned"?[5] Or must it make the paradoxical attempt to review also what might be overlooked, what is invisible, what was not envisioned to avoid mistaking the contingencies of a tradition handed down to us in the form of physical pictures for the many facets and factors of the history of the image?

Before All Eyes

Arrayed in room 17 of the Ipha were devices designed to equate the human eye to the photographic camera. In total, forty-four[6] apparatuses built at the Royal Academy of Graphic Arts and Bookcraft in Leipzig under the direction of Emanuel Goldberg[7] were set up in individual cabinets, beckoning visitors to engage in "self-guided experimentation"; even the official catalogue acknowledged that they had an air of *bricolage* rather than technical perfection.[8] The correspondent of a photographic trade journal reports of their construction: "The framework consists of wooden shipping boxes such as may be had cheaply, and almost all components of the apparatuses are mass-manufactured products of the watchmaking, toy, and lighting industries; for the peepholes, for

2 "Von der Dresdner Ausstellung I," *Die Photographische Industrie [7]*, no. 19 (1909): 582.

3 Ulrich Pohlmann, "Das 'Historische Lehrmuseum für Photographie' von Hermann Krone: Der museale Blick auf die Fotografie im 19. Jahrhundert," in *Photographie und Apparatur: Der Photopionier Hermann Krone; Bildkultur und Phototechnik im 19. Jahrhundert*, ed. Wolfgang Hesse and Timm Starl (Marburg: Jonas, 1998), 208.

4 Jonathan Crary, *Suspensions of Perception: Attention, Spectacle, and Modern Culture* (Cambridge, MA: MIT Press, 2001), 2.

5 Georges Didi-Huberman, *Confronting Images: Questioning the Ends of a Certain History of Art*, trans. John Goodman (University Park: Pennsylvania State University Press, 2005), 3.

6 See *Offizieller Katalog der Internationalen Photographischen Ausstellung Dresden 1909, ausgegeben im Juni 1909*, 2nd illustrated ed. (Dresden: Baensch, 1909), 53–64.

7 In addition to Goldberg, a few sources mention Dr. Scheffer at Carl Zeiss, in Jena, as well as the Institute of Photography at the Royal Technical College, Dresden, as responsible for the section. See "Internationale Photographische Ausstellung Dresden 1909," *Zeitschrift für wissenschaftliche Photographie, Photophysik und Photochemie* 7, no. 5 (1909): 180; Michael Buckland, *Emanuel Goldberg and His Knowledge Machine: Information, Invention, and Political Forces* (Westport, CT: Libraries Unlimited, 2006), 46.

8 *Offizieller Katalog*, 53.

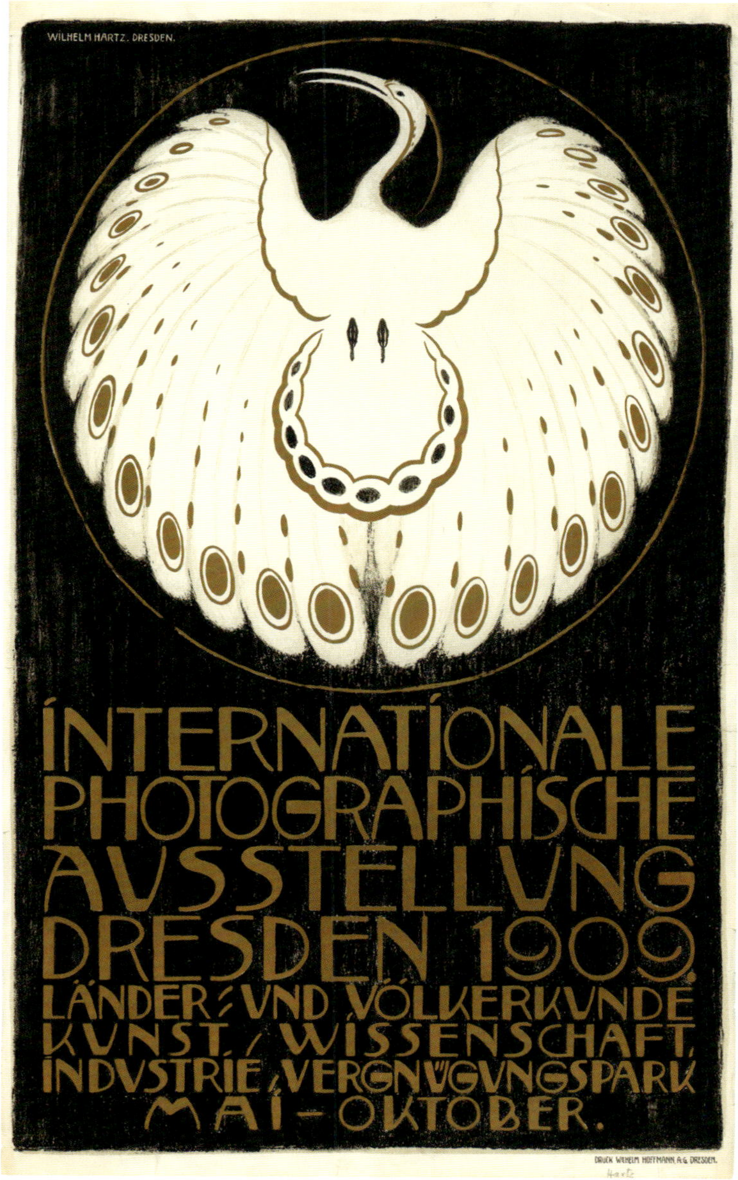

FIG. 3: Wilhelm Hartz, poster for the *Internationale Photographische Ausstellung Dresden 1909*, printed by Wilhelm Hoffmann AG, Dresden, two-color lithograph, 97.3 × 60.3 cm (motif), 100.2 × 62.2 cm (folio), ca. 1909. Kupferstich-Kabinett, Staatliche Kunstsammlungen Dresden. Photographer: Herbert Boswank.

example, the constructors used porcelain sockets for light bulbs, the axes and bearings are cheap mass products, the light relays are operated with buttons for home telegraphs and electromagnets for doorbells. The entire arrangement offers useful indications on how similar demonstration objects may be cheaply mass-manufactured for schools, exhibitions, Urania-style endeavors, and the like."[9]

Built from simple parts, these apparatuses afforded the beholders an opportunity to probe the limits of their own perceptive capacity by operating the equipment and experiencing the resulting visual effects for themselves. Goldberg's devices featured experiments about optical accommodation, irradiation, and chromatic aberration[10] as well as the eye's lens system and the limitations of its visual acuity, among other

9 [K. W.] W[olf]-Cz[apek], "Von der Dresdner Ausstellung XV," *Die Photographische Industrie* [7], no. 35 (1909): 1165.

10 *Accommodation* describes the eye's ability to adapt dynamically its refraction to the distance of a nearby or remote object; in optics, *irradiation* designates an optical illusion that makes light objects on a dark background appear larger, and dark objects on a light background appear smaller, than they actually are; *chromatic aberrations* are image defects caused by lenses and the different wavelengths of light in different colors, such as blurring or red and green edges.

phenomena. Many a visitor attempting, for example, to see the "blind spot" may have despaired of the eye's inadequacy when it came to detecting its own inadequacy; as much is suggested by the catalogue entry for apparatus no. 8, which explains that the "photosensitive plate" of the human eye suffers from a defect of which its owner remains unaware, whence the experiment requires some practice and concentration.[11] The experiment designed to illustrate this systemic flaw of the physiology of perception proceeded as follows: "Experiment: Fix the rigid right eye on the white circle while slowly turning the crank. At a certain point, without averting the eye from the circle, you may see that the little square disappears. If you continue to slowly turn the crank, however, the square will reappear. If the experiment fails, repeat it with a slight inclination of the head."[12] A similar challenge to perceive the "carnal density" of one's own sense of vision[13] was posed by the exploration of the eye's focal length and its adjustment of visual acuity, for which Goldberg's class had developed the following self-experiment, conducted with apparatus no. 2: "Experiment: Place the eye as close as possible to the peephole and regard the letters inscribed on the inside of the box. (Press the 'Light' button.) The veil stretched across the center of the box is then virtually imperceptible and blurred. With some effort, you may also see the veil clearly; but then the writing disappears. In the elderly, the mechanism of lens switching is heavily impaired and the experiment will not work."[14] The apparatus thus made the fact manifest—enabled the user to experience for himself—that there are always objects in his field of vision that are literally located outside the focus of perception.

Physics Lessons around 1900 and the Urania Model

The conception of Goldberg's devices patently referred to the promise photography held for the sciences; since the 1870s, this promise had found formulaic expression in the trope of the photosensitive plate as the researcher's new retina.[15] During the same period, pedagogical theory and practice ascribed growing relevance to experiments conducted by high school students. If the study of formulas and facts, or else frontal demonstrations by teachers, had hitherto been the predominant form of natural-science instruction, experiments—also called student exercises—now started to appear in physics lessons. As a photograph taken at an advanced vocational school in Hamburg and published in 1905 illustrates, laboratories were set up in which physical phenomena—in this instance, the "heat of water vaporization"[16]—could be produced and observed (fig. 4). As Hermann Hahn, himself a senior teacher, explained, practical exercises of this sort allowed the students to shed their "blindness to nature" and "learn to use their senses correctly, or more precisely, to interpret their sense perceptions correctly."[17] One consequence of this didactic innovation was that teachers, for their part, needed to be trained in how to guide students in such experiments. One expert who rose to particular prominence in this connection was Hermann Schäffer, who knew how "to build vividly illustrative apparatuses with few means"[18] that he employed in so-called vacation classes. The Jena-based physicist and mathematician's "physica pauperum" was an important material predecessor to Goldberg's devices, as were the seminars offered by the Urania, in Berlin, which likewise held courses to build the pertinent skills in teachers.

11 *Offizieller Katalog*, 55.

12 Ibid.

13 Cf. Jonathan Crary, *Techniques of the Observer: On Vision and Modernity in the Nineteenth Century* (Cambridge, MA: MIT Press, 1992), 150.

14 *Offizieller Katalog*, 54.

15 The origins of this metaphor have been dated to the year 1877; see Francoise Launay, "Jules Janssen et la Photographie," in *Dans le Champ des Étoiles: Les Photographes et le Ciel, 1850–2000*, ed. Marie Dominique de Teneuille, exhibition catalogue, Musée d'Orsay, Paris; Staatsgalerie Stuttgart (Paris: Éditions de la Réunion des musées nationaux, 2000), 26; on the metaphor of the retina, see also Peter Geimer, *Bilder aus Versehen: Eine Geschichte fotografischer Erscheinungen* (Hamburg: Philo Fine Arts, 2010), 233–43; André Gunthert, "La rétine du savant: La fonction heuristique de la photographie," *Études photographiques*, no. 7 (2000), http://etudesphotographiques.revues.org/index205.html; Christoph Hoffmann, "Zwei Schichten: Netzhaut und Fotografie, 1860/1890," *Fotogeschichte: Beiträge zur Geschichte und Ästhetik der Fotografie* 21, no. 81 (2001): 21–38; Franziska Brons, "Das Versprechen der Retina: Zur Mikrofotografie Robert Kochs," in *Instrumente des Sehens*, ed. Angela Fischel, *Bildwelten des Wissens: Kunsthistorisches Jahrbuch für Bildkritik*, vol. 2, no. 2 (Berlin: Akademie, 2004), 19–28.

16 Karl T. Fischer, *Der naturwissenschaftliche Unterricht—insbesondere in Physik und Chemie—bei uns und im Auslande* (Berlin: Springer, 1905), 20–21.

17 Hermann Hahn, *Wie sind die physikalischen Schülerübungen praktisch zu gestalten?* (Berlin: Springer, 1905), 7; on the didactic culture of experimentation, see Peter Heering and Roland Wittje, eds., *Learning by Doing: Experiments and Instruments in the History of Science Teaching* (Stuttgart: Steiner, 2011); Viola van Beek, *"Man lasse doch diese Dinge selber einmal sprechen." Experimentierkästen, Experimentalanleitungen und Erzählungen zwischen 1870 und 1930*, Preprint, vol. 365 (Berlin: Max-Planck-Institut für Wissenschaftsgeschichte, 2009).

18 Massinger, "Professor Schäffer in Jena und seine 'Physica pauperum,'" *Südwestdeutsche Schulblätter* 8, no. 11 (1896): 261. For his apparatuses, see Heinrich Bohn, *Physikalische Apparate und Versuche einfacher Art aus dem Schäffermuseum* (Berlin: Salle, 1902).

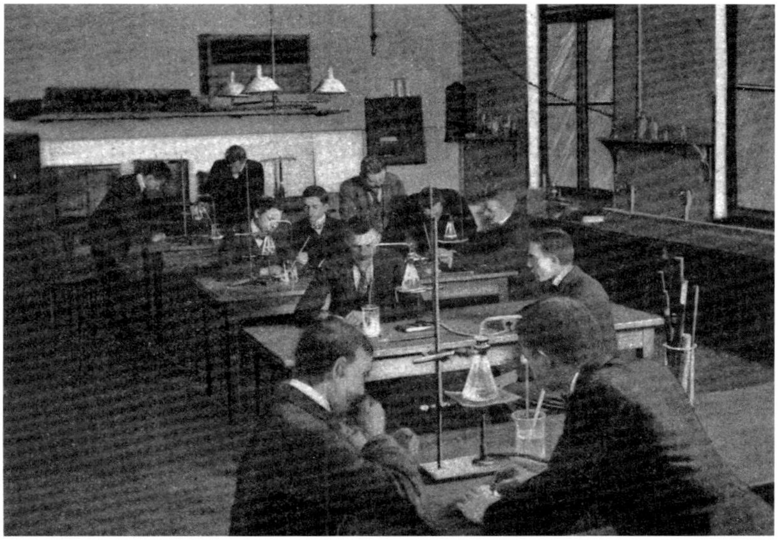

FIG. 4: Student laboratory at the Oberrealschule auf der Uhlenhorst higher vocational school, Hamburg, ca. 1905. Karl Tobias Fischer, *Der naturwissenschaftliche Unterricht—insbesondere in Physik und Chemie—bei uns und im Auslande* (Berlin: Springer, 1905), 23.

Several historians have highlighted the part the Urania played in the popularization of the natural sciences and their images.[19] Franz Goerke, who was the institution's acting director in 1909, had been commissioned to lead the "Education and Entertainment" section in Dresden. The push-button experiments the physicist Eugen Goldstein had designed for the Urania that can be seen in an interior shot from around 1890 (fig. 5) served as a model for Goldberg's Ipha contribution.[20] The exposition at the Urania had redefined the role of the beholder with exhibits that elicited his active involvement: the devices allowed him to explore scientific phenomena in hands-on experiments. "Each experimental arrangement has been meticulously prepared so that, pressing a button, the visitor may trigger the production of the natural spectacle at his own accord. A brief and clearly framed description, moreover, explains the significance of the process,"[21] the journal *Gartenlaube* informs its readers in 1896; the accompanying series of pictures shows visitors interacting with the apparatuses designed to visualize physical facts (fig. 6). Instead of walking past a sequence

19 See Andreas W. Daum, *Wissenschaftspopularisierung im 19. Jahrhundert: Bürgerliche Kultur, naturwissenschaftliche Bildung und die deutsche Öffentlichkeit 1848–1914,* 2nd ed. (Munich: Oldenbourg, 2002), 178–83; Ole Molvig, "The Berlin Urania, Humboldtian Cosmology and the Public," in *The Heavens on Earth: Observatories and Astronomy in Nineteenth-Century Science and Culture*, ed. David Aubin, Charlotte Bigg, and H. Otto Sibum (Durham, NC: Duke University Press, 2010), 340.

20 *Offizieller Katalog*, 52. As a site of the popularization of knowledge, the Urania was a crucial model not only for the Dresden world exhibition. The push-button experiments at the Deutsches Museum in Munich likewise drew on preliminary work Goldstein had done at the Urania. See "Zum 25-jährigen Bestehen der Gesellschaft Urania in Berlin: Die Festsitzung der Urania am 29. April 1913," *Himmel und Erde* 25, no. 9 (1913): 391.

21 Franz Bendt, "Die neue Berliner 'Urania,'" *Die Gartenlaube: Illustriertes Familienblatt*, no. 38 (1896): 634.

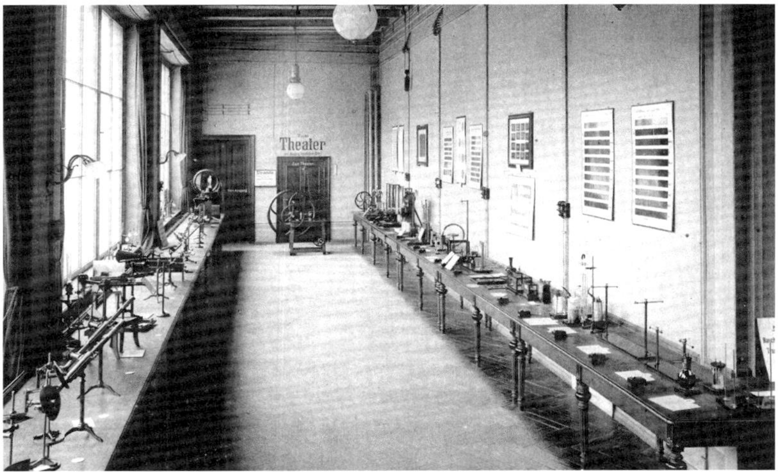

FIG. 5: View of the physics hall, Urania, near Lehrter Bahnhof, Landes-Ausstellungspark, Berlin, eastern division, ca. 1890. M. Wilhelm Meyer, "Die Urania nach ihrer Fertigstellung," *Himmel und Erde: Illustrierte naturwissenschaftliche Monatsschrift*, no. 5 (1890): 238.

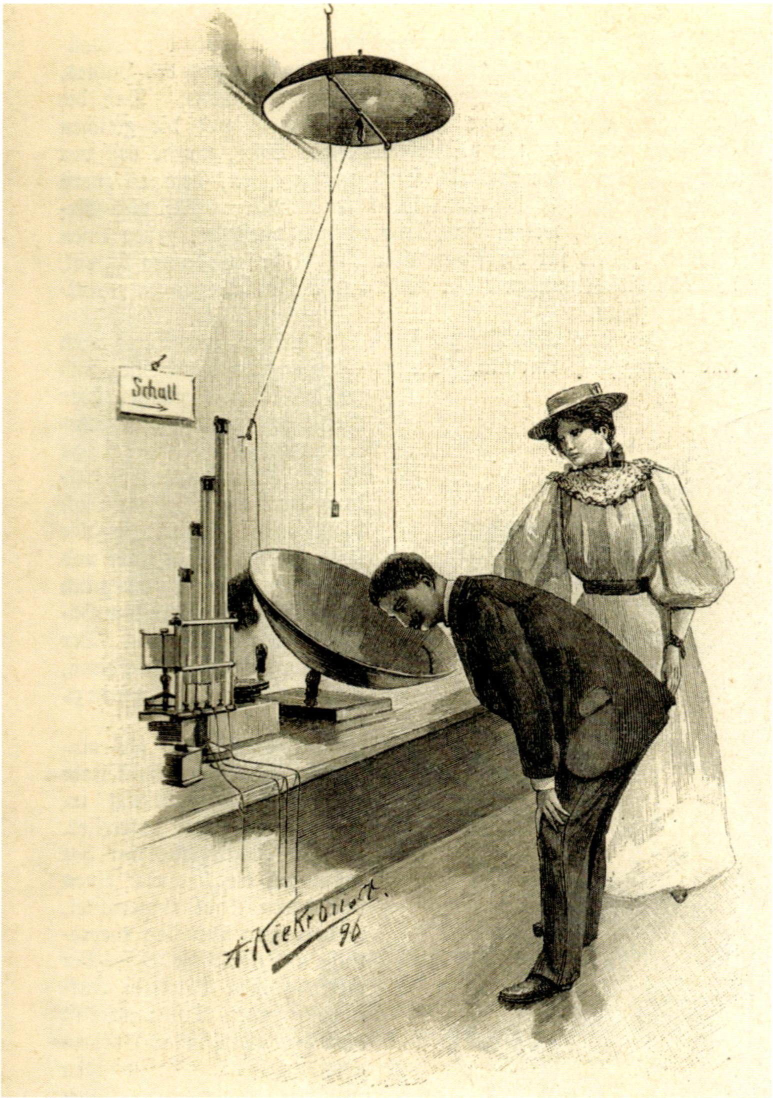

FIG. 6: Albert Kiekebusch, "sonic mirror," Urania, Taubenstraße, Berlin, 1896. Franz Bendt, "Die Neue Berliner 'Urania.'" *Die Gartenlaube. Illustriertes Familienblatt,* no. 38 (1896): 636.

of exhibits, the visitor to the Urania conducted programmed "show experiments,"[22] implementing the idea of "self-teaching by means of physical 'automata.'"[23]

Of Retina and Camera

The devices Goldberg and his students conceived and built for the Dresden exhibition presented the structure of the human eye, which is to say, the physical and physiological laws of light and the process of seeing—or in Max Frank's words, "the most basic facts of the effects of vision"—by way of analogy with the photographic camera.[24] The comparison between optical instruments and the human eye had influenced the study of human vision since the invention of the camera obscura.[25] Hermann von Helmholtz, for instance, used the relation between the oculus and the image-generating apparatus as the principle structuring his argument in an 1868 lecture series on "Recent Progress in the Theory of Vision."[26]

22 "The physical department is especially interesting to visitors, and this is largely due to the arrangement of the apparatus. [...] In this [department], each piece is so arranged that the visitor can set it in motion by touching an electric button near it." Edward S. Holden, "The Urania Institute of Berlin," *The Engineering Magazine* 1 (1891): 784.

23 Otto Reichenheim, "Eugen Goldstein," *Die Naturwissenschaften* 8, no. 36 (1920): 719.

24 Max Frank, "Die Gruppe 'Unterhaltung und Belehrung' auf der Internationalen Photographischen Ausstellung zu Dresden," *Photographische Chronik und Allgemeine Photographen-Zeitung* 16, no. 92 (1909): 565. The correspondent of the *British Journal of Photography* describes the section as presenting "phenomena of the eye," "elementary facts of light and colour," and "basic fact[s] of photography or vision." "The Dresden Exhibition [I]," *British Journal of Photography* 56, no. 2560 (1909): 418.

25 Crary, *Techniques of the Observer,* 40–50; see also Brons, "Das Versprechen der Retina," 22–23; Birgit Schneider and Vera Dünkel, "Rundbild und Augenblick," in *Der Ball ist rund: Kreis Kugel Kosmos,* exhibition catalogue (Berlin: Staatliche Museen zu Berlin, Stiftung Preußischer Kulturbesitz, 2006), 110.

26 Hermann von Helmholtz, "Die neueren Fortschritte in der Theorie des Sehens" [1868], in *Vorträge und Reden,* vol. 1, 4th ed. (Braunschweig: Vieweg, 1896), 267–365, esp. 271–94.

As early as 1855, he had declared that "the eye is an optical instrument formed by nature, a natural camera obscura."[27] Such theoretical considerations transposed physiological data onto the elements of the camera construction. The physicist's goal in taking this approach, however, was not to give an apologia of the human sense of sight: in Helmholtz's view, the performance of the eye, when compared to that of photography, was first and foremost deficient, given such phenomena as blurring and fatigue.

The first exhibit in the experimentation room arranged under Goldberg's leadership was all about this equation between the physical process of seeing and technical mechanisms. The visitor encountered a chart explaining the apparatuses that showed a cross section of a human eye disassembled into camera components (fig. 7):[28] the section opened with the attempt to prove that nature itself relied on a mechanical modus operandi. The catalogue note accordingly observed that "the human eye bears an extraordinary resemblance to the photographic camera. Like the latter, the eye has a lens, a photosensitive plate, an aperture, adjustment mechanisms, etc."[29] By this logic, the eye's cornea corresponded to the lens cap, the conjunctiva to the camera body, and the blind spot to a defect of the plate.[30] The "ocular objective" had a "focal length." In accommodation, the eye actuated a "mechanism of lens switching." The eye's aperture system was a protection against "overexposure."[31] The Ipha turned the experimenting visitor into a living camera. Operating the forty-four devices, he successively became conscious of his inherently photographic gaze. Although the "ocular objective" suffered from "pretty much all the defects a bad and cheap objective might have,"[32] it still was an objective and hence sufficiently qualified to act as an illustrative stand-in for photographic principles. The experiments of the professor from Leipzig were thus fundamentally committed to the idea that "we knew nothing about our senses until media provided models and metaphors."[33]

Impression and Insight

The discussion surrounding the experiments expanded on this rhetorical fusion of eye and camera. The insight conveyed in the exhibition that the eye, too, was "adjustable in several ways"[34] led the visitor to understand based on his own visual experience what, by converse implication, a photograph was not supposed to look like. Be it the absence of the "photosensitive layer" in the "blind spot" or the difficulty of adjusting the eye's focus: the same phenomena that were to be eliminated in the domain of scientific photography, such as plate defects or blurs, were physiological facts of the sense of vision. Goldberg's apparatuses sought to leave a lasting impression of this deficiency in the eye of the beholder.

Incessantly turning cranks, flipping switches, and pressing buttons, surrounded by the din and clatter of the machinery, the visitor, briefed by a signboard about the use and purpose of a particular experiment,[35] probed the process of his own vision. The practical instruction, that is to say, took the form of the student's own observations because, as the correspondent of the *Photographisches Wochenblatt*, among others, highlighted, visual impressions promised to be much more lasting.[36] Walter Benjamin wrote about this form of visual insight made possible by expositions in the full sense of the term: "The masses do not wish to

27 Hermann von Helmholtz, "Über das Sehen des Menschen" [1855], in *Vorträge und Reden*, vol. 1, 90.

28 In 1913, Goldberg published a cross section of an eye that fits the description in the exhibition catalogue perfectly. See Emanuel Goldberg, "Das Auge des Menschen," in *Farbenphotographie: Eine Sammlung von Aufnahmen in natürlichen Farben*, vol. 10, ed. Fritz Schmidt (Leipzig: Seemann, 1913), 78.

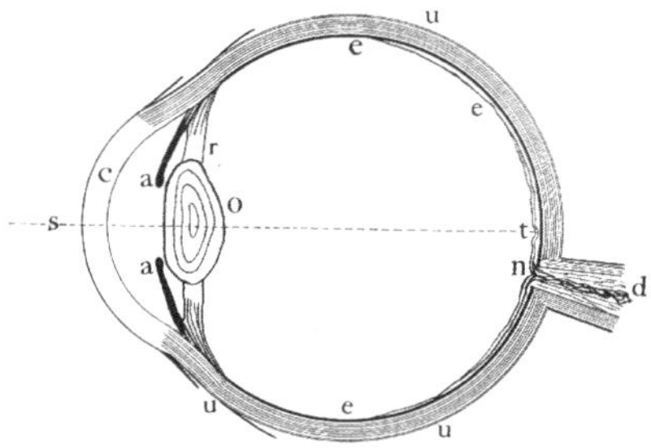

FIG. 7: Schematic cross section of the human eye as a camera, ca. 1909. Emanuel Goldberg, "Das Auge des Menschen," *Farbenphotographie: Eine Sammlung von Aufnahmen in natürlichen Farben*, ed. Fritz Schmidt (Leipzig: Seemann, 1913), 78.

29 *Offizieller Katalog*, 53.

30 Frank, "Die Gruppe 'Unterhaltung und Belehrung,'" 565.

31 *Offizieller Katalog*, 53–54.

32 Ibid., 55.

33 Friedrich Kittler, *Optical Media*, trans. Anthony Enns (Cambridge: Polity Press, 2010), 34.

34 Frank, "Die Gruppe 'Unterhaltung und Belehrung,'" 565.

35 "The Dresden Exhibition [I]," 418; cf. Vanessa Rocco, "Pictorialism and Modernism at the Dresden *Internationale Photographische Ausstellung*," *History of Photography* 33, no. 4 (2009): 392–93.

36 "Die internationale photographische Ausstellung in Dresden 1909, III," *Photographisches Wochenblatt* 35, no. 33 (1909): 322.

be 'instructed.' They can absorb knowledge only if it is accompanied by the slight shock that nails down inwardly what has been experienced."[37] In this perspective, the defamiliarization of the sense of vision Goldberg's apparatuses effected was in fact of lasting epistemic value, since "at a single stroke, it elucidated a whole series of important chapters of photography."[38] The Ipha, it would seem, was indeed one of "the most advanced outposts in the realm of display methods," as Benjamin described exhibitions of his time.[39]

Examining the Traces of Ephemeral Events

In 1909, Goldberg presented a "science center" *avant la lettre* to wide audiences that involved the visitor by means of what the late twentieth century would call "hands-on exhibits." The Goldbergian concept of "'press-button' science-teaching"[40] drew international attention[41] and became a source of crucial impulses for museums of technology in the early twentieth century: the Ipha prompted Oskar von Miller, the founder of the Deutsches Museum in Munich, to recruit Goldberg to design the planetarium he planned to build.[42] In October 1909, the "apparatuses for the popular exposition of the foundations of vision and photography"[43] built for the Dresden exhibition were presented to the German Book Trade Association and permanently installed in the machine hall of its headquarters in Leipzig.[44] The materials published in connection with the association's celebration of its twenty-fifth anniversary offer the last verifiable indication concerning the fate of the devices.[45]

Five years later, the *International Exhibition for the Book Trade and Graphic Arts* (Bugra) in Leipzig featured a section on "Scientific Photography"; set into an eighty-foot wall studded with peepholes, cranks, and handles, Goldberg presented thirty apparatuses whose handling and thematic orientation resembled those of the experimental arrangements he had exhibited in 1909. Visitors to the 1914 show thus enjoyed another opportunity to survey "all fields of photographic optics"[46] almost in the blink of an eye (fig. 8).[47] A 1915 report mentions that exhibits shown at the Ipha as well as the Bugra would form the core collection of a museum of photography to come.[48] Yet the plan never came to fruition. As Michael Buckland surmises, the unique devices made exclusively for the Bugra by Carl Zeiss, in Jena,[49] probably passed into

37 Walter Benjamin, "Food Fair: Epilogue to the Berlin Food Exhibition," in *Selected Writings*, vol. 2: *1927–1930*, ed. Michael W. Jennings, Howard Eiland, and Gary Smith (Cambridge, MA: Harvard University Press, 2005), 136. On the connection between shock and exposition in Benjamin, see Gottfried Korff, "*Omnibusprinzip* und *Schaufensterqualität*: Module und Motive der Dynamisierung des Musealen im 20. Jahrhundert," in *Geschichte und Emanzipation: Festschrift für Reinhard Rürup*, ed. Michael Grüttner, Rüdiger Hachtmann, and Heinz-Gerhard Haupt (Frankfurt am Main: Campus, 1999), 732.

38 "Aus der Gruppe 'Belehrung' der Dresdener Ausstellung," *Photographische Mitteilungen* 46 (1909): 271.

39 Benjamin, "Food Fair," 135; cf. Korff, "*Omnibusprinzip* und *Schaufensterqualität*," 732.

40 "The Dresden Exhibition [I]," 418.

41 Buckland, *Emanuel Goldberg and His Knowledge Machine*, 47; cf. Rolf Sachsse, "Beginnen wir! Die photographischen Abteilungen der Hochschule für Graphik und Buchkunst in Leipzig zwischen 1890 und 1950," in *Fotografie: Arbeiten von Absolventen und Studenten 1980–93; 100 Jahre Fotografie an der Hochschule für Grafik und Buchkunst Leipzig* (Leipzig: Hochschule für Grafik und Buchkunst Leipzig, 1993), 10.

42 Buckland, *Emanuel Goldberg and His Knowledge Machine*, 47, 284. On the influence that world exhibitions had on museums of technology, see the early survey in Eugene S. Ferguson, "Technical Museums and International Exhibitions," *Technology and Culture* 6, no. 1 (1965): 30–46.

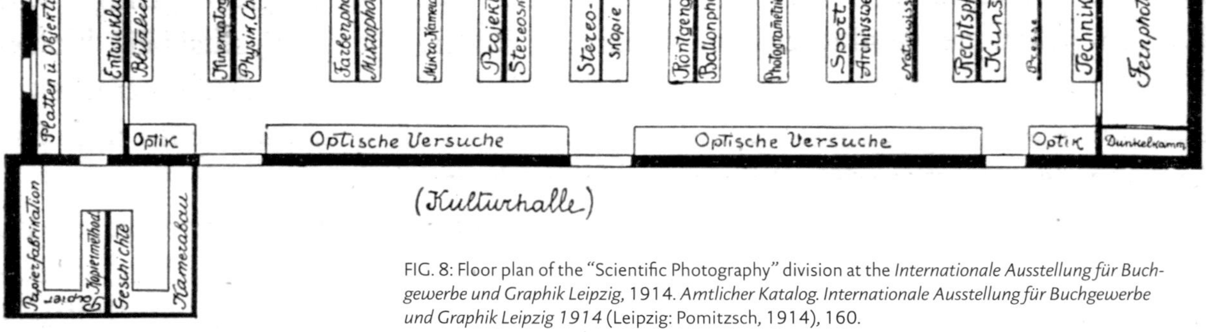

FIG. 8: Floor plan of the "Scientific Photography" division at the *Internationale Ausstellung für Buchgewerbe und Graphik Leipzig,* 1914. *Amtlicher Katalog. Internationale Ausstellung für Buchgewerbe und Graphik Leipzig 1914* (Leipzig: Pomitzsch, 1914), 160.

possession of the German Museum of Books and Writing,[50] so that the latter at some point must have had a collection of seventy-four apparatuses exploring issues concerning the sense of vision, color, optics, and photography. Several devices were presented again at the *International Photographic Exhibition of the Union of German Amateur Photographers' Associations* held in Leipzig in 1932.[51] Then the trail goes cold.[52]

"To build awareness of the most variegated phenomena in the field of vision"[53] was the Ipha's stated goal and the genuine opportunity it offered. In room 17, the experience of seeing, the prerequisite on which any exhibition rests, became the explicit thematic object of the Dresden exposition of photography. Yet vision itself is irreproducible. Exhibitions are ephemeral events. Anything not documented in pictures or descriptions is lost and cannot be recovered. What remains are fragments. To approach an exhibition from a distance of more than a hundred years means to perform a "historiographical operation,"[54] locating surviving pictures, descriptions, documents, and objects and devising an interpretive narrative. Still, not everything can be put into words, as Michel de Certeau has argued—tellingly enough, he illustrates his point by comparing the historian to a museum guide—and even when exhibits that were once parts of an exposition may be embedded in a shared narrative, the necessity remains of pointing to the sheer fact of their visible existence.[55] In the instance of Goldberg's apparatuses and the phenomena they allowed their users to see, we lack even that option. "An exhibition wants to be seen, not read,"[56] an anonymous correspondent put it with a view to the Ipha in August 1909. The retrospective engagement with the apparatuses in room 17 must face up to this contradiction. Yet the meticulous description of what is now absent may turn out to be no less productive—indeed, perhaps especially productive—for the analysis of the relationship between technical images and ideas of what seeing is: the sediments the history of photography has deposited consist not only of shots and prints but also of the traces that devices and optical impressions have left behind in the documents of an exhibition and the recorded experiences of their beholders. Such afterimages, in the physiological as well as the figural sense, are indispensable objects for an art history that seeks to set the nexus between technically generated images, media, technology, science, and aesthetics before all eyes and preserve its recollection.

43 *Bericht der Königlichen Akademie für Graphische Künste und Buchgewerbe zu Leipzig, Schuljahre 1908–1910* (Leipzig: Breitkopf & Härtel, n.d.), 49.

44 Ibid., 54–55.

45 See "Die Jubelfeier des Deutschen Buchgewerbevereins aus Anlaß seines fünfundzwanzigjährigen Bestehens," *Archiv für Buchgewerbe* 46, nos. 11–12 (1909): 325.

46 Emanuel Goldberg, "Die wissenschaftliche Photographie auf der Internationalen Ausstellung für Buchgewerbe und Graphik," *Photographische Rundschau und Mitteilungen* 51 (1914): 219.

47 Ibid., 217.

48 "Gründung eines photographischen Museums," *Photographische Rundschau und Mitteilungen* 52 (1915): 72.

49 Goldberg, "Die wissenschaftliche Photographie," 217.

50 Buckland, *Emanuel Goldberg and His Knowledge Machine*, 49.

51 Sachsse, "Beginnen wir," 10.

52 A bombing raid on December 4, 1943, destroyed roughly 90 percent of the museum's holdings and files. The author is grateful to Wolfgang Hohensee at the Cultural History and History of Paper Collections of the German Museum of Books and Writing (German National Library) in Leipzig for this information; e-mail message to the author, September 2, 2010.

53 Frank, "Die Gruppe 'Unterhaltung und Belehrung,'" 568.

54 Michel de Certeau, *The Writing of History*, trans. Tom Conley (New York: Columbia University Press, 1988), 56–113.

55 Ibid., 99–100; cf. Wim Weymans, "Michel de Certeau and the Limits of Historical Representation," *History and Theory* 43, no. 2 (2004): 176.

56 "Die internationale photographische Ausstellung in Dresden 1909, III," *Photographisches Wochenblatt* 35, no. 33 (1909): 321.

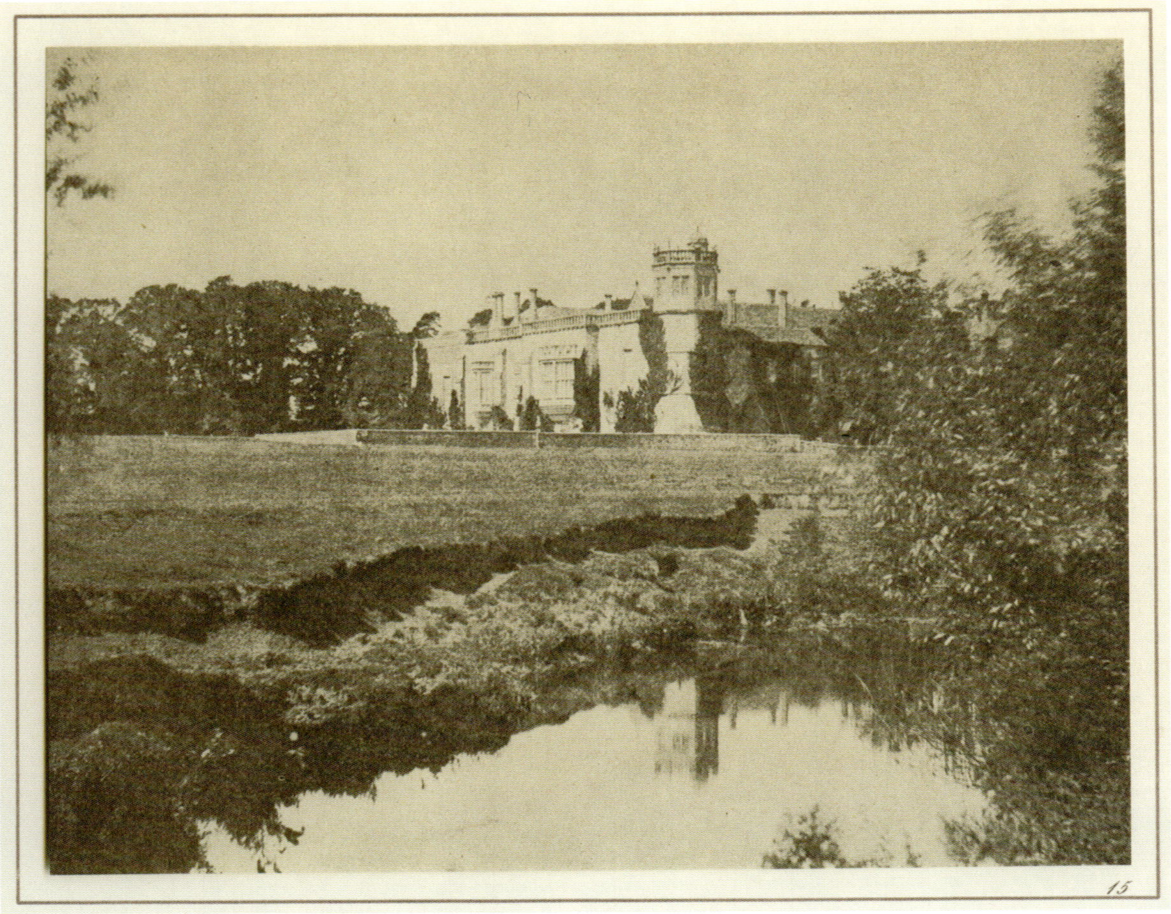

OBJECTIVITY AND EVIDENCE

Objectivity describes science's claim to represent the truth. Accordingly, it constitutes an overarching category in the sense of a scientific ideal that is mutable and so must be regarded in a historical perspective, as Lorraine Daston and Peter Galison have pointed out in a seminal study on the history of science (Daston and Galison 1992, 2007). Historical scrutiny reveals what is regarded as "objective" at any given time to be a convention. Objectivity, that is to say, is the result of agreements, negotiation processes, and norms set by the representatives and actors of a certain scientific community that can claim to be binding within that community (Fleck 1935).

As Daston and Galison have shown with reference to the use of images representing scientific objects, the word *objectivity* came into fashion in the nineteenth century in connection with the invention and application of image-producing technology such as photography. The ideal of "mechanical objectivity" is associated with the notion of a technical recording or inscription of scientific phenomena that eliminates any subjective intervention (Nagel 1986). This ideal is evident in Talbot's appraisal of photography as the "pencil of nature" (Talbot 1844; fig. 1) but also in the use of prints and measuring methods for collecting forensic data (figs. 2 and 3). At the same time, use of the term often proves to be a trope that is current and effective even today, employed to bolster the truth and authenticity of what represented (fig. 4). Behind this trope stand the construction of the equipment and the techniques of image generation, processing, and staging that are inherent to the production and presentation of images. In an analysis of scientific images, the focus must therefore be not only on written ascriptions, but also—beyond what Daston and Galison

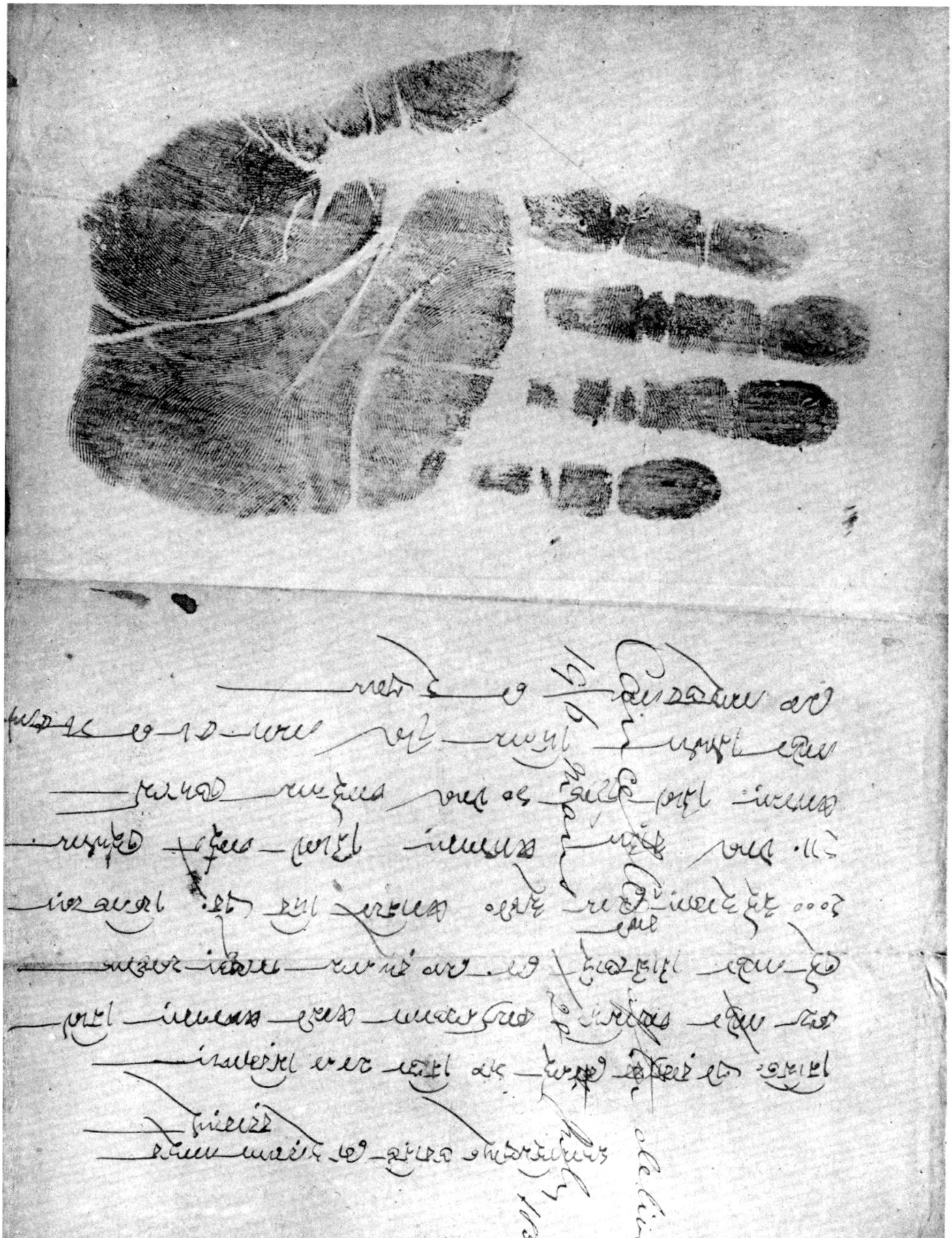

FIG. 1 (facing): "Lacock Abbey in Wiltshire." According to William Henry Fox Talbot, this building, which appears in plate 15 of his book *The Pencil of Nature,* was the first "that was ever yet known to have drawn its own picture." Talbot described his photographic images as "self-representations" of the things represented, thereby founding a rhetoric of objectivity that is still current today. William Henry Fox Talbot, *The Pencil of Nature* (New York: Da Capo Press, 1969), plate 15.

FIG. 2 (above): Print of the right hand of Rajyadhar Konai on a contract concluded on July 28, 1858, in Hugli-Chunchura, India. Sir William Herschel, a member of the British Civil Service in India, was the first to use handprints and, later, fingerprints to identify signatories to contracts unambiguously and so participated in the establishing of dactyloscopy. Karl Pearson, *The Life, Letters and Labours of Sir Francis Galton,* vol. 3a: *Correlation, Personal Identification and Eugenics* (Cambridge: Cambridge University Press, 1930), plate 5.

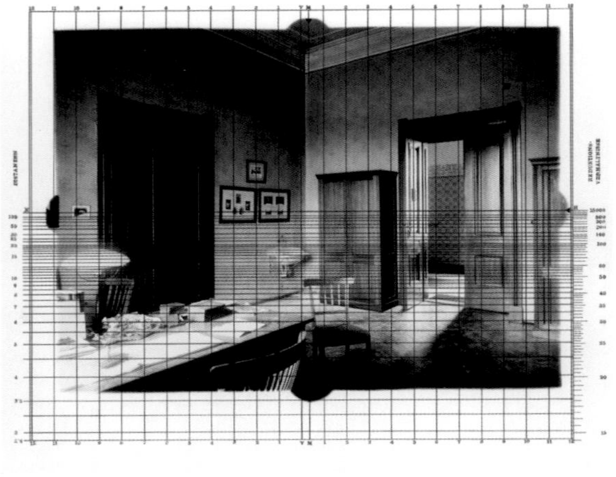

FIG. 3: Franz Eichberg, photogrammetric picture, Vienna, ca. 1920. Photogrammetry, a method for measuring objects represented in a photograph by photographic means, was developed in the nineteenth century, based on the observation that a photographic image records images in perspective. Using various processes, information about the dimensions of spaces, architecture, and other objects could be derived from the image alone. The method requires precise knowledge of the conditions under which the photograph was taken or the inclusion of yardsticks in the picture. In Vienna, Franz Eichberg invented a photographic apparatus containing a thin steel grid, which, in combination with the photograph, was designed to enable all the elements of housing interiors to be completely reconstructed so as to establish the facts in criminal cases. In this way, plans of a crime scene could be made that would then provide an objective basis for reconstructing the circumstances. Archive of the Technisches Museum Wien, Vienna, Austria. © TMW-Archiv 2008.

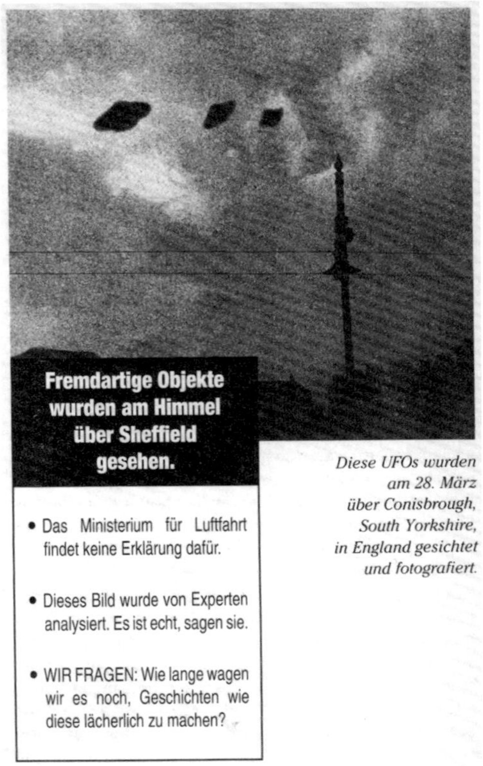

FIG. 4: UFOs over Sheffield (anonymous and undated image). The slightly blurry appearance of saucer-shaped objects in the sky, combined with captions in the style of newspaper reports, support the authors' claim to veracity. Nigel Blundell and Roger Boar, *The World's Greatest UFO Mysteries* (London: Hamlyn, 1983), 33. Fortean Picture Library.

have pointed out—on the contexts and conditions of production and the practices and ways of engaging with images. Precise case studies have decisively qualified the notion of the "image of objectivity" by building critical awareness of the constructed nature of technical images in scientific contexts (e.g., Soojung-Kim Pang 1997; Hoffmann 2002; Brons 2004).

Terms such as *evidence* and *demonstration* are closely associated with the concept of objectivity as a scientific ideal. Images have turned out to be a constitutive element of scientific argumentation (fig. 5) because they are seen as especially suited to convey unquestionable, obvious, and immediately manifest truth. In its original philosophical sense, the Latin word *evidentia* refers quite literally to seeing: what is evident is "directly present to the eye." Diagrams, for example, are predicated on their claim to make something manifest (fig. 6). Here, too, issues such as how evidence is produced in specific contexts, how arguments are made using images, and how an image could possibly function as proof must be investigated. Jennifer Tucker, Tal Golan, Miloš Vec, and others have undertaken studies in this area (Vec 2002; Golan 2004; Tucker 2005). On the one hand, such analysis must consider the image itself with a view to its formal characteristics—that is to say, the specific pictorial operations by which a visual creation of meaning functions—and embed the image in a wider visual history. On the other hand, it must take into account the contexts of production and the claims, attributed meanings, expectations, and practices related to the image. —VD

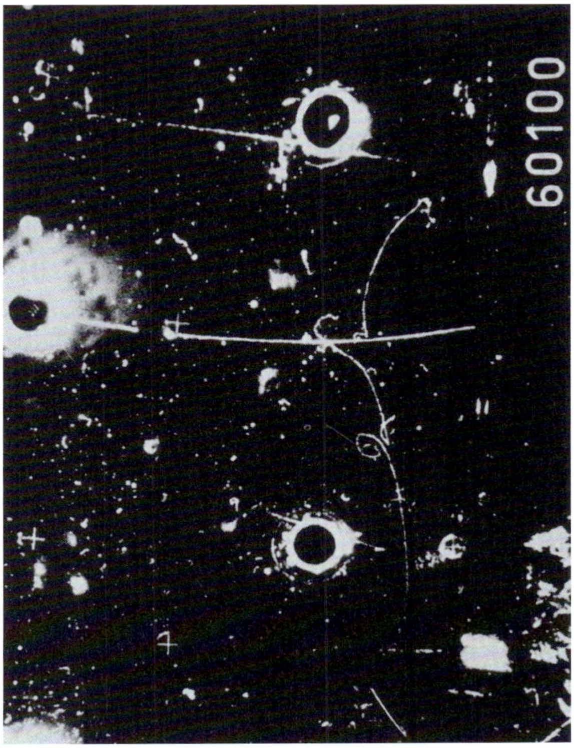

FIG. 5: "Golden Event." This image provided convincing proof of neutral currents to the wider community of physicists in 1973. Peter Galison describes the potential of a single image to furnish such cogent evidence as the fundamental characteristic of an image-based research tradition in physics, which—in contrast to a "tradition of logic" that emphasizes the statistical evaluation of masses of data—aspires to successfully record a "golden event" so as to then base an argument on the image (see Peter Galison, *Image and Logic* [Chicago: University of Chicago Press, 1997]). F. J. Hasert et al. "Search for Elastic Muon-Neutrino Electron Scattering," *Physics Letters* 46B (1973): 122, fig. 1, with permission from Elsevier.

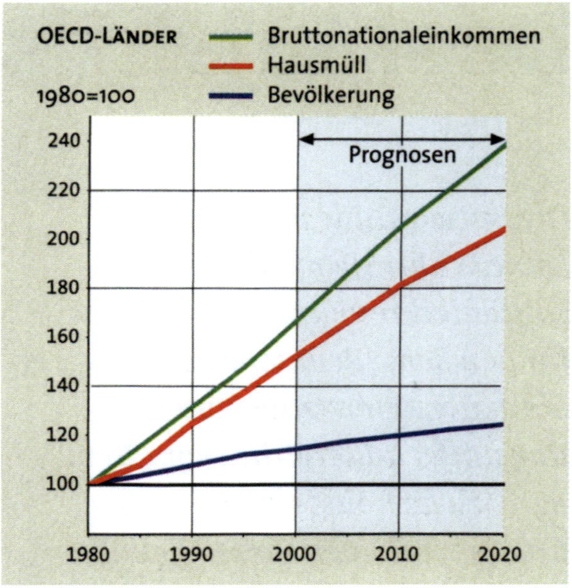

FIG. 6: "Mit der Wirtschaft wächst der Müll" (As the economy grows, so does the garbage output): Statistical curve by Philippe Rekacewicz from his *Atlas der Globalisierung*, 2006. The correlation of the three curves demonstrates the author's contention that increases in household waste depend not on population developments but on economic growth. *Atlas der Globalisierung. Die neuen Daten und Fakten zur Lage der Welt*, ed. Le monde diplomatique, Dietmar Bartz et al. (Berlin: Taz, 2006), 27.

LITERATURE

Brons, Franziska, "Sachverständige Fotografie: Der Mikrokosmos vor Gericht." *Fotogeschichte* 24, no. 94 (2004): 15–29.

Chandler, James, Arnold I. Davidson, and Harry D. Harootunian, eds. *Questions of Evidence: Proof, Practice, and Persuasion across the Disciplines*. Chicago: University of Chicago Press, 1994.

Daston, Lorraine, and Peter Galison. "The Image of Objectivity." *Representations*, no. 40 (1992): 81–128.

Daston, Lorraine, and Peter Galison. *Objectivity*. Boston: Zone Books, 2007.

Fleck, Ludwik. *Genesis and Development of a Scientific Fact* [1935]. Chicago: University of Chicago Press, 1979.

Gillispie, Charles C. *The Edge of Objectivity: An Essay in the History of Scientific Ideas*. Princeton, NJ: Princeton University Press, 1960.

Golan, Tal. *Laws of Men and Laws of Nature: The History of Scientific Expert Testimony in England and America*. Cambridge, MA: Harvard University Press, 2004.

Hoffmann, Christoph. "Die Dauer eines Moments: Zu Ernst Machs und Peter Salchers ballistisch-fotografischen Versuchen 1886/87." In *Ordnungen der Sichtbarkeit: Fotografie in Wissenschaft, Kunst und Technologie*, ed. Peter Geimer, 342–77. Frankfurt am Main: Suhrkamp, 2002.

Nagel, Thomas. *The View from Nowhere*. New York: Oxford University Press, 1986.

Pang, Alex Soojung-Kim, "'Stars Should Henceforth Register Themselves': Astrophotography at the Early Lick Observatory." *British Journal for the History of Science* 30, no. 2 (1997): 177–201.

Snyder, Joel. "Res Ipsa Loquitur." In *Things That Talk: Object Lessons from Art and Science*, ed. Lorraine Daston, 195–221. New York: Zone Books, 2004.

Talbot, William Henry Fox. *The Pencil of Nature*. London: Longman, Brown, Green and Longmans, 1844.

Tucker, Jennifer. *Nature Exposed: Photography as Eyewitness in Victorian Science*. Baltimore: Johns Hopkins University Press, 2005.

Vec, Miloš. *Die Spur des Täters: Methoden der Identifikation in der Kriminalistik (1879–1933)*. Baden-Baden: Nomos-Verlag, 2002.

X-Ray Vision and Shadow Image: On the Specificity of Early Radiographs and Their Interpretations around 1900

Vera Dünkel

The discovery of X-rays in late 1895 led to the invention of an imaging technique whose domains of application were at first far from settled. Its initial openness to interpretation and the immense fascination of the images it generated fostered a sometimes playful experimentation with the new technique that expanded into a wide variety of fields in the sciences as well as in popular culture. The goal of such experimentation was to produce images and explore areas of investigation in which the technique might be fruitfully applied. A number of questions remained open in this early phase: Which categories of visual culture and which concepts ought to be applied to understanding the peculiar appearance of these images? What did they show and what should one call this form of depiction? How might what was novel about them be shown while also tying them back to familiar forms of representation?

The following pages draw on contemporary journalistic writings and tracts as well as the first radiographic atlases to present different interpretations of the new technology and the images it produced. The aim is to show how witnesses to the emergence of radiography were torn between euphoric belief in the possibilities of what would be known as X-ray vision and the critical awareness that these images were constructions, comparable, perhaps, to silhouettes. In order to gain a better understanding of the specific ways in which these images generated meaning in different contexts, we examine both their formal qualities and the meanings that were assigned to them by those who engaged with them and harnessed their evidentiary force.

The Hand with the Floating Rings

On January 23, 1896, Wilhelm Conrad Röntgen, a professor of physics, delivered a lecture before the Physical and Medical Society at the University of Würzburg on a "new kind of rays."[1] During the meeting, he made a picture of the hand of one attendee, the professor of anatomy Albert von Kölliker (fig. 1); the audience greeted the image with applause.[2]

Set before a diffuse monochrome backdrop, the hand stretches into the field of view from the bottom left; it is fully visible above the wrist. Its image is composed of three strata: from the lighter outline of the hand, the bones emerge in darker tones, and superimposed above these two strata, or so it seems, we see the sharply delimited oval shapes of two rings. The clear outline or silhouette of the hand recalls its familiar appearance; by contrast, the prominent emergence of the bones, as the innermost stratum, produces an utterly novel sight. The way they are

1 Wilhelm Conrad Röntgen, *Über eine neue Art von Strahlen (Vorläufige Mittheilung): Sonderdruck, aus den Sitzungsberichten der Würzburger Physik.-medic. Gesellschaft* (Würzburg: Stahel, 1895); see Wilhelm Conrad Röntgen, "On a New Kind of Rays," trans. Arthur Stanton, *Nature* 53, no. 1369 (1896): 274–76. On January 1, 1896, Röntgen sent a copy of his report together with a number of images to colleagues in Germany and abroad. See Otto Glasser, *Wilhelm Conrad Röntgen und die Geschichte der Röntgenstrahlen* [1931] (Berlin: Springer, 1959), 24.

2 On Röntgen's lecture, see Glasser, *Wilhelm Conrad Röntgen und die Geschichte der Röntgenstrahlen*, 37–40.

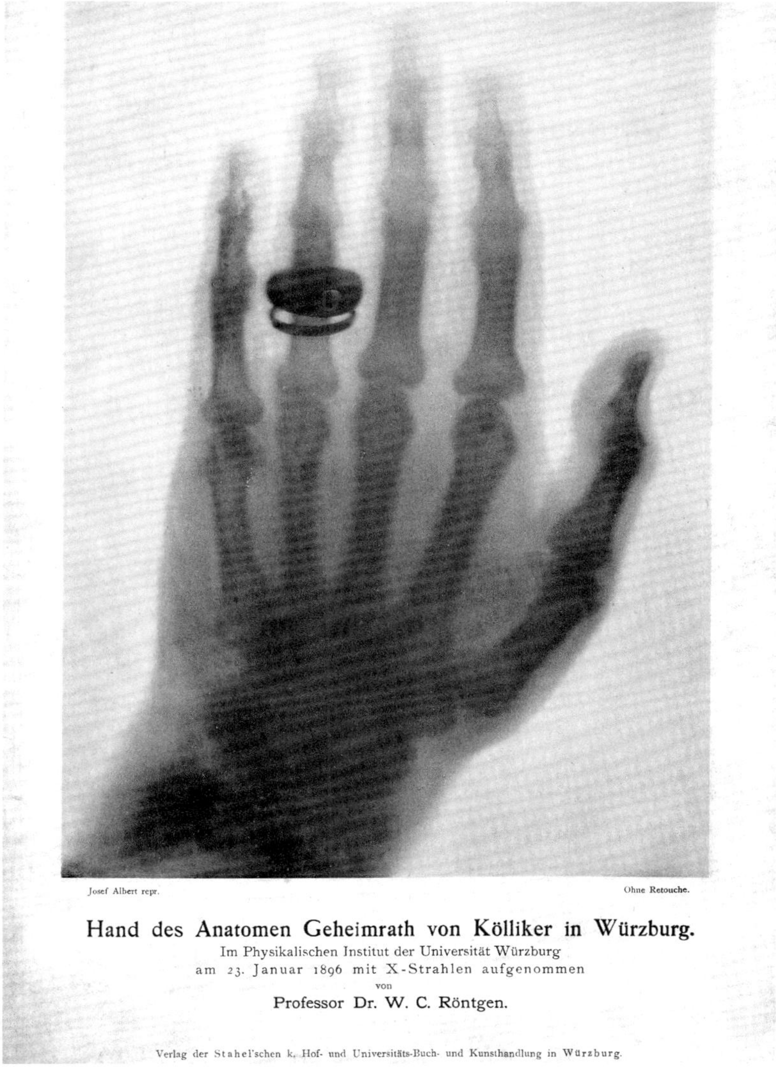

FIG. 1: Wilhelm Conrad Röntgen, radiograph of the hand of the anatomist Albert von Kölliker, taken at the Institute of Physics of the University of Würzburg, January 23, 1896. Archive of Deutsches Röntgen-Museum, Remscheid-Lennep, Germany.

loosely embedded in the hand's outline heightens the impression that we are looking through a veil-like screen at something below, as though the very flesh of the hand had become transparent. The bones appear curiously suspended within the silhouette of the hand. What further lends them a certain illusion of prominence is that the interior is in some places lighter than their dark contours. A third stratum consists of the rings, which seem to hover above the rest. Their appearance is flat, but they also evince foreshortening. The distance that separates them from the bones, as though the visceral outline of the hand had pushed them outward, suggests their position on the finger, while their similarly dark tone associates them more closely with the bones, which they nonetheless do not touch.

An early newspaper report described the disconcerting relationship between the different parts of the body by noting that the rings "floated around the fingers"; the wording bespeaks the bewilderment this sight caused.[3] On the one hand, the image suggests a gaze into a body, a gaze that penetrates its outermost layer to fall upon something that is usually

3 "Eine sensationelle Entdeckung," *Die Presse*, Vienna, January 5, 1896. A facsimile of the original report can be found in Klaus Hübner, *Die zwei ersten Zeitungsberichte über Röntgens Entdeckung* (Berlin: ERS, 2000), 36.

FIG. 2: Advertisement of Stahel's university bookstore and art gallery, Würzburg, for Wilhelm Conrad Röntgen's radiograph of the hand of the anatomist Albert von Kölliker, 1896. Archive of Deutsches Röntgen-Museum, Remscheid-Lennep, Germany.

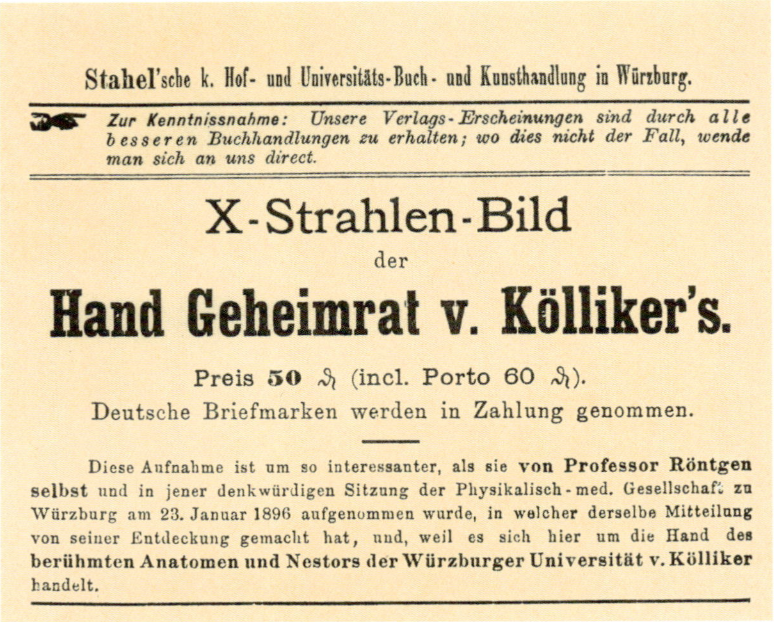

Stahel'sche k. Hof- und Universitäts-Buch- und Kunsthandlung in Würzburg.

Zur Kenntnissnahme: Unsere Verlags-Erscheinungen sind durch alle besseren Buchhandlungen zu erhalten; wo dies nicht der Fall, wende man sich an uns direct.

X-Strahlen-Bild

der

Hand Geheimrat v. Kölliker's.

Preis **50** ₰ (incl. Porto 60 ₰).

Deutsche Briefmarken werden in Zahlung genommen.

Diese Aufnahme ist um so interessanter, als sie **von Professor Röntgen selbst** und in jener denkwürdigen Sitzung der Physikalisch-med. Gesellschaft zu Würzburg am 23. Januar 1896 aufgenommen wurde, in welcher derselbe Mitteilung von seiner Entdeckung gemacht hat, und, weil es sich hier um die Hand des **berühmten Anatomen und Nestors der Würzburger Universität v. Kölliker** handelt.

concealed, although the beholder is cognizant of its existence. Because the hand remains discernible as such, the beholder's perception wavers between recognition and disconcertment. The image has been inverted—it is a positive, not a negative—promoting the reading that recognizes a familiar format in it: the image as the gaze at something. The impression of spatial depth produced by the varying shades of light and darkness likewise allows us to draw on past visual experience. On the other hand, this ostensibly concrete spatial perception of a body remains fairly tenuous when it comes to precisely locating the individual elements in the picture in relation to each other: Where exactly on the finger is the ring? Are the bones in, above, or beneath the shadow of the hand? Setting empirical knowledge about hands aside, the precise arrangement of the body parts and their relations to each other remain, quite literally, in suspense. A positively surreal spectacle comes into view: rings, bones, and hand overlap and interpenetrate. The jewelry, placed on the outside of the body, attaches to its innermost part, the bone, after what lay between them—muscles, tendons, blood vessels—has been dissolved. The inside and outside of the body coincide in the image plane.

A process rapidly ensued in which this image became an auratic object: soon after Röntgen's lecture, original prints, laminated to a matte and furnished with a printed title and a brief account describing the genesis of the image, went on sale at Stahel's university bookstore (fig. 2). A notice advertising the sale praised the image as follows: "This picture is the more interesting since it was taken *by Professor Röntgen himself* and during that memorable meeting [...] and because it shows the hand of the *famous anatomist and nestor of the university, von Kölliker*."[4] The caption lends the image the aura of a piece of evidence as well as a memorial, calling to mind the event of the "memorable meeting" and engaging the anatomist's renown to attest that the presentation actually took place. The reference to the living anatomist (whose bones are now

4 The advertisement appears in the back flyleaf of Röntgen's second report, published on March 9, 1896, by Stahel's university bookstore and art gallery in Würzburg. See Wilhelm Conrad Röntgen, *Eine neue Art von Strahlen. II. Mittheilung: Sonderdruck, aus den Sitzungsberichten der Würzburger Physik.-medic. Gesellschaft* (Würzburg: Stahel, 1896).

in turn revealed to our eyes by the new technique!) imbues the image of the hand with significance and turns it into a commemorative object of sorts. At the same time, the witness's gesture the anatomist performed during the lecture by holding his beringed hand steady on the plate for a few minutes—a technical requirement of the production of the image—is charged with meaning as well (see fig. 4). That such witnessing is taking place is also reflected in the formal and upright appearance of the hand in the image: the anatomist's hand and, by extension, his public character vouch for the authenticity of the method.[5] The caption underlines this iconic content.

Designating the image as a "hand," moreover, generates a tension that proceeds from the difference between the universally known original object, the hand, and the novel phenomenon that appears in the image, the bone skeleton: the latter is associated with death, but the caption emphasizes that this is a living person's hand. The label "hand" calls up a familiar picture, which then makes for a peculiar contrast with the phenomenon in the image. By engendering this difference, the framing inscription enables the beholder to experience the tension between recognition and astonishment, comparing the familiar and novel appearances and deriving an understanding of the particular quality of this sort of image.

Initial Interpretations: X-Ray Vision . . .

Early newspaper reports and tracts about the new technique bear titles such as "photography of the invisible," "new photography," and "photographie à travers les corps opaques." These descriptions would seem to be expedients that serve to capture the novelty of the technique while also allowing beholders to draw connections to what they are familiar with. Describing the technique as a form of photography placed it in a category that modeled the coexistence of the fantastic with scientific objectivity: the nineteenth century regarded photography as a medium that allowed for the accurate and detailed depiction of reality while also transcending the perception of the naked eye. Pointing to contemporary technical achievements such as the telephone and the photographic snapshot, an early newspaper report accordingly located the new "photographic" technique halfway between science and fiction, referring to "fantastic-futuristic speculation in the style of Jules Verne" while also emphasizing that "serious men of learning take the matter seriously" and that "the photographic evidence for this discovery has so far seemed to stand up to the scrutiny of serious critics."[6]

Photography and works of literary fiction prepared the ground for the euphoric reception of the new technique and established the framework within which the earliest attempts to classify and interpret it took place. One of the first handbooks on X-ray imaging, published in 1897, even quoted a "medical fairy-tale" from 1892: "If only there were a means to make man transparent as a jellyfish!"[7] The ancient medical dream of the transparent human being articulated in the fantastic tale seemed to have become a reality with the new technique, which was seen as the answer to the "efforts of the physicians to lay the interior of the body open to visual inspection without any dissection of tissue."[8] The jellyfish as a metaphor for transparency and pellucidity framed

5 As the organ of action, the hand may stand in for the entire person and his or her intentions; legal custom establishes placing one's hand on something as a gesture of oath-taking and witnessing. See Hanns Bächtold-Stäubli, ed., *Handwörterbuch des deutschen Aberglaubens*, vol. 3 (Berlin: de Gruyter, 1930/1931), s.vv. "Hand" and "Handauflegen," 1379–98 and 1398–1402, and *Reallexikon für Antike und Christentum*, vol. 13 (Stuttgart: Hiersemann, 1986), s.v. "Hand II (ikonographisch)," 445. On the motif of the hand, its expressive range, and its symbolism in evidentiary contexts, see Jennifer Blessing, ed., *Speaking with Hands: Photographs from The Buhl Collection*, exhibition catalogue (New York: Solomon R. Guggenheim Museum, 2004); Bernd Evers, ed., *Sprechende Hände*, exhibition catalogue (Berlin: Staatliche Museen zu Berlin, Stiftung Preußischer Kulturbesitz, 2006), 22–25.

6 "Eine sensationelle Entdeckung," 37.

7 Oskar Büttner and Kurt Müller, *Technik und Verwerthung der Röntgen'schen Strahlen im Dienste der ärztlichen Praxis und Wissenschaft* (Halle an der Saale: Knapp, 1897); Philander, "Elektra: Ein physikalisch-diagnostisches Märchen aus dem zwanzigsten Jahrhundert," in *Medizinische Märchen* (Stuttgart: Levy & Müller, 1892), 186–98.

8 Büttner and Müller, *Technik und Verwerthung der Röntgen'schen Strahlen*, 3.

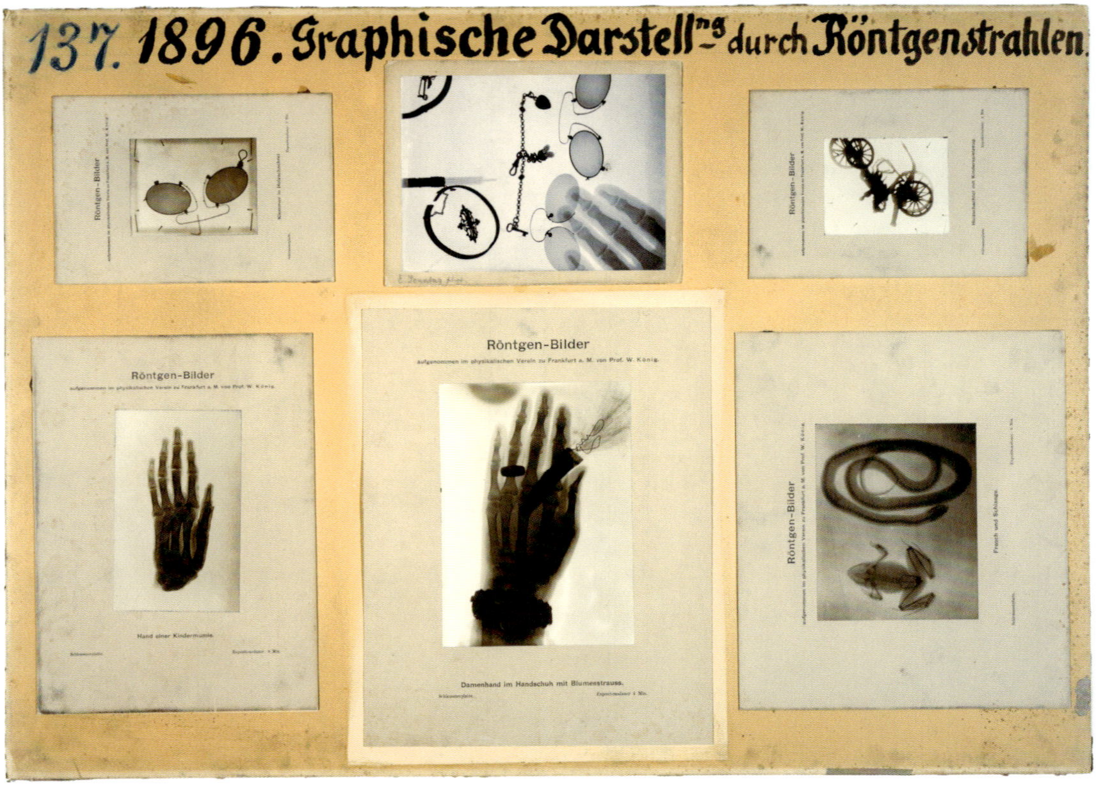

FIG. 3: "Graphical depiction using X-rays," 1896. Plate 137 from Hermann Krone's *Historisches Lehrmuseum für Photographie.* © Hermann-Krone-Sammlung, Institut für Angewandte Photophysik / Technische Universität Dresden.

by the fairy-tale penned five years earlier became emblematic of the radiographic method, which permitted what seemed to be unlimited insight into the living body, as distinct from anatomical techniques, which were predicated on the death of the body under examination.[9] The new images suggested this distanced gaze into the body. In this context, they were primarily read not as shadows of the body but as images of a body become transparent. This reading gave rise to the idea of "X-ray vision,"[10] which is also symbolic of the physician's power over the healthy body controlled by public authorities—thirty years later, the "transparent man," the anatomical model made of clear synthetic material, would be the ideal image of that body.[11]

The euphoria of this idea inspired all sorts of experiments and application trials, from the visualization of the interior of rare species to determining the authenticity of mummies and diamonds, to the inspection of passengers and luggage at railway-station customs offices.[12] Numerous experiments tested the ability of the new technique to generate imagery, producing a prodigious variety of motifs. A plate from Hermann Krone's *Historisches Lehrmuseum für Photographie*, or "Historical-Didactic Museum of Photography" (fig. 3), is paradigmatic of this variety; it also demonstrates how the classification of X-ray imaging as a species of photography was seen as self-evident.[13] The prints assembled for this plate present a sample of typical motifs in the early era of radiography, such as hands (fig. 3, bottom center and left), small animals like snakes and frogs

9 See Monika Dommann, "Die magische Büchse der Elektra: Röntgenstrahlen und ihre Wahrnehmung um 1900," in *Licht und Leitung*, ed. Lorenz Engell, Joseph Vogel, and Bernhard Siegert (Weimar: Universitätsverlag, 2002), 35; Monika Dommann, *Durchsicht, Einsicht, Vorsicht: Eine Geschichte der Röntgenstrahlen 1896–1963* (Zurich: Chronos, 2003), 263.

10 I use the term to refer to the utopian idea of superhuman eyesight, which will be the special power, for instance, of the 1930s comic-strip hero Superman; see the earliest Superman shorts directed by Dave Fleischer and produced by Fleischer Studios starting in 1941.

11 The first transparent man was manufactured at the Deutsches Hygiene-Museum, Dresden, in 1928. For the genesis and significance of the transparent man in the context of measures of social and health policy, see Rosmarie Beier and Martin Roth, eds., *Der gläserne Mensch—eine Sensation: Zur Kulturgeschichte eines Ausstellungsobjekts* (Stuttgart: Hatje, 1990).

(bottom right), and everyday objects such as small cases for trinkets or toys (top, left to right).

. . . and Shadow-Image

Matching the bewildering diversity of motifs, the wide variety of terms used to describe these images attests to the lack of agreement on how exactly one should characterize them. For more than a decade, terminological chaos reigned; proposed names for the technique and the images it generated included *diagraphy* (or transparent image), *skiagraphy* (shadow-image), *radiography* (ray-image), and *pyknography* (density-image), illustrating how bewildering this form of representation seemed to contemporaries.[14] These terms at the same time also reflect the deliberations about what the new images depicted and how their epistemological benefits should be assessed. The debate, moreover, closely examined the particularities of the process of their production: "An X-ray image differs from a photographic image in two respects. First, it is not rays of light but X-rays that disintegrate the light-sensitive layer, i.e., the silver bromide gelatin coating of the recording plate. Second, photography is a light-image of the object to be depicted; the X-ray exposure, its shadow-image."[15] This observation in an early newspaper report names crucial properties of the radiograph that flow from the process of its production and characterizes it by distinguishing it from conventional photography. It is not visible light but invisible rays artificially generated using a spark inductor, voltage, and a vacuum tube that produce the image (fig. 4). And although these rays strike conventional silver bromide gelatin plates like those used to make

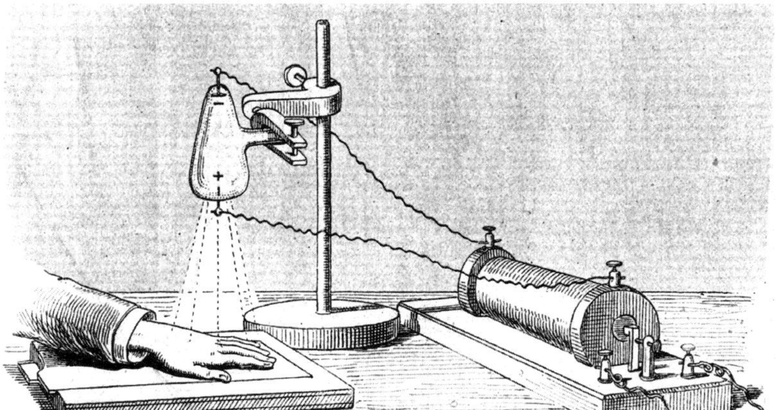

FIG. 4: Experimental arrangement: taking the radiograph of a hand. Popular illustration, late nineteenth century. *Cosmos* 45, no. 576 (1896): 301.

photographic negatives, this is a photographic process that takes place without the mediation of optical equipment such as a camera or lens, since the new rays were not amenable to refraction by the lenses of the era. The rays traverse the objects placed on the plate and project shadows of their parts with their varying densities onto the photographic layer.

To some observers, the fact that radiographs are shadow-images seemed to call their scientific value into question. An article in the medical journal *The Lancet* noted: "We possess no power of refracting

12 See Vera Dünkel, "Das Auge der Radiographie: Zur Wahrnehmung einer neuen Art von Bildern," in *Modernisierung des Sehens: Sehweisen zwischen Künsten und Medien*, ed. Matthias Bruhn and Kai-Uwe Hemken (Bielefeld: Transcript, 2008), 207–20.

13 See Vera Dünkel, "Die Fotografie mit Röntgenstrahlen: Hermann Krones Rezeption des Mappenwerks von Walter König und die Ikonographie der frühen Röntgenbilder," in *Wahr-Zeichen: Fotografie und Wissenschaft*, ed. Andreas Krase and Agnes Matthias, exhibition catalogue (Dresden: Technische Sammlungen, 2006), 45–46.

14 As Monika Dommann has shown, these terms varied with the contexts in which the authors worked and emphasized different qualities of the technique. See Dommann, *Durchsicht, Einsicht, Vorsicht*, 330–32; Monika Dommann, "'Das Röntgen-Sehen muss im Schweisse der Beobachtung gelernt werden': Zur Semiotik von Schattenbildern," *Traverse: Zeitschrift für Geschichte* 6, no. 3 (1999): 114–30.

15 Max Wildermann, "Über Entstehung und Verwendung der X-Strahlen nach dem heutigen Stande ihrer Erforschung," *Alte und neue Welt* 33, no. 10 (1898–99): 594–95.

the x rays so that we are utterly unable to obtain anything of the nature of a 'true image,' the results are nothing more than photographic prints of shadows."[16] The wording hints at the negative connotations—a secondary, incorporeal, dark image, a mere semblance—that have clung to the shadow ever since Plato's allegory of the cave.[17] Throughout a long history, images were described as devoid of truth and objectivity, being "nothing but shadows."[18]

16 "Roentgen Rays," *The Lancet* 148, no. 3826 (December 26, 1896): 1832.

17 See Hanns Bächtold-Stäubli, ed., *Handwörterbuch des deutschen Aberglaubens*, vol. 9 (Berlin: de Gruyter, 1938/41), s.v. "Schatte(n)," 126–42; Victor I. Stoichita, *A Short History of the Shadow*, trans. Anne-Marie Glasheen (London: Reaktion Books, 1997).

18 See Barron H. Lerner, "The Perils of 'X-ray Vision': How Radiographic Images Have Historically Influenced Perception," *Perspectives in Biology and Vision* 35, no. 3 (Spring 1992): 382–97; on the negative connotations that attach to the X-ray shadow, see 385–87.

As a flat shadow-image in which the different layers of which a body is composed appeared in inextricable superimposition and that nonetheless conveyed a sense of—however confusing—spatial depth (fig. 1), the radiograph indeed posed considerable challenges of interpretation to those who employed the new technique. Several potential solutions were subsequently proposed that aimed to improve the legibility of the images with a view to specific applications. Highly expensive processes designed to spatialize radiographs, such as radiostereoscopy, once more illustrate the desire for an unlimited X-ray vision. Other approaches attest to an acute awareness of the limitations constraining such a gaze and of the constructed nature of the X-ray image. In fact, its shadow-like quality was not always grounds for a negative assessment; in some instances, it became a crucial element in a positive critique of X-ray images that examined the problems of their interpretation with a view to the translation from flat image to physical and spatial reality.

Critique of the Shadow-Image: Learning to Read Shadows

One prominent advocate of this view was the surgeon Rudolf Grashey, the author of one of the earliest radiographic atlases, which began to appear in print around 1900. In the preface to his *Atlas typischer Röntgenbilder vom normalen Menschen* (Atlas of typical radiographic images of the normal human body, 1905), Grashey points out the fundamental difficulties the interpretation of radiographs encounters as a result of the difference between naked-eye perception and the perception of images—which is to say, of the real added value of the new technique: "Some of the silhouettes and lines that emerge in the radiographic image do not correspond to any self-contained anatomical concept. [...] That is why resolving and disentangling the lines compressed into the image plane is often a laborious task."[19] According to Grashey, the physician interpreting these images should draw on his knowledge of anatomy, which will nonetheless be only of limited help when it comes to making a diagnosis. This clearly calls the application of findings derived from the image to the body into question. In his atlas, Grashey places radiographs of individual body parts in their anatomical contexts by juxtaposing them with anatomical illustrations of the bones that bring out their three-dimensional shapes, schematic line drawings, and explanations in writing (fig. 5).[20] Only the synopsis of these materials renders the radiograph intelligible; at the same time, it is what constitutes the epistemic object of the atlas in the first place.[21] Even before publishing his "atlas of the normal body," Grashey had pointed out that there were further prerequisites to reading radiographs and classifying them according to categories such as "normal" and "pathological": "One cannot regard too many normal images, comparing them again in each instance, in order to gain a firm grasp of the notion of the normal."[22] The atlas therefore rested on the production

19 Rudolf Grashey, *Atlas typischer Röntgenbilder vom normalen Menschen* [1905], 2nd ed. (Munich: Lehmann, 1912), preface to the first edition of 1905, iii.

20 This plate and accompanying text appear in the second edition of the atlas, published in 1912, which was much enlarged in terms of both imagery and text.

21 See Vera Dünkel, "Vergleichendes Röntgensehen: Lenkungen und Schulungen des Blicks angesichts einer neuen Art von Bildern," in *Vergleichendes Sehen*, ed. Lena Bader, Martin Gaier, and Falk Wolf (Munich: Fink, 2010), 368–75.

22 Rudolf Grashey, "Fehlerquellen und diagnostische Schwierigkeiten beim Röntgenverfahren," *Münchener Medizinische Wochenschrift* 52, no. 1 (1905): 810.

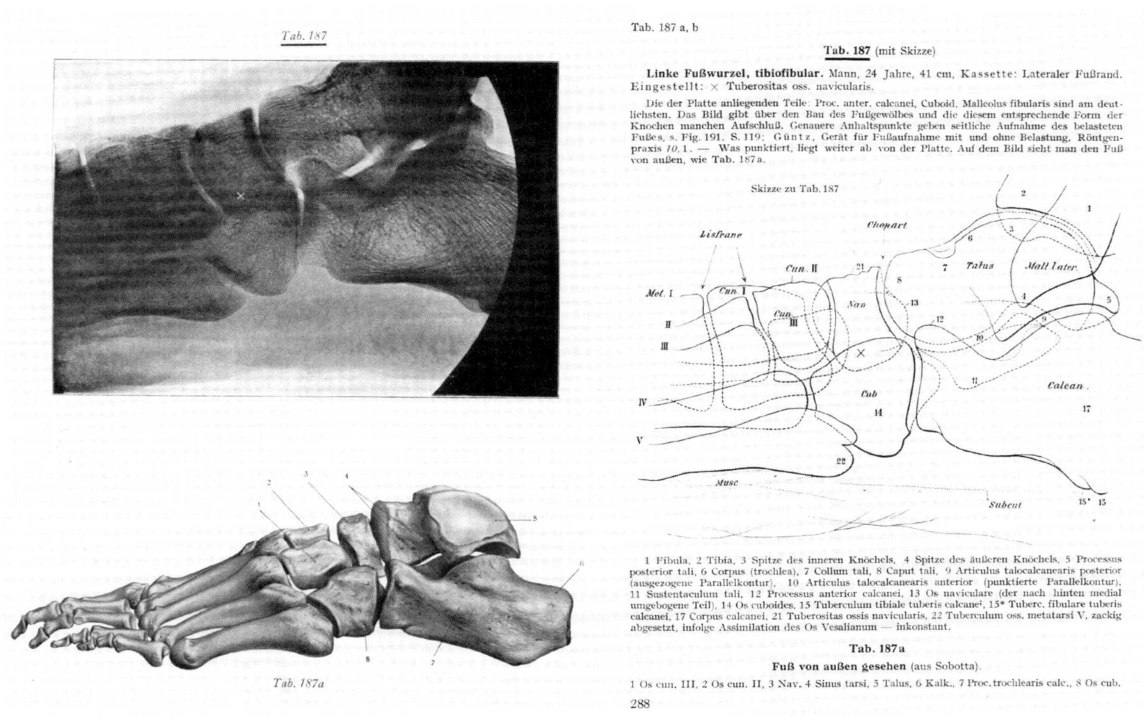
Tab. 187

Tab. 187a

Tab. 187 a, b

Tab. 187 (mit Skizze)

Linke Fußwurzel, tibiofibular. Mann, 24 Jahre, 41 cm, Kassette: Lateraler Fußrand. Eingestellt: × Tuberositas oss. navicularis.

Die der Platte anliegenden Teile: Proc. anter. calcanei, Cuboid, Malleolus fibularis sind am deutlichsten. Das Bild gibt über den Bau des Fußgewölbes und die diesem entsprechende Form der Knochen manchen Aufschluß. Genauere Anhaltspunkte geben seitliche Aufnahme des belasteten Fußes, s. Fig. 191, S. 119; Güntz, Gerät für Fußaufnahme mit und ohne Belastung, Röntgenpraxis *10*, 1. — Was punktiert, liegt weiter ab von der Platte. Auf dem Bild sieht man den Fuß von außen, wie Tab. 187a.

Skizze zu Tab. 187

1 Fibula, 2 Tibia, 3 Spitze des inneren Knöchels, 4 Spitze des äußeren Knöchels, 5 Processus posterior tali, 6 Corpus (trochlea), 7 Collum tali, 8 Caput tali, 9 Articulus talocalcanearis posterior (ausgezogene Parallelkontur), 10 Articulus talocalcanearis anterior (punktierte Parallelkontur), 11 Sustentaculum tali, 12 Processus anterior calcanei, 13 Os naviculare (der nach hinten medial umgebogene Teil), 14 Os cuboides, 15 Tuberculum tibiale tuberis calcanei, 15* Tuberc. fibulare tuberis calcanei, 17 Corpus calcanei, 21 Tuberositas ossis navicularis, 22 Tuberculum oss. metatarsi V, zackig abgesetzt, infolge Assimilation des Os Vesalianum — inkonstant.

Tab. 187a

Fuß von außen gesehen (aus Sobotta).

1 Os cun. III, 2 Os cun. II, 3 Nav. 4 Sinus tarsi, 5 Talus, 6 Kalk., 7 Proc. trochlearis calc., 8 Os cub.

288

FIG. 5: Two pages from Rudolf Grashey's *Atlas typischer Röntgenbilder vom normalen Menschen* (Munich, 1912), plates 187, 187a, and accompanying text.

of an indefinite series of images: "The more normal images of a region we have, the broader the basis on which we build our assessment."[23] The images needed to be produced in accordance with uniform standards of exposure, Grashey argued, so that "pictures of the same region in different individuals become as similar to one another as possible and large series are built that facilitate comparative study."[24] Such comparison was in fact the prerequisite for the emergence of a concept of the "normal" and the construction of a medical iconography of the radiographic image. Still, the radiograph can ever only depict a particular individual case, so the normal remains a category that must comprehend a series of deviations that is hard to delimit; Grashey calls them "varieties" and exhorts his readers to "hunt" for them whenever the opportunity arises.[25] Grashey's "normal atlas" cannot offer more than orientation amid a plethora of images, each of which shows a singular case.

The atlas thus provides a basis also for the interpretation of pathological images, which are to be taken in accordance with the same standards.[26] Yet the status of the images becomes especially precarious when they record pathological phenomena, as Grashey's *Atlas chirurgisch-pathologischer Röntgenbilder* (Atlas of radiographic images from surgical pathology),[27] published three years later, illustrates (fig. 6). The problem of the singular case is even more prominent here, since the variability of the ailments leads to a virtually infinite diversity of possible outcomes. In order to present the greatest number of cases, Grashey selected details that brought out the relevant "pathology" of each individual, emphasizing the exemplary nature of the images, which show obscure and abstract shapes that are inscrutable to the lay reader. Set close together, these pictures—now without accompanying anatomical

23 Grashey, *Atlas typischer Röntgenbilder vom normalen Menschen*, iii.

24 Ibid.

25 Ibid.

26 Grashey, "Fehlerquellen und diagnostische Schwierigkeiten," 810: "Images of pathological cases must be taken precisely in accordance with the model presented by the normal types."

27 Rudolf Grashey, *Atlas chirurgisch-pathologischer Röntgenbilder* (Munich: Lehmann, 1908).

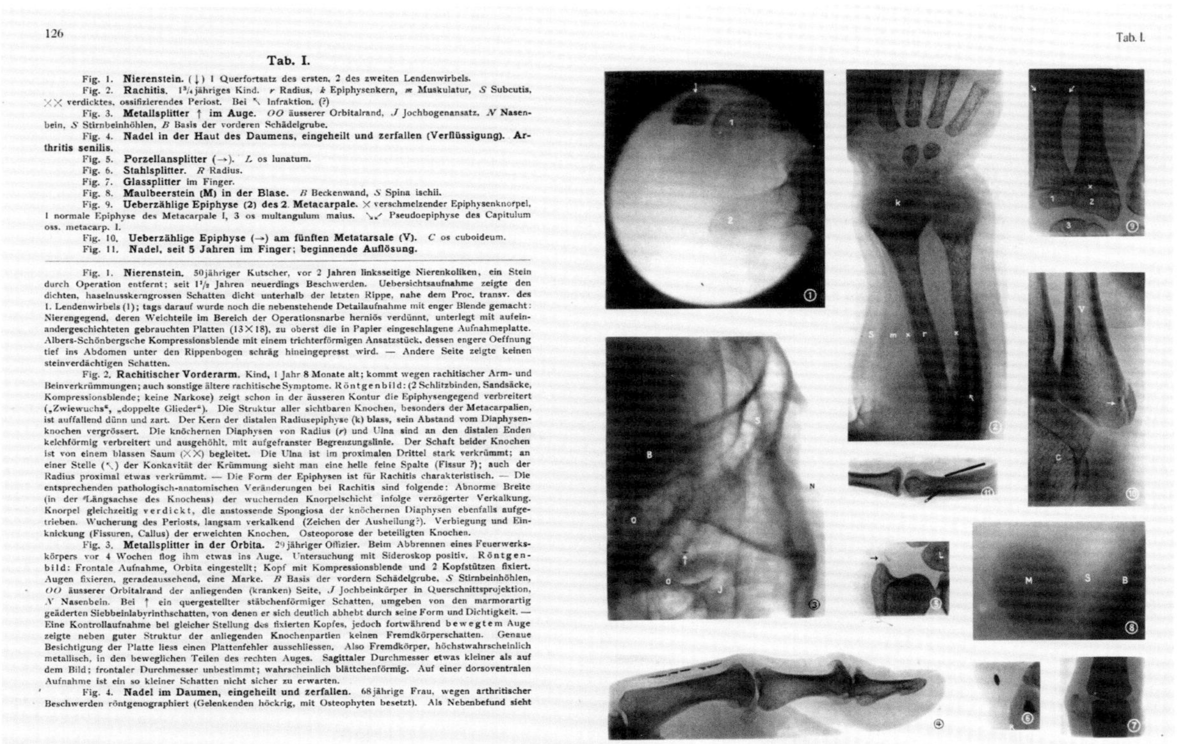

126

Tab. I.

Fig. 1. **Nierenstein.** (↓) 1 Querfortsatz des ersten, 2 des zweiten Lendenwirbels.

Fig. 2. **Rachitis.** 1¾jähriges Kind. *r* Radius, *k* Epiphysenkern, *m* Muskulatur, *S* Subcutis, ✕✕ verdicktes, ossifizierendes Periost. Bei ↖ Infraktion. (?)

Fig. 3. **Metallsplitter ↑ im Auge.** *OO* äusserer Orbitalrand, *J* Jochbogenansatz, *N* Nasenbein, *S* Stirnbeinhöhlen, *B* Basis der vorderen Schädelgrube.

Fig. 4. **Nadel in der Haut des Daumens, eingeheilt und zerfallen (Verflüssigung). Arthritis senilis.**

Fig. 5. **Porzellansplitter** (→). *L* os lunatum.

Fig. 6. **Stahlsplitter.** *R* Radius.

Fig. 7. **Glassplitter** im Finger.

Fig. 8. **Maulbeerstein (M) in der Blase.** *B* Beckenwand, *S* Spina ischii.

Fig. 9. **Ueberzählige Epiphyse (2) des 2. Metacarpale.** ✕ verschmelzender Epiphysenknorpel, 1 normale Epiphyse des Metacarpale I, 3 os multangulum maius. ↘↙ Pseudoepiphyse des Capitulum oss. metacarp. I.

Fig. 10. **Ueberzählige Epiphyse (→) am fünften Metatarsale (V).** *C* os cuboideum.

Fig. 11. **Nadel, seit 5 Jahren im Finger; beginnende Auflösung.**

Fig. 1. **Nierenstein.** 50jähriger Kutscher, vor 2 Jahren linksseitige Nierenkoliken, ein Stein durch Operation entfernt; seit 1½ Jahren neuerdings Beschwerden. Uebersichtsaufnahme zeigte den dichten, haselnusskerngrossen Schatten dicht unterhalb der letzten Rippe, nahe dem Proc. transv. des 1. Lendenwirbels (1); tags darauf wurde noch die nebenstehende Detailaufnahme mit enger Blende gemacht: Nierengegend, deren Weichteile im Bereich der Operationsnarbe herniös verdünnt, unterlegt mit aufeinandergeschichteten gebrauchten Platten (13×18), zu oberst die in Papier eingeschlagene Aufnahmeplatte. Albers-Schönbergsche Kompressionsblende mit einem trichterförmigen Ansatzstück, dessen engere Oeffnung tief ins Abdomen unter den Rippenbogen schräg hineingepresst wird. — Andere Seite zeigte keinen steinverdächtigen Schatten.

Fig. 2. **Rachitischer Vorderarm.** Kind, 1 Jahr 8 Monate alt; kommt wegen rachitischer Arm- und Beinverkrümmungen; auch sonstige ältere rachitische Symptome. Röntgenbild: (2 Schlitzbinden, Sandsäcke, Kompressionsblende; keine Narkose) zeigt schon in der äusseren Kontur die Epiphysengegend verbreitert („Zwiewuchs", „doppelte Glieder"). Die Struktur aller sichtbaren Knochen, besonders der Metacarpalien, ist auffallend dünn und zart. Der Kern der distalen Radiusepiphyse (k) blass, sein Abstand vom Diaphysenknochen vergrössert. Die knöchernen Diaphysen von Radius (*r*) und Ulna sind an den distalen Enden kelchförmig verbreitert und ausgehöhlt, mit aufgefranster Begrenzungslinie. Der Schaft beider Knochen ist von einem blassen Saum (✕✕) begleitet. Die Ulna ist im proximalen Drittel stark verkrümmt; an einer Stelle (↖) der Konkavität der Krümmung sieht man eine helle feine Spalte (Fissur ?); auch der Radius proximal etwas verkrümmt. — Die Form der Epiphysen ist für Rachitis charakteristisch. — Die entsprechenden pathologisch-anatomischen Veränderungen bei Rachitis sind folgende: Abnorme Breite (in der Längsachse des Knochens) der wuchernden Knorpelschicht infolge verzögerter Verkalkung. Knorpel gleichzeitig verdickt, die anstossende Spongiosa der knöchernen Diaphysen ebenfalls aufgetrieben. Wucherung des Periosts, langsam verkalkend (Zeichen der Ausheilung?). Verbiegung und Einknickung (Fissuren, Callus) der erweichten Knochen. Osteoporose der beteiligten Knochen.

Fig. 3. **Metallsplitter in der Orbita.** 29jähriger Offizier. Beim Abbrennen eines Feuerwerkskörpers vor 4 Wochen flog ihm etwas ins Auge. Untersuchung mit Sideroskop positiv. Röntgenbild: Frontale Aufnahme, Orbita eingestellt; Kopf mit Kompressionsblende und 2 Kopfstützen fixiert. Augen fixieren, geradeaussehend, eine Marke. *B* Basis der vordern Schädelgrube, *S* Stirnbeinhöhlen, *OO* äusserer Orbitalrand der anliegenden (kranken) Seite, *J* Jochbeinkörper in Querschnittsprojektion, *N* Nasenbein. Bei ↑ ein quergestellter stäbchenförmiger Schatten, umgeben von den marmorartig geäderten Siebbeinlabyrinthschatten, von denen er sich deutlich abhebt durch seine Form und Dichtigkeit. — Eine Kontrollaufnahme bei gleicher Stellung des fixierten Kopfes, jedoch fortwährend bewegtem Auge zeigte neben guter Struktur der anliegenden Knochenpartien keinen Fremdkörperschatten. Genaue Besichtigung der Platte liess einen Plattenfehler ausschliessen. Also Fremdkörper, höchstwahrscheinlich metallisch, in den beweglichen Teilen des rechten Auges. Sagittaler Durchmesser etwas kleiner als auf dem Bild; frontaler Durchmesser unbestimmt; wahrscheinlich blättchenförmig. Auf einer dorsoventralen Aufnahme ist ein so kleiner Schatten nicht sicher zu erwarten.

Fig. 4. **Nadel im Daumen, eingeheilt und zerfallen.** 68jährige Frau, wegen arthritischer Beschwerden röntgenographiert (Gelenkenden höckrig, mit Osteophyten besetzt). Als Nebenbefund sieht

FIG. 6: Two pages from Rudolf Grashey's *Atlas chirurgisch-pathologischer Röntgenbilder* (Munich, 1908), plate i and accompanying text.

illustrations—fill the plates, dotted by numbers, letters, and small arrows that refer to the explanatory text. In order to demonstrate the limitations of radiography as a technique of visualization, Grashey deliberately included images of questionable diagnostic value in his atlas: "I initially intended to give only cases in which the diagnoses had been anatomically ascertained; on the other hand, it seemed important to me to include questionable cases in order to show to the novice that a question mark must be put on many diagnoses based on radiographs."[28]

Picture 11 in figure 6 (center) shows pieces of a needle embedded in a finger close to the bone. The report notes that the foreign body has "begun to disintegrate": "Semicircular indentations may be seen in several places along the needle, to each of which a drop-shaped and accordingly prominent lighter shadow attaches (partial liquefaction)."[29] The description attests to a keen observing eye, but the confidence with which it is offered suggests that experience with other cases played a role as well. In an article, Grashey had referred to an earlier medical report published in 1897 about a case of putative pieces of a needle in a girl's forearm;[30] even though the patient did not complain of any symptoms, the mother requested that X-ray images of the arm be taken and then demanded that the pieces be surgically removed, "since it was now known where they were and they were so clearly visible."[31] The surgeon had been unable to find the foreign bodies, leading the physicians to undertake a systematic study of the process of dissolution metal objects undergo when embedded in tissue. It was found that iron rust "produced the same shadow a solid needle did."[32]

28 Ibid., iv.

29 Ibid., 127.

30 Grashey, "Fehlerquellen und diagnostische Schwierigkeiten," 809.

31 Quoted in Dommann, *Durchsicht, Einsicht, Vorsicht*, 8.

32 Grashey, "Fehlerquellen und diagnostische Schwierigkeiten," 809.

This case provides clear evidence of the difference that may intervene between the image and the object it depicts; Grashey, obviously cognizant of it, accorded particular importance to the diagnostically questionable cases. These images visualize something, but what exactly it is they show is not always unequivocally clear. They engender a reality of their own whose reference is sometimes in question, and the beholder must be aware of that possibility if the radiograph is to be of diagnostic value. That is why every image ought to be placed in an anatomical and clinical context and read only in the light of familiarity with the peculiar traits of the individual patient's illness. What is most important, however, is that every diagnosis must be supported by a visual memory trained over a long and time-consuming course of comparative seeing—and even then false conclusions are possible: "I hope that the atlas will contribute to raising the reputation of X-rays as a diagnostic tool, but also demarcate the boundaries within which this peculiar method operates."[33] The greatest accomplishment the images in Grashey's atlas are capable of is accordingly "to serve as a signpost";[34] as the author emphasizes in the preface, the book aims to "help the student improve his discernment" and "exercise and train the eye in the peculiar task of perceiving subtle nuances of shading."[35] Rudolf Grashey thus advocates a conception of objectivity that assigns the scientist assessing a case a crucial role in interpreting the visual evidence and intervening as necessary.[36]

His use of the radiographic technique emerges as an element in a positive history of the shadow in which critics emphasize as well as qualify its diagnostic value.[37] By explicitly addressing the limitations and defects of radiography, Grashey points up what is even now the fundamental problem of these images: explicit design choices and interpretive decisions must be made to unlock their scientific and diagnostic potential. At the same time, Grashey paves the way for an application of the critique of imagery to the products of radiography; we must seek to understand the specific qualities of these images—their strengths *and* their weaknesses—by analyzing their historical contexts, genesis, and effects.

33 Grashey, *Atlas chirurgisch-pathologischer Röntgenbilder*, iv.

34 See Lorraine Daston and Peter Galison, "The Image of Objectivity," in "Seeing Science," special issue, *Representations*, no. 40 (Autumn 1992): 81–128, 107.

35 Grashey, *Atlas chirurgisch-pathologischer Röntgenbilder*, iv.

36 See Peter Galison, "Judgment against Objectivity," in *Picturing Science, Producing Art*, ed. Caroline A. Jones and Peter Galison (New York: Routledge, 1998), 327–59.

37 For the positive history of the shadow, see Stoichita, *Short History of the Shadow*, 153–85, where Stoichita discusses the silhouette as an instrument of physiognomic insight in Lavater, as well as the shadow as the soul and essential part of man in the example of Adalbert von Chamisso's "Peter Schlemihls wundersame Geschichte" (1839).

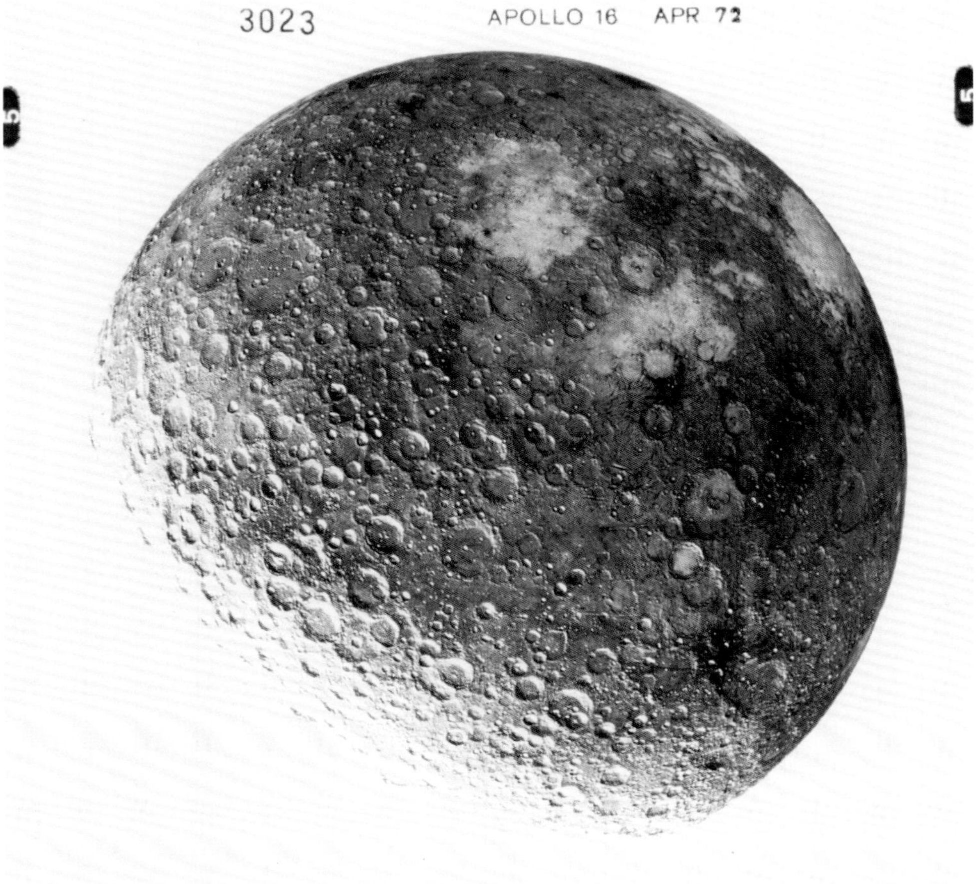

FIG. 1: Photograph of the moon taken with a survey camera by Kenneth Mattingly from Apollo 16, April 1972. It was taken during a return flight from the moon from a distance of around 1,600 km. The boundary between the moon's near side and the far side, which cannot be seen from the Earth, runs through the center of the picture. NASA.

VISUALITY, VISUALIZING, IMAGING

In science and technology, numerous forms of graphical information are often referred to quite generally as "representations" of something, in contrast or addition to alternative terms such as *depiction* or *reproduction* that suggest a direct resemblance between (or a mimetic approximation of) an image and the object it seems to substitute for. On the other hand, the ambiguous term *representation* carries diverse (artistic, political, theatrical) connotations and so is not altogether suitable to the specific problems of research in the natural sciences (Elkins 2008). Since the 1980s, it has increasingly been questioned and replaced by alternative expressions such as *visualization* or *rendering visible*. In contradistinction to the notion of a passive representation of things and the resemblance it entails, these terms are meant to describe the productive and constructive aspect of imaging in scientific practice (Rheinberger, Hagner, and Wahrig-Schmidt 1997; Huber and Heller 1999; Haupt and Stadler 2006; Elkins 2008), where images appearing on the computer screen are usually translations of measurements into models that give visual access to the object of research. The creation of such images depends on basic research and knowledge in fields such as physics or biochemistry and demands considerable technical and financial efforts. As an essential and sometimes exclusive means of observation and operation, such images also stand for a fundamental epistemological shift and a new paradigm of *visuality* (a term coined by visual culture studies) in technology-based societies (Elkins 1999).

In the literal sense, *visualization* could refer to any kind of man-made visual pattern, ranging from diagrams to radiographs, that renders an object or a fact perceptible to the human eye. As already stated by pioneers of modern and abstract art, the aesthetic manifestation

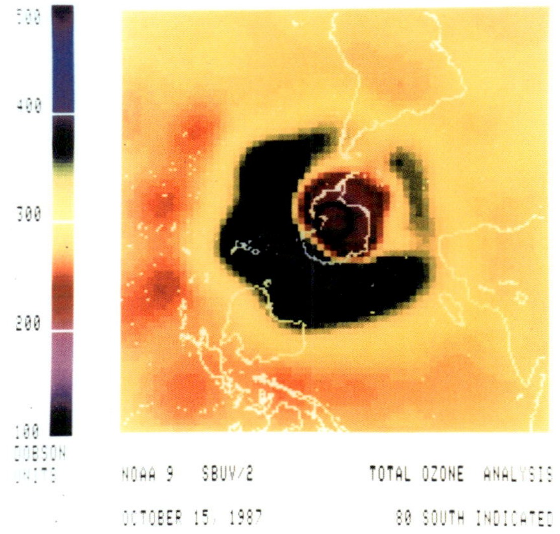

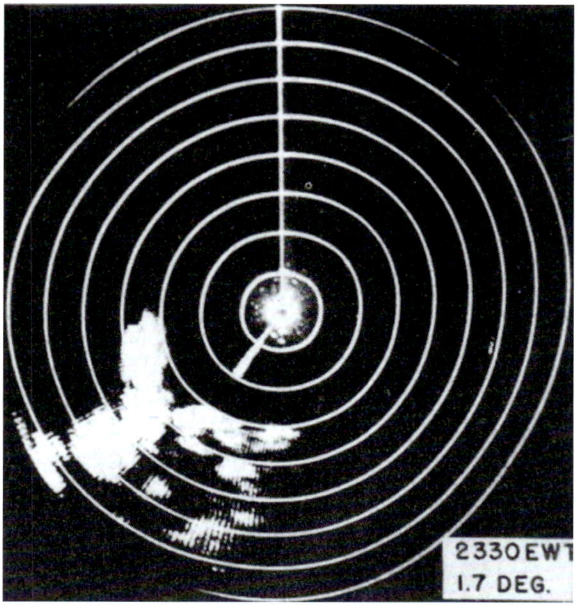

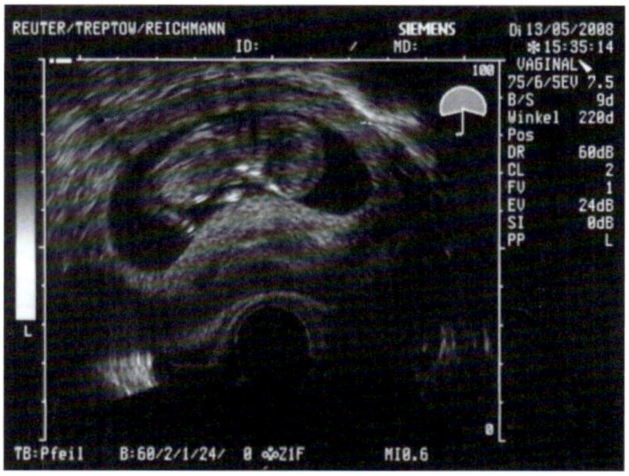

FIG. 2 (above): An analysis of the ozone distribution over Antarctica recorded using the Solar Backscatter Ultraviolet (SBUV/2) instrument of the U.S. National Oceanic and Atmospheric Administration and showing the hole in the ozone layer on October 15, 1987.

FIG. 3 (top right): Radar image of a heavy storm southwest of Spring Lake, New Jersey, July 27, 1944. From the "AAF Manual 105-101-2 Radar Storm Detection" issued by U.S. Army Air Forces Headquarters, August 1945.

FIG. 4 (bottom right): Ultrasound picture of a fetus during the first half of pregnancy, 2008. Sonography enables gynecologists to make a non-invasive diagnosis and examine the development of the child's body and organs while it is still in the womb. Private archive of Vera Dünkel.

of emotions, ideas, or imaginings might likewise be regarded as a kind of visualization of the unseen, comparable to scientific practices. Enhancing the eye by using optical instruments such as a microscope or a telescope reveals new visible worlds (fig. 1); photography and film capture time and motion in a variety of forms and thereby define them (cf. Snyder 1998). Such a wide interpretation would be supported by the fact that the term *visualization* is also used for design solutions and screen-based media of all kinds for the virtual modeling and simulation of constructions and surfaces in architecture, in industrial design, or in the games industry, where three-dimensional figures are usually created out of large quantities of numerical data (Elkins 2007; Lima 2011; Wilson 2002).

In contrast with a broad interpretation of the term *visualization*, *imaging* is often and explicitly used in medical and scientific contexts to describe particular procedures and equipment and the problems involved with their application in these areas. Here, images may be the result of mechanical phenomena recorded graphically (e.g., blood pressure, seismic events) or of extended electromagnetic wave spectra made visible (fig. 2); they may also be based on the active emission of pulses (figs. 3 and 4). Imaging includes the electronic registration of objects in scanning tunneling microscopy, which uses reference values to create an interaction with the object, the summation of radiological data in medicine, and comparable tomographic means of observation, e.g., in brain research (fig. 5). Subsequent to the success of electronic media and the possibility of displaying data measurements on a computer screen, terms like *imaging* have become primarily associated with digital media. However, *imaging* can as well remain entirely analog; see, for instance, the tracing of sound curves by oscilloscope (fig. 6) or the

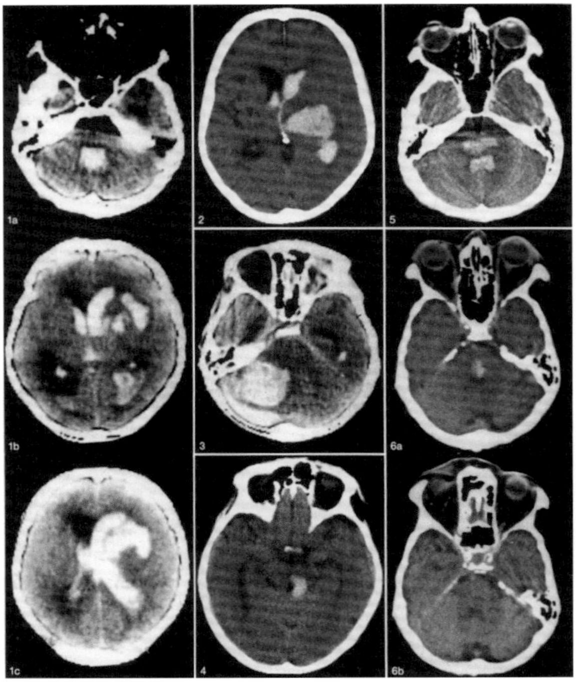

FIG. 5: Plate from a computer tomography textbook: examples of CT images showing axial sections of six patients with intracerebral hemorrhage. The hematomas are recognizable as lighter areas in the tomogram. Sebastian Lange et al., *Cerebral and Spinal Computerized Tomography*, 2nd rev. ed. (New York: Karger, 1989), 95, plate 38. S. Karger AG, Basel.

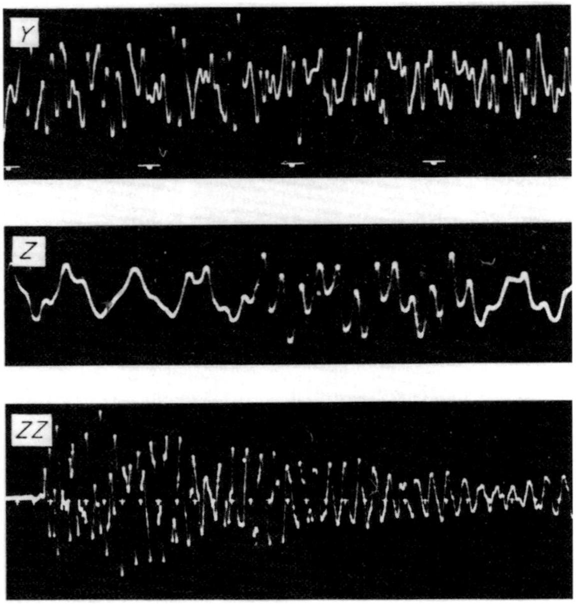

FIG. 6: Acoustic wave recordings made using the early oscillograph of Dayton Clarence Miller, called the Phonodeik, around 1910. From top to bottom: brass band music, a bell ringing, and the sound of a rocket. V. J. Phillips, *Waveforms: A History of Early Oscillography* (Bristol: A. Hilger, 1987), 58, fig. 2.53.

"sound figures" of Ernst Chladni, who visualized acoustic oscillations as patterns in sand in the late eighteenth century (fig. 7).

When the alleged object under investigation is brought into the register of the visible, screens, prints, or algorithms themselves become a field of operations. Classifications, experiments, and changes to parameters turn the visible image into a product of interventions that refers to the object only indirectly, through the "inscription" of measurements (Latour and Woolgar 1986). This has raised fundamental questions concerning the interpretation of results based on visual tools (Hacking 1983). The philosopher of science Hans-Jörg Rheinberger has regarded such images, in a systematic way, as the product of a constructive process of "performing" knowledge (Rheinberger, Hagner, and Wahrig-Schmidt 1997, 2001). Owing to their artificial character, images created in this manner are subject to the respective rules of representation, e.g., in terms of coloring, spatiality, perspective, scale, and speed; deliberately or unintentionally, their producers abide by these rules so as to provide better means of analysis and evaluation (cf. Lynch and Edgerton 1988).

Visualization and *imaging* are thus in many ways synonymous; the latter has become the established term of art, especially in the medical equipment industry, where images can become the basis of vital decisions, diagnosis, and treatment. Nevertheless, since visualized data is increasingly present in the "operation theater," there are still intersections with a performative notion of "representation." —MB

LITERATURE

Elkins, James. *The Domain of Images*. Ithaca, NY: Cornell University Press, 1999.

Elkins, James. *Six Stories from the End of Representation: Images in Painting, Photography, Astronomy, Microscopy, Particle Physics, and Quantum Mechanics, 1980–2000*. Stanford, CA: Stanford University Press, 2008.

Elkins, James. *Visual Practices across the University*. Munich: Wilhelm Fink Verlag, 2007.

Foster, Hal, ed. Vision and Visuality. Seattle: Bay Press, 1988.

Hacking, Ian. *Representing and Intervening*. Cambridge: Cambridge University Press, 1983.

Haupt, Sabine, and Ulrich Stadler, eds. *Das Unsichtbare sehen: Bildzauber, optische Medien und Literatur*. Zurich: Edition Voldemeer, 2006.

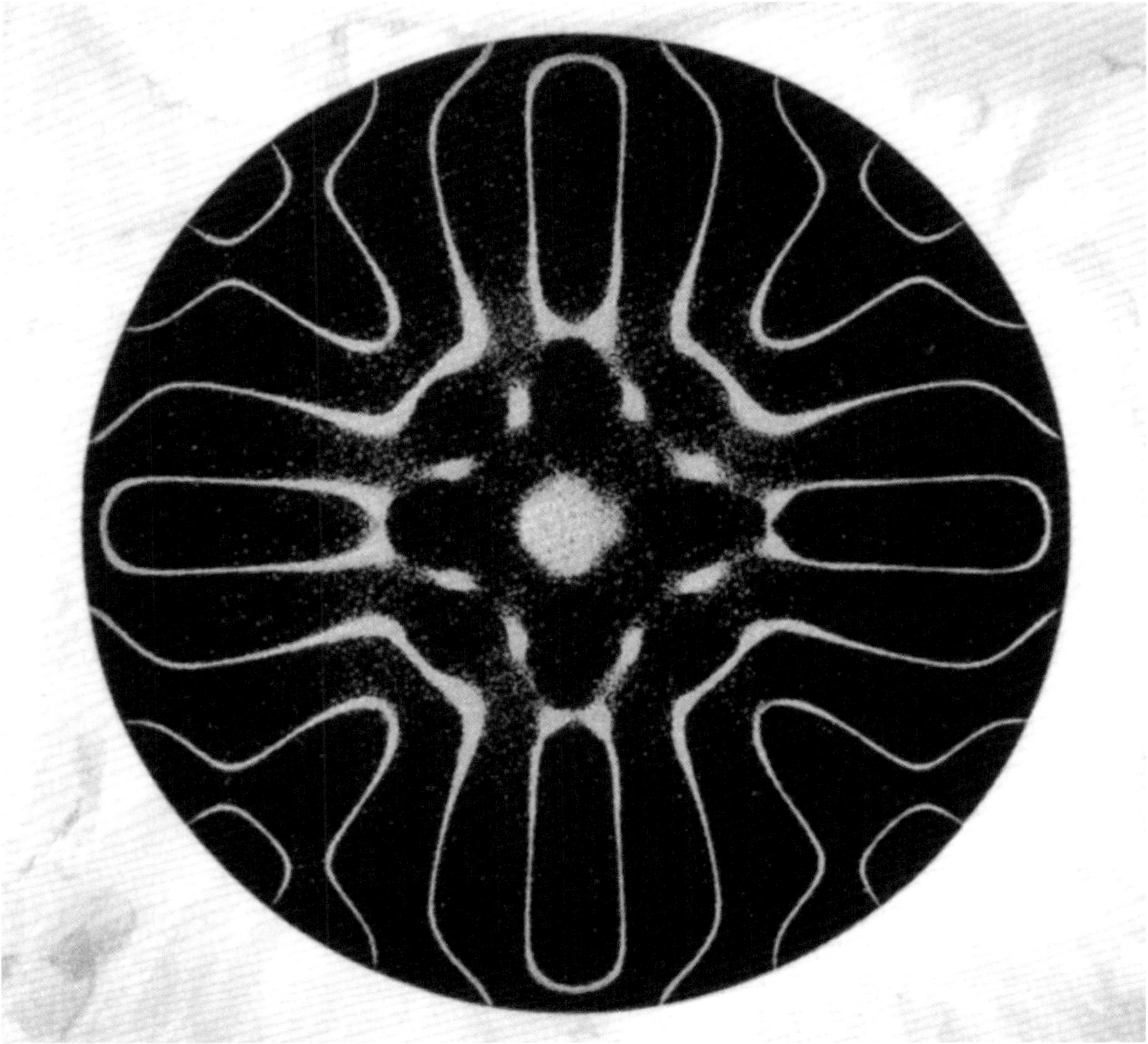

FIG. 7: Sound figure created using Ernst Chladni's method of 1787. Oscillations produced on a sand-covered metal plate using violin bows or loudspeakers create frequency-dependent patterns on the plate. Alexander Lauterwasser, *Wasser Klang Bilder. Die "schöpferische Musik des Weltalls,"* 2nd ed. (Aarau: AT Verlag, 2003), 43.

Huber, Jörg, and Martin Heller, eds. *Konstruktionen Sichtbarkeiten*. Vienna: Springer, 1999.

Latour, Bruno, and Steve Woolgar. *Laboratory Life: The Construction of Scientific Facts*. Princeton, NJ: Princeton University Press, 1986.

Lima, Manuel. *Visual Complexity: Mapping Patterns of Information*. New York: Princeton Architectural Press, 2011.

Lynch, Michael, and Samuel Y. Edgerton. "Aesthetics and Digital Image Processing: Representational Craft in Contemporary Astronomy." In *Picturing Power: Visual Depiction and Social Relations*, ed. Gordon Fyfe and John Law, 184–220. London: Routledge, 1988.

Lynch, Michael, and Steve Woolgar, eds. *Representation in Scientific Practice*. Cambridge, MA: MIT Press, 1990.

Rheinberger, Hans-Jörg. "Objekt und Repräsentation." In *Mit dem Auge denken: Strategien der Sichtbarmachung in wissenschaftlichen und virtuellen Welten*, ed. Bettina Heintz and Jörg Huber, 55–61. Zurich: Edition Voldemeer, 2001.

Rheinberger, Hans-Jörg. *Towards a History of Epistemic Things: Synthesizing Proteins in the Test Tube*. Palo Alto, CA: Stanford University Press, 1977.

Rheinberger, Hans-Jörg, Michael Hagner, and Bettina Wahrig-Schmidt, eds. *Räume des Wissens: Repräsentation, Codierung, Spur*. Berlin: Akademie Verlag, 1997.

Snyder, Joel. "Visualization and Visibility." In *Picturing Science, Producing Art*, ed. Peter Galison and Caroline A. Jones, 379–97. London: Routledge, 1998.

Wilson, Stephen. *Information Arts: Intersections of Art, Science, and Technology*. Cambridge, MA: MIT Press, 2002.

Wise, Norton. "Making Visible." *Isis* 97 (2006): 75–82.

Instrument-Aided Vision and the Imagination: The Migration of Worms and Dragons in Early Microscopy*

Stefan Ditzen

* This essay is an abridged version of my "Der Satyr auf dem Larvenrücken: Zum Verhältnis von instrumentellem Sehen und Bildtraditionen," in *Konstruierte Sichtbarkeiten: Wissenschafts- und Technikbilder seit der frühen Neuzeit*, ed. Martina Heßler (Munich: Fink, 2006), 41–56. See also Stefan Ditzen, *Kunstformen instrumenteller Sichtbarkeit: Etappen einer Bildgeschichte des Mikroskops* (Aachen: Shaker, 2008).

At the dawn of microscopy, the view through the instrument was clouded by doubts concerning the quality of the optical equipment. This skepticism concerned the capacity of the lenses as well as the constructions combining several glasses: the instruments were not subject to any standardization. As a consequence, each microscope was unique, offering distinctive glimpses of the microcosm, which made it difficult to compare observations and build consensus over findings. The identification of minute structures, that is to say, was embedded in a network of dependencies. On the one hand, instrumental conditions were crucial to the results a microscopist was able to present. On the other hand, he would make sense of what he saw by looking for phenomena he was familiar with from reading compendia of microscopy, which helped him counter his doubts concerning his own instrument-aided eye. Pictorial traditions emerged that, as it were, prefigured the microscopic vision.

Instrument and Visibility

To begin with, the convex lenses manufactured for magnifying and reading glasses were insufficient for the observation of microscopic structures. Instead of these conventional products, small lenses in particular needed to be manufactured in a way that allowed for greater focal lengths. For a long time, however, the manufacture of such lenses presented considerable difficulties.

Among the best microscopes of the seventeenth century were those made by Antoni van Leeuwenhoek (1632–1723). Their construction was perfectly simple; in contemporary parlance, given the way they were used, they would be called magnifying glasses rather than microscopes, properly speaking. Leeuwenhoek's biconvex lenses were no larger than a twentieth of an inch, allowing him to achieve magnifications by a factor of at least 270.[1] The downside of these instruments was that they absorbed a great deal of light, making it hard to arrive at reliable observations. Owing to their small size, they were also exceptionally difficult to handle. Moreover, not all microscopists enjoyed the opportunity to use such high-quality lenses.[2] For a long time, the designs and quality of the instruments varied widely.

Besides the sort of microscopes Leeuwenhoek built, the era's scholars also used so-called compound microscopes. Arranging several lenses in series made it possible to multiply their magnifying effects. These instruments, too, had a drawback: they produced optical effects. For example, when the points in which the rays intersected did not coincide, the images

1 Estimates of the resolution based on the results of his observations range up to a magnification of 770. See Edward G. Ruestow, *The Microscope in the Dutch Republic: The Shaping of Discovery* (Cambridge: Cambridge University Press, 1996), 2.

2 Leeuwenhoek himself did not sell any of his instruments, nor did he divulge the method he used to manufacture the lenses. See ibid., 152, 290–91.

would be out of focus; in other instances, fringes of color would appear in the image, sometimes strong enough to produce artifacts that defied the attempt to decide whether they originated in a phenomenon in the object being examined or the instrumental conditions of examination.[3]

Both types of microscopes, moreover, were incapable of resolving a fundamental problem that beset the transfer of observations to paper: the involvement of *memoria*—the intermediate storage of what the scholar saw in his memory—with the process of recording them was regarded as another source of error. In practice, the challenge the microscopist faced was to use one eye to look through the lens while keeping the other eye on the drawing. In addition, the instrument could always only be set to focus on one detail; in making a drawing, the scholar "scanned" the specimen piece by piece. Historical atlases of microscopy allow us to trace how microscopic visualization operated between the poles of imagination, contemporary theory, pictorial precedent, and macroscopic views.

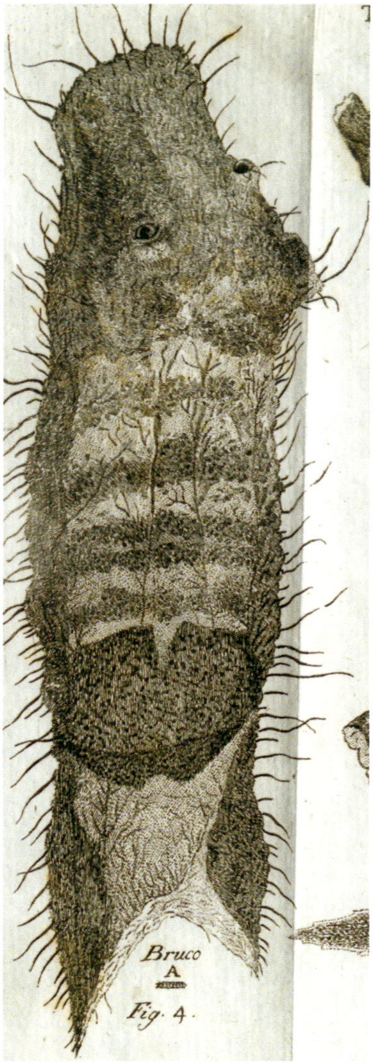

FIG. 1: Giovanni Maria della Torre, enlarged depiction of a caterpillar, from *Nuove osservazioni microscopiche* (Naples, 1776). © Staatsbibliothek zu Berlin—PK.

Fabulous Creatures, Dragons, and Microscopic Worms

One example of the interaction between observation and a preexisting figment of the imagination appears in the *Nuove osservazioni microscopiche* that Giovanni Maria della Torre (1710–1782) published in 1776. The third plate of these "New Microscopic Observations" features a caterpillar, *un picciolo Bruco*, depicted both in actual size and magnified by a factor of fifty (fig. 1).[4] The image illustrates della Torre's view of the specimen as a fabulous creature that combines an elongated hairy body with a dog's or bear's head and a fishlike tail end. The figure suggests that uncertainty concerning the scholar's own observations and its liability to imprecision left room for the transfer of macroscopic visual habits. To the extent that he was unable to unequivocally determine the nature of the head, he resorted to familiar physiognomies.

A second example illustrates how such transfers did not necessarily originate in the realm of real visible objects; scholars might also interweave figments of an imagination untethered to reality with what they saw through the microscope. In the *System of Medicine* published anonymously in 1726,[5] ideas concerning the causes of the plague that raged in Toulon and Marseille in 1721 gave rise to peculiar visions. The book describes ninety-one illnesses said to be brought on by tiny insects that were given names such as "faint-maker, body-pincher, boil-causer, tear-duct-fistulator, lust-arouser, the-runs-causer."[6] The liberties the illustrators took in presenting the appearance of the alleged pathogens are apparent in a little animal said to be found in patients with rheumatic pain. The magnified view shows an animal with a wide-open maw and large eyes (fig. 2). Mounted on its back is a long serrated fin that may be read as a sort of sting or possibly a wing. The prone outstretched body with bent legs suggests a crocodile, but the tail, unlike a crocodile's, is forked. The microscopic animal illustrates what the draftsman imagined a dragon looked like, rather than anything derived from the observation of reality.[7]

FIG. 2: Pathogen causing rheumatic pain, from Anonymous (Nicolas Andry?), *Systême d'un Medecin Anglois sur la cause de Toutes les especes de maladies [...]* (Paris: Mesnier, 1726). © Staatsbibliothek zu Berlin—PK.

Meanwhile, the graphic representations of the pathogens themselves test the limits of visibility. Some of them measure no more than a few millimeters, essentially amounting to spots of printer's ink on the paper. Others evoke associations of animals such as fishes or larger insects. As the book's full title indicated, "le moyen d'un bon Microscope" was supposedly what allowed the scholar to see these animals, which conversely meant that if a reader failed to see them, his own instrument was not up to standard. But precisely because the microscopic images were virtually impossible to falsify at this level of detail, the images represented visions rather than visual evidence—embodiments of fantasies about what causes illnesses that attest not to the acuity of the author's eye but rather to an imagination fueled by the severity of various illnesses.

Yet we should not just dismiss such images as, in today's perspective, "purely non-scientific." Instead, they must be regarded as part of a certain "thought style," to use a concept the medical scholar Ludwik Fleck (1896–1961) introduced.[8] Fleck's theory of thought styles and thought collectives seeks to identify the core of knowledge and scientific insight in a closely knit network of social groups, possibilities and conditions of the transfer of knowledge, and scientific procedures. Considered in this perspective, even monstrous visions and images play a constructive part in the multifaceted process of scientific advancement. In the present instance, for example, the images may have helped focus subsequent research on living infectious agents, leading scientists to abandon earlier models such as that of a disruption of the equilibrium of humors for good.

Not only did the problematic conditions under which microscopic observation took place invite the interference of the *imaginatio*, engendering images today's beholder finds difficult to comprehend. The capacity of the microscopist's equipment has changed so fundamentally that we also cannot study the products of early microscopy without considering the original instruments as well. For a particularly illustrative example of this nexus, we may look to the history of the detection of wormlike microscopic animals, which Bruno Zanobio and Luigi Belloni worked to retrace in the 1960s.[9]

In 1658, Athanasius Kircher had pointed out the existence of such *animalcula* that were invisible to the "unarmed" eye, identifying them as the pathogens causing the plague, among other illnesses.[10] Over the decades that followed, microscopy positively teemed with wormlike creatures. Theodor Kerckring (1639–1693) discovered them in intestines, livers, and the parenchyma of all other viscera;[11] Thomas Willis (1621–1675) used his microscope to observe them in the visceral pleura.[12]

Spermatozoa, which were discovered around the time these studies were undertaken, were likewise regarded as wormlike animalcules. In 1677, Leeuwenhoek confirmed the discovery of the spermatozoon made by Johan Ham (1651–1723), a student at the medical school in Leiden, who brought him a semen sample from a man suffering from gonorrhea that contained tailed animalcules.[13] Ham believed that the little animals were products of putrefaction, but Leeuwenhoek regarded them as a normal component of semen.[14] The Dutch instrument maker Nicolas Hartsoeker (1656–1725) also claimed prior discovery, arguing that his amazement had discouraged him from publishing what he had seen.[15] The studies of the spermatozoon by Ham, Hartsoeker, and Leeuwenhoek showed tiny organisms that were henceforth understood to be "seed-animals." Man, it seemed, resembled a worm at the beginning of his ontogeny; the logical next step was the belief that he was pervaded by worms from cradle to grave. The ultimate conclusion that might be drawn

3 Saville Bradbury, "The Quality of the Image Produced by the Compound Microscope: 1700–1840," *Proceedings of the Royal Microscopical Society* 2 (1967): 151–73.

4 Giovanni Maria della Torre, *Nuove osservazioni microscopiche* (Naples, 1776), 51.

5 The full title reads *Systême d'un Medecin Anglois sur la cause de Toutes les especes de maladies, avec les surprenantes configurations des differentes especes de petits Insectes, qu'on voit par le moyen d'un bon Microscope dans le Sang & dans les Urines des differens Malades, & même de tous ceux que doivent le devenir* (Paris: Mesnier, 1726). The book was probably written by the French professor Nicolas Andry (1658–1742).

6 These terms appear in the brief discussion of the *Systême* in Christian Gottfried Ehrenberg, *Die Infusionsthierchen als vollkommene Organismen: Ein Blick in das tiefere organische Leben der Natur. Nebst einem Atlas von vierundsechszig colorierten Kupfertafeln, gezeichnet vom Verfasser* (Leipzig: Voss, 1838).

7 *Systême d'un Medecin Anglois*, 10.

8 Ludwik Fleck, *Genesis and Development of a Scientific Fact*, trans. Fred Bradley, ed. Thaddeus J. Trenn and Robert K. Merton (Chicago: University of Chicago Press, 1979).

9 Bruno Zanobio, "L'immagine filamentoso-reticolare nell'anatomia microscopica dal XVII al XIX secolo," *Physis: rivista internazionale di storia della scienza* 2 (1960): 299–317; Luigi Belloni, "I capillari sanguini nelle tavole del Malpighi," *Physis: rivista internazionale di storia della scienza* 5 (1963): 70–77; Luigi Belloni, "The Repetition of Experiments and Observations: Its Value in Studying the History of Medicine (and Science)," *Journal of the History of Medicine and Allied Sciences* 25, no. 2 (1970): 158–67; Luigi Belloni, "Athanasius Kircher: Seine Mikroskopie, die Animalcula und die Pestwürmer," *Medizinhistorisches Journal* 20 (1985): 58–65.

10 Athanasius Kircher, *Scrutinium physico-medicum Contagiosae Luis, quae Pestis dicitur* (Rome: Mascardi, 1658). See also Belloni, "Athanasius Kircher," 61.

11 Theodor Kerckring, "Observatio XVIII: Per Microscopia incertum in Anatomia judicium," in *Spicilegium anatomicum* (Amsterdam: Frisius, 1670): "Intestina scilicet, hepar, ceteraque viscerum parenchymata infinitis scatere minutis-

from this conception was that all beings on earth had existed from the first day of creation, layered into one another in an edifice of worms of vastly varying sizes. This view allowed the idea of the "seed-animalcule" in preformationism to be linked to other observations and imaginings.

These webs of interconnected ideas might reach as far as the Christian history of creation. That is what we find, for example, in the *Jobi physica sacra* (1721) of the Swiss physicotheologist Johan Jacob Scheuchzer (1672–1733); he invokes Leeuwenhoek's findings in his discussion of man as existing, as a compound of "little worms," from God's creation and until the postmortem decomposition of his body effected by worms.[16] This theory relates the wormlike appearance of spermatozoa to the maggots consuming the decaying corpse. The latter were thought to arise from the dead body itself. An important observation corroborating these views was that such wormlike phenomena were found in virtually all microscopic samples.

In the 1960s, Bruno Zanobio's examination of the resolving capacity of early microscopes and microphotographs he took using the original instruments allowed him to demonstrate that what these old microscopes showed actually resembled what the microscopists had identified as a tangle of worms. That was where the world of the imagination allied itself with real observations. As a consequence, such web-like ensembles of fibers appear in various treatises on microanatomy.[17] Yet the interwoven filaments that may also be seen in Zanobio's reconstructions of historical microscopic images in fact resulted from a deceptive interplay of aberrations and diffraction lines in the compound microscopes.[18] As a result, "various theories" advanced at the time "concerning the elementary structure of the organism apparently rested on different interpretations of the same deceptive image."[19]

It was not until 1783 that the Scottish anatomist Alexander Monro II (1733–1817) determined, in his *Observations on the Structure and Functions of the Nervous System*, that such wormlike phenomena were to be found in virtually all minute structures he studied, and the more so the higher the resolutions of the lenses being used became. He found them in nerve pathways, lymph nodes, and hair roots no less than in microbes (fig. 3) or the microscopic structures of vegetables.[20] So the great number and variety of worms and filamentous animalcules observed in historical microscopy is in fact evidence that the latter deserves to be taken seriously as a science based on actual instrument-aided observation.

Knowledge of the Microcosm and the Contribution of a Pictorial Tradition

That is not to say that we can identify physical explanations for all microscopic imagery that strikes us as odd today. Rather, the image comes into being in the interplay between what the particular instrument being used allows the microscopist to see and what he is in fact willing to see—what, guided by a theory, he is trying to see.

So the imagery itself is another factor shaping the process of perception. In order to gain confidence in producing an image of what they saw through the microscope, draftsmen resorted to information about, and representations of, the object under examination provided by other microscopists; in some instances, however, interference would result. For

simis animalculis," quoted in Belloni, "Athanasius Kircher," 62.

12 Thomas Willis, *Pharmaceutica rationalis, sive, Diatriba de medicamentorum operationibus in humano corpore* (London: Scott, 1674); see also Belloni, "Athanasius Kircher," 63.

13 The possibility remains that Ham had already studied other samples containing healthy spermatozoa; see Ruestow, *Microscope in the Dutch Republic*, 188–89.

14 Stuart Howards, "Antoine van Leeuwenhoek and the Discovery of Sperm," *Fertility and Sterility* 67, no. 1 (January 1997): 16.

15 Catherine Wilson, "Leibniz and the Animalcula," in *Studies in Seventeenth-Century European Philosophy*, ed. Alexander Stewart (Oxford: Clarendon, 1997), 153–75.

16 See Johannes Jacob Scheuchzer, *Jobi Physica sacra, oder Hiobs Natur-wissenschafft vergliechen mit der Heutigen* (Zurich: Bodmer, 1721), 152–53.

17 Zanobio, "L'immagine filamentoso-reticolare."

18 Ibid.

19 Belloni, "Athanasius Kircher," 64.

20 Alexander Monro, *Observations on the Structure and Functions of the Nervous System* (Edinburgh: Creech, 1783), 67–68.

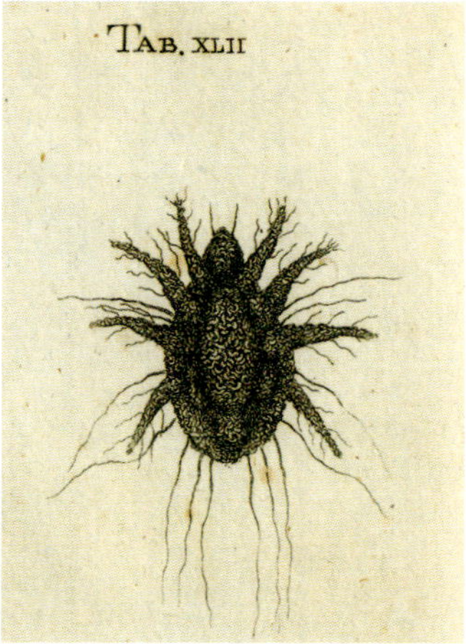

FIG. 3: Alexander Monro, wormlike structures on the body of a mite, from *Observations on the Structure and Functions of the Nervous System* (Edinburgh: Creech, 1783). © Staatsbibliothek zu Berlin—PK.

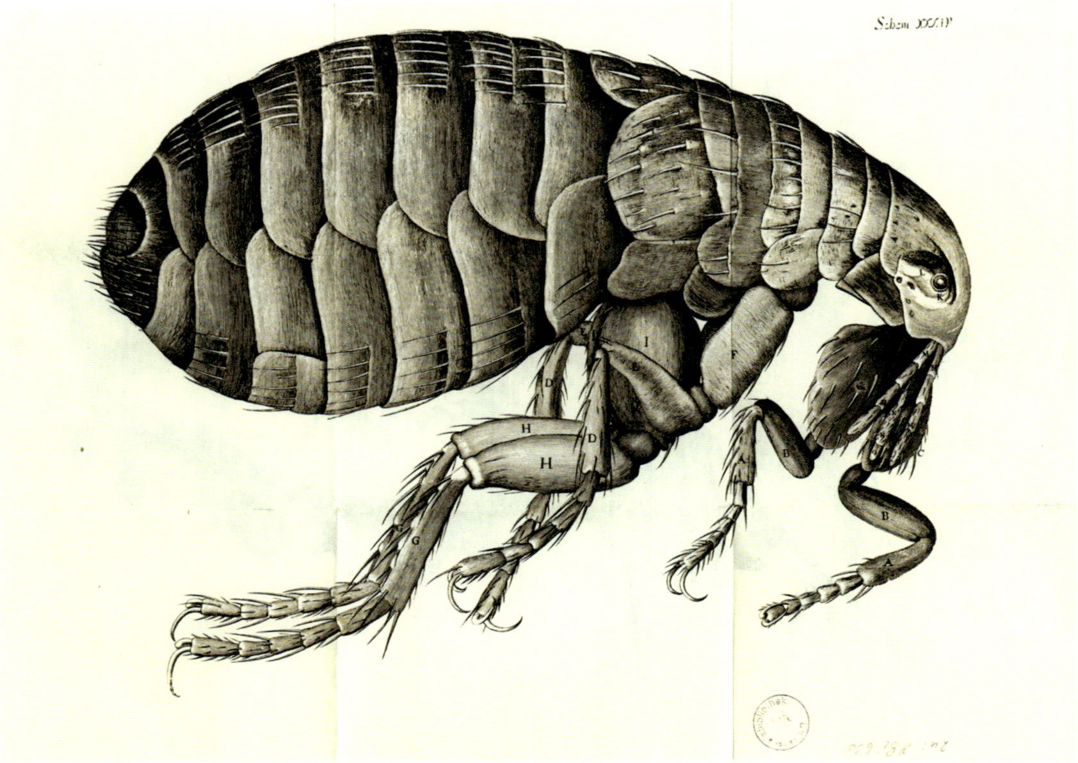

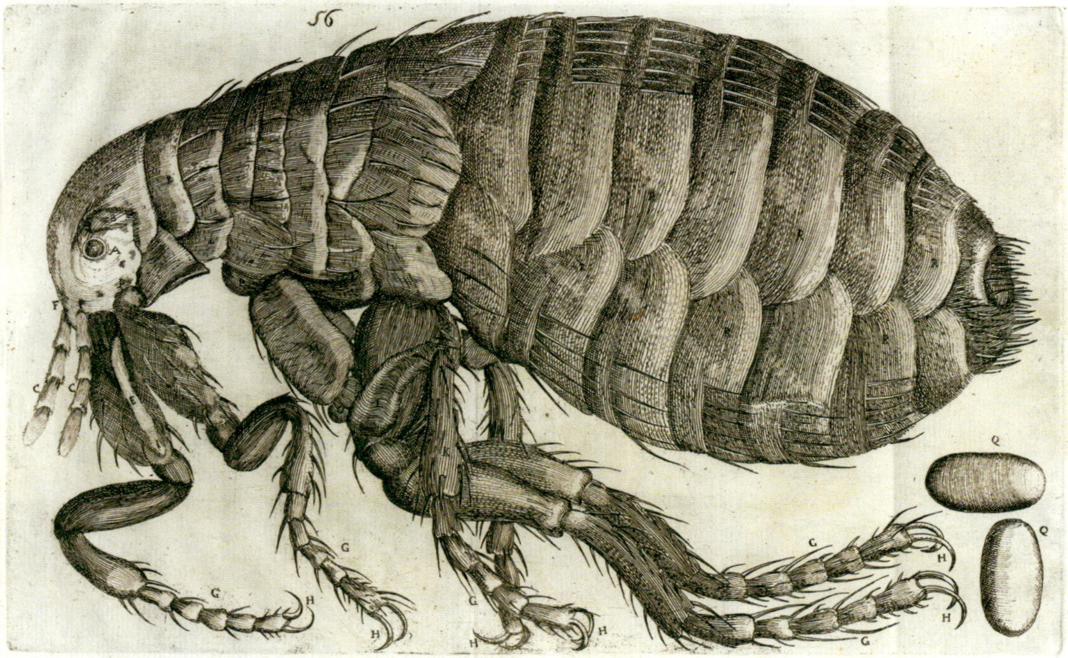

FIGS. 4a, 4b: Comparison of depictions of a flea; top: Robert Hooke, *Micrographia, or Some Physiological Descriptions of Minute Bodies, Made by Magnifying Glasses: with Observations and Inquiries thereupon* (London: Martyn & Allestry, 1665); bottom: Filippo Bonanni, *Observationes circa viventia, quae in rebus non viventibus reperiuntur: cum micrographia curiosa* (Rome: Hercules, 1691). 4a, © Rare Book and Manuscript Library University of Pennsylvania. 4b: © Staatsbibliothek zu Berlin—PK.

example, Robert Hooke (1635–1703) was a trained artist, so the plates in his *Micrographia* (1665) were of such quality that they came to constitute the standard for all later microscopists; especially when the latter doubted their own observations and drawings—whether because of defects of their own equipment or because their graphic skills were inferior—it made sense to rely on Hooke's images and, indeed, to copy them.[21] The latter possessed the authenticity that derives from attention to detail, so scholars had every reason to regard them as "correct." They effectively represented a truth that had eluded the skills of other microscopists.

Filippo Bonanni (1638–1723) first published his *Micrographia curiosa*[22] twenty-six years after Hooke's *Micrographia* had come out. Comparison between the images of a flea in both books shows that Bonanni's is a direct copy (fig. 4): in other words, he renders not some flea he observed with his own eyes but a revision of the specific image in Hooke's work. The animal, however, is now laterally reversed, the legs are in a slightly different position, and Bonanni has added views of the flea's eggs. Comparing this image to the other depictions in Bonanni's work suggests that the author decided to resort to copying due to misgivings concerning his own graphic skills. The engravings based on his own observations show highly simplified forms; their quality cannot match that of Hooke's illustrations.[23] In other words, it was not the engraver's abilities that constituted the limiting factor; he would have been able to make much more precise copperplates than Bonanni's own drawings necessitated. Moreover, the latter used copies irrespective of whether an image showed a detail of a specimen selected for its representative quality or an individual view. In copying Hooke's depiction of the fly's compound eye, for instance, Bonanni even retained the reflections of the windows that appeared on the surface of each facet (fig. 5).

Examples like these allow us to trace, even today, how certain aspects are transmitted and handed down through copying. When not directly copying a precedent image, the draftsman might nonetheless rely on illustrations he had seen, and like a preconceived theory, such depictions would influence his composition and imagination. The history of microscopy offers diverse examples of how the conglomerate of mental images based on pictures a microscopist had seen influenced the design of his own visualizations.

The *Descriptions et usages de plusieurs nouveaux microscopes* published by the mathematician Louis Joblot (1645–1723)[24] in 1718 consists of two parts: a description of various types of microscopes and an account of Joblot's own microscopic studies. The observations of infusoria in the second part include the first published drawing of a hydrachnid larva. Its particular feature is the recording of what Joblot thought he had discerned on the animal's dorsal plate: a distinctive maculation showing the likeness of a human face, a man furrowing his brows as he is glancing sideways (fig. 6, top).[25] The detailed depiction of the face includes precisely drawn nasal wings, lips, and a moustache.

This satyr's face on the back of a tiny animal proved so impressive that, like Robert Hooke's illustrations, it was adopted by the producers of other micrographies. It resurfaces, for example, in the *Micrographia illustrata* of George Adams Sr. (1704–1772).[26] In the first edition of this work, which came out in 1746, the animal with the face maculation

21 See Stefan Ditzen, "Brechen, Schleifen, Brennen: Aspekte instrumenteller Bedingungen in den Bildern der frühen Mikroskopie," in *Instrumente in Kunst und Wissenschaft: Zur Architektonik kultureller Grenzen im 17. Jahrhundert*, Theatrum Scientiarum, vol. 2, ed. Helmar Schramm, Ludger Schwarte, and Jan Lazardzig (Berlin: de Gruyter, 2006), 362–76.

22 Filippo Bonanni, *Observationes circa viventia, quae in rebus non viventibus reperiuntur: cum micrographia curiosa* (Rome: Hercules, 1691).

23 Hooke's illustrations are so precise that it was even possible to identify a particular fossil in the pictures he made of it. In 1990, Tozer compared the image of an ammonoid fossil (fig. 1, table 5) from Hooke's *Posthumous Works* (London: Smith and Walford, 1705) with the fossil now held by the British Museum and made a positive identification. See E. T. Tozer, "Discovery of an Ammonoid Specimen Described by Robert Hooke," *Notes and Records of the Royal Society of London* 44, no. 1 (January 1990): 3–12.

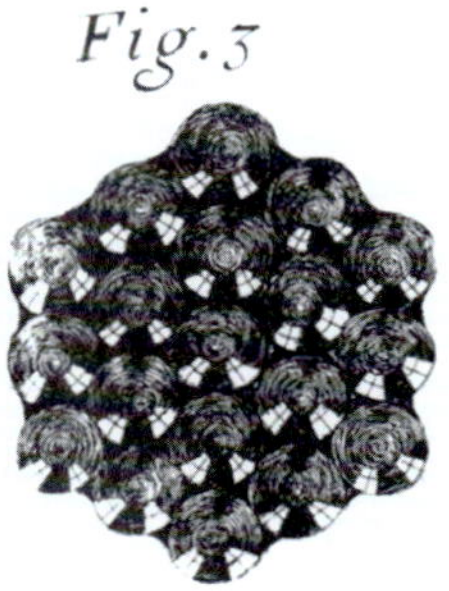

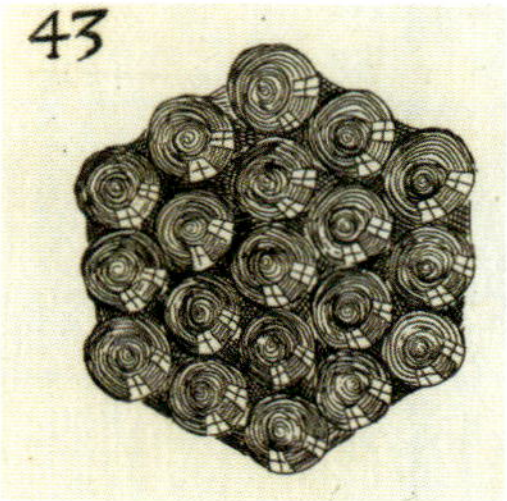

FIGS. 5a (top), 5b (bottom): Comparison of depictions of the compound eye of a fly. (a) Robert Hooke, *Micrographia, or Some Physiological Descriptions of Minute Bodies, Made by Magnifying Glasses: with Observations and Inquiries thereupon* (London: Martyn & Allestry, 1665). (b) Filippo Bonanni, *Micrographia curiosa* (Rome: Hercules, 1691). 5a: © Rare Book and Manuscript Library University of Pennsylvania. 5b: © Staatsbibliothek zu Berlin—PK.

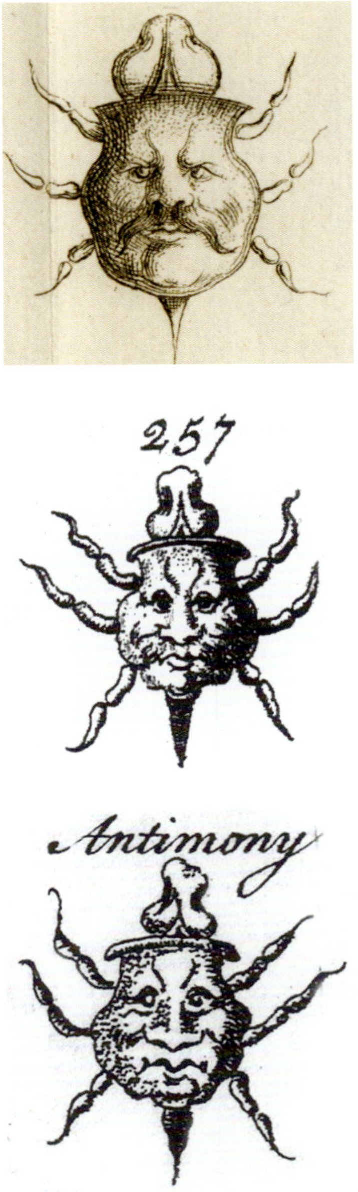

FIGS. 6a (top), 6b (center), 6c (bottom): Comparison of depictions of a hydrachnid larva showing the features of a human face. (a): Louis Joblot, *Descriptions et usages de plusieurs nouveaux microscopes, tant simples que composez* (Paris: Collombat, 1718). © Staatsbibliothek zu Berlin - PK. (b): George Adams, *Micrographia Illustrata, or the Microscope Explained [...]* [1746] (London, 1771). © The British Library Board. General Reference Collection 1651/1154. (c): Francis Watkins, *L'exercise du microscope, contenant un abregé de tout ce qui a été écrit par les meilleurs autheurs [...]* (London, 1754). © The British Library Board. General Reference Collection T.446.(4).

appears in plate 34; by the second edition, published in 1771, it had continued its "migration" to plate 40 (fig. 6, center). The images in the *Micrographia illustrata* also served to advertise the quality of the instruments Adams had developed. An image such as the depiction of the larva, that is to say, was testament to the fact that instruments constructed according to Adams's designs allowed the user to observe what Louis Joblot had discovered with his own microscopes.[27]

Adams made several alterations to the face on the animal's back. The large dark eyes now look straight at the beholder; no more than a patch of fuzz remains of the moustache. Yet the new image retains enough of the original's features to identify it as a copy. In this picture, too, the intricate face-like dorsal maculation contrasts strikingly with the paucity of detail in the rest of the animal's body: the draftsman focused his energies on capturing the human facial features, whereas the depiction of the larva is otherwise fairly desultory. The image makes it look like a vase-shaped head to which the animal's actual head and extremities have been loosely attached.

Whether it is indeed a copy or instead based on microscopic observation inspired by the foregoing atlases, the image of the face on the larva's dorsal plate also shows up in *L'exercise du microscope* (1754) by Francis Watkins (1723–1782),[28] where it appears amid "other Animalcules" the author had harvested from infusions of lemon flowers, oak bark, or antimony (fig. 6, bottom). In his sketchy visualization, Watkins adopted the basic shape and primary features of the face, though the beard has now completely vanished. Once again, details of the depiction of the face such as the wide nasal wings and the two furrows creasing the forehead suggest the earlier images no less than the basic shape of the animal.

The Pictorial Mycelium

The study of historical atlases of microscopy suggests that the scholar's gaze through the instrument is rarely, if ever, unaffected by traditions. We should not assume that a practitioner of microscopy has had no visual information concerning the insights the device will enable him to gain when he first peers into the instrument. On the contrary, such images play a major role in generating interest in the microscope in the first place; once assimilated, they form a considerable part of the scholar's competence. The more "invisible" an object is, the more difficult it is to identify it repeatedly and uniformly, the more room it leaves for a tradition of specific microscopic images. The discovery of something new, that is to say, is based on previous familiarity with what is already known. That has consequences both for the need to normalize instruments and for the normative definition of standardized procedures of preparation and imaging techniques.

The imagery of a research area is accordingly a product not only of the capacities of the instruments and the prerequisites and limitations of their employment but also, and in connection with these tools, of theoretical predispositions, as the interplay of worms seen in observation, theoretical premises, and deceptive optical phenomena illustrates. Individual images, moreover, actively intervene in observations and the transmission of knowledge; they may make structures emerge from the realm of the invisible—like a face on a larva's back—that prove strikingly durable. The concrete image in the natural sciences, then, must not be considered in isolation; we must conceive it instead as a sort of node in which these many factors intersect, imparting to it the dynamism of the various fields that conditioned its genesis.

Natural-scientific visualization, we might say, *embodies* the totality of these formative conditions. The preparation procedures and the processes of consensus-building that tie a community of scientists together and ultimately lead to the emergence of doctrines and the creation of atlases are likewise an elementary constituent of their form. All the fields I have mentioned must therefore be regarded as part of the concrete visual product. Taken together, these factors are what should properly be addressed as the actual composite pictorial body. The concrete image, conversely, is a "surrogate," filling the blank of the composite body that cannot be represented as such, which we may call the *pictorial mycelium* (fig. 7). I borrow this term from biology, where it describes the organization of fungal organisms: the fungus, properly speaking, grows underground and initially remains invisible before forming the fruiting bodies that, considered separately, are commonly known as "mushrooms."

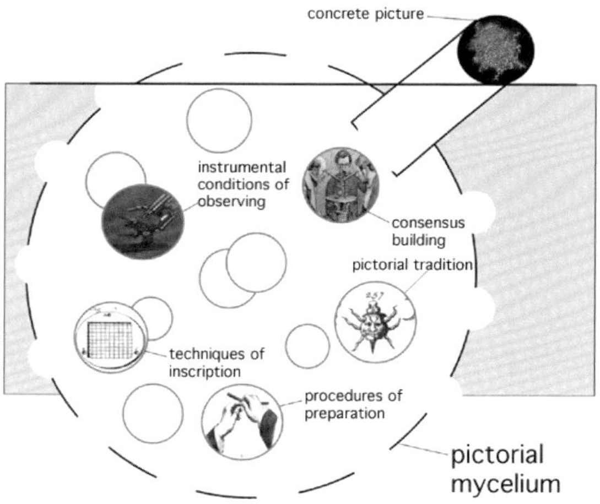

FIG. 7: Schematic illustration of the pictorial mycelium. Illustration by the author. © Stefan Ditzen.

The *pictorial mycelium* similarly encompasses the totality of the conditions determining the production of an artificial image. That includes the conditions of instrument-aided visibility as well as the phenomena of shared perception, and hence the procedures a scientific community relies on to arrive at consensus on insights to be integrated into the canon. The chains of images discussed in these pages likewise constitute an attractor in this ensemble, one whose emergence is interwoven with the totality of the other conditions influencing the genesis of imagery. The concrete image, that is to say, is two things at once: a concrete graphic product that may have its own claim to physical presence and an index of the underlying pictorial body, a composite that comprises the conditions of its genesis in their entirety.

24 Joblot was among the first Frenchmen to devote sustained attention to microscopy. A particular influence in rousing his interest in this branch of science may have been Christiaan Huygens (1629–1695), who appeared before the Académie des Sciences in 1678 to demonstrate the use of the microscopes he had brought from Holland. See Gerard L'Estrange Turner, *Collecting Microscopes* (New York: Mayflower, 1981), 52.

25 Louis Joblot, *Descriptions et usages de plusieurs nouveaux microscopes, tant simples que composez* (Paris: Collombat, 1718). See also John H. Hammond and Jil Austin, *The Camera Lucida in Art and Science* (Bristol: Hilger, 1987), 114.

26 George Adams, *Micrographia Illustrata, or the Microscope Explained [...]* (London, 1771).

27 See John Millburn, *Adams of Fleet Street, Instrument Makers to King George III* (Oxford: Ashgate, 2000), 31–36.

28 Francis Watkins, *L'exercise du microscope, contenant un abregé de tout ce qui a été écrit par les meilleurs autheurs [...]* (London, 1754).

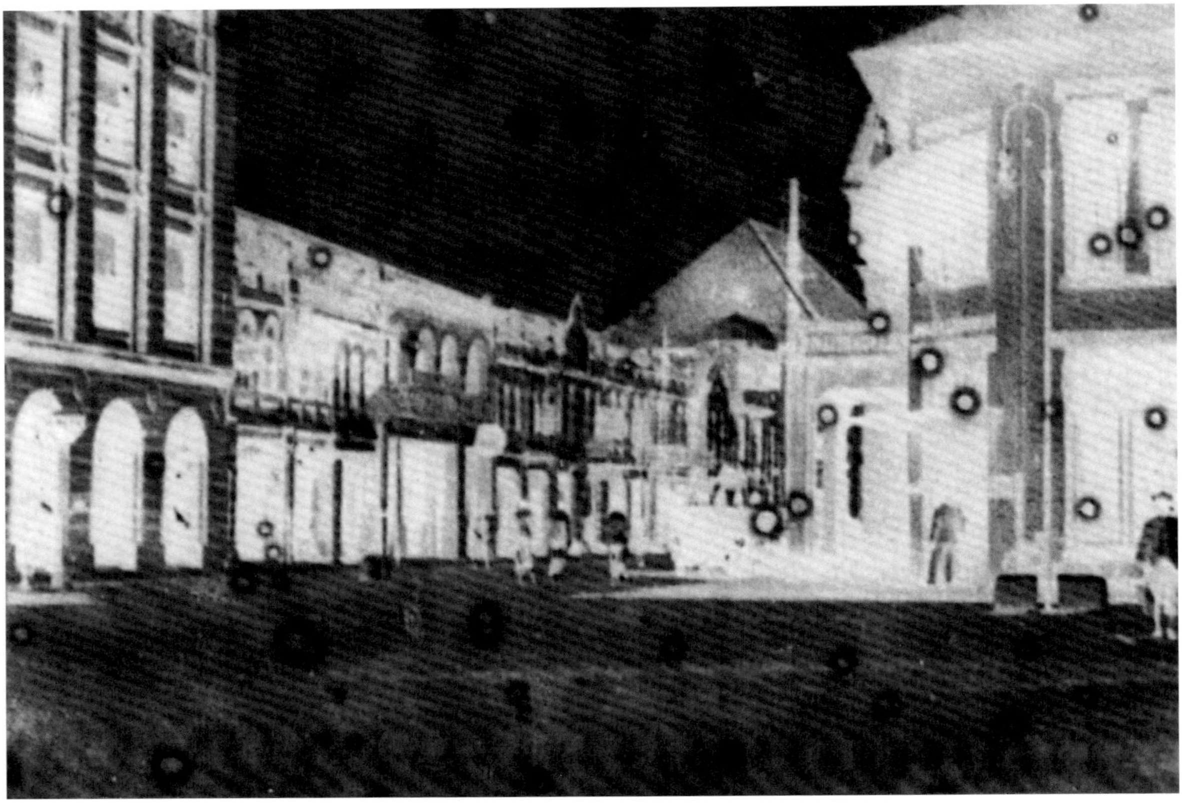

FIG. 1: "Bacterial erosion," from Kurt Fritsche's 1969 *Fotofehler-Buch* (Book of photographic errors). "Small, light depressions in the gelatin layer suggest erosion by bacteria. Insects, such as ants, also like to nibble the soft gelatin and sometimes cause similar-looking damage. [. . .] To distinguish bacterial or insect damage from other lesions in the photosensitive layer that often look very similar, compare 'holes in the layer' with 'lesions in the layer.'" Kurt Fritsche, *Das große Fotofehler-Buch*, 6th ed. (Leipzig: Fotokinoverlag, 1969), 97, fig. 50.

IMAGE NOISE

In the history of technical images, the term *noise* refers to the unforeseen intrusion of a medium's material or technical conditions, which are supposed to remain invisible, in the graphical transmission of a phenomenon or visual presentation of a message. Ideally, the medium used to render the object visible is present only as an absence; image noise occurs when it becomes visible.

The effects of such noise can include streaks, hazes, blurring, ablution of image layers, falsifications of perspective, scratches, blotches, dots (fig. 1), false color values, interference, the moiré effect (fig. 2), and even the disappearance of the image, to mention just a few.

Image noise thwarts the intentions of the image producers and confuses the beholder of the resulting image. Image noise is by definition uncalculated and uncontrollable; historically, it has usually been treated as producing a defective reject. This is how Josef Maria Eder regards the matter in his photography handbook *Ausführliches Handbuch der Photographie* (1930), a compilation of photographic mishaps. His aim in classifying and describing them in analogy with successful images is not an aesthetic appreciation of image noise but an analysis of the sources of errors that will help the reader avoid them in future. Such systematic cataloguing continues today in "books of photographic errors" or in scanning instructions, which seek to counteract unintended visual effects based on a logic of cause and effect (Fritsche 1969).

According to the dictionary definition of the term, *image noise* is generally synonymous with the flickering of the TV picture; in some instances, *signal noise*, in the sense of "snow" (fig. 3), is regarded as the most visible trait of image interference (Brockhaus 2006). During other transmission errors, the test pattern appears instead of the noise (fig. 4); although transmitted

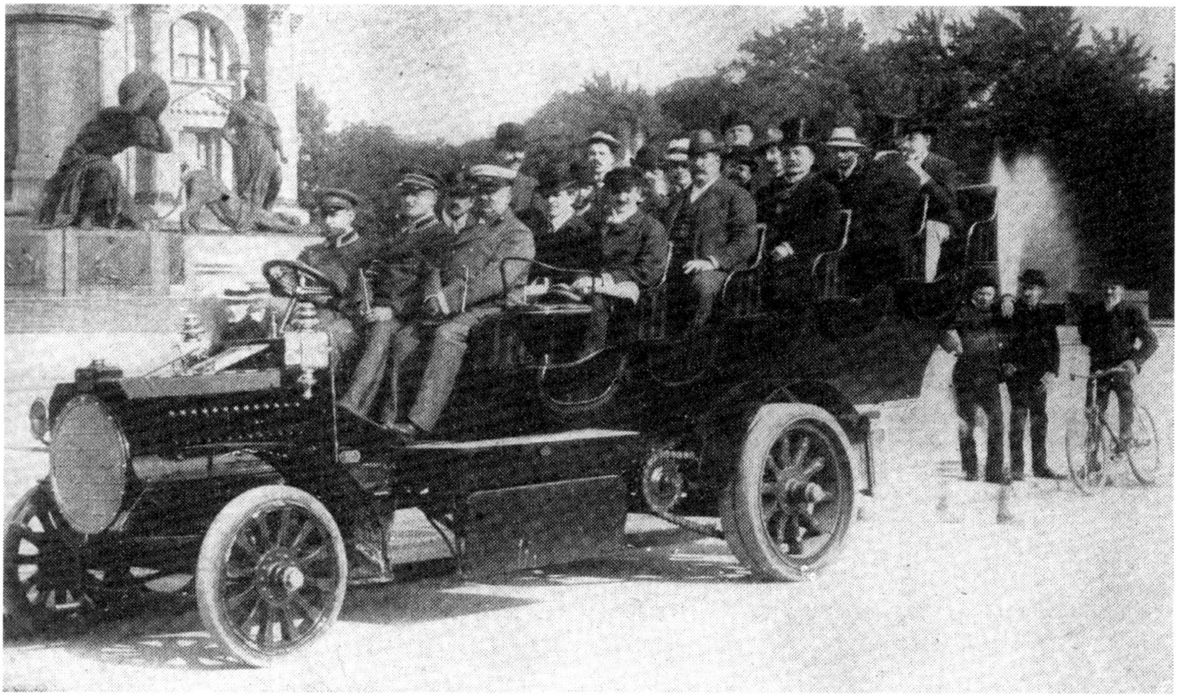

FIG. 2: Moiré effect in a book of photographs. A grid is used to reproduce photographs for halftone printing; similar grids also appear in scanning. The image error occurs when two grids are superimposed, producing vibrant patterns. Egon Jameson, *Berlin, so wie es war,* 8th ed. (Düsseldorf: Droste, 1988), 87.

without interference, it is perceived as disruptive as well (Schneider 2002). In the area of film, Joachim Paech has described the breaking of film as the "appearance of disappearance in film. [. . .] In the image of its destruction, the film becomes visible as a material medium and may be observed in a form which substitutes itself, as the form of the medium of film, for those formulations to which in their turn only the transparency, the invisibility of the medium has contributed in favor of the represented form" (Paech 1999, 125–26). Friedrich A. Kittler, following Marshall McLuhan and Claude Elwood Shannon's model of communication, also gave signal noise, interference in the transmission of information from sender to recipient, a central position in his media theory (Kittler 1993), taking up Michel Foucault's hypothesis that a medium's message is inalienably bound up with the asemantics of (image) noise. "But in fact there must, if there is to be a 'message' at all, [. . .] be 'noise'" (Foucault 1999, 140). In his study *The Parasite,* Michel Serres established the related figure of the third man, who, both troublemaker and mediator, makes long-established dichotomous principles of order and dualistic semantics precarious (Serres 1982).

In this perspective, it would seem worthwhile to analyze not only the incunabula of the histories of art and science but also the forms of technically impaired images. Work on the production and use of images in the area of scanning tunneling microscopy in the 1980s and 1990s, for example, has shown that images produced in the early stages of a new technology may well be subject to interference and that such noise is even explicitly addressed. Yet as such techniques become established, noise, although it is a fundamental part of laboratory practice, tends to disappear from publications (Hennig 2006).

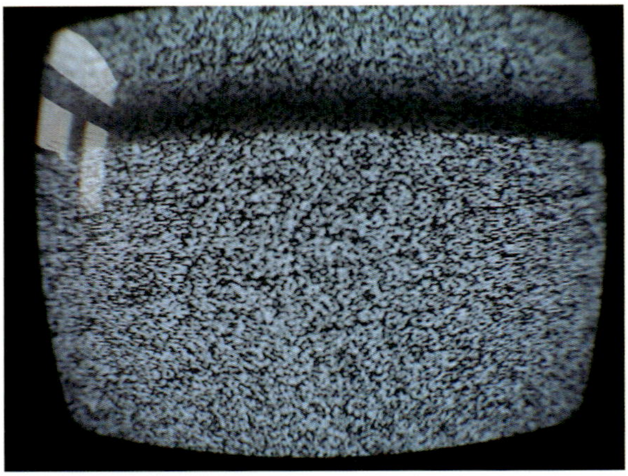

FIG. 3: Signal noise or snow on a television screen. Physics defines signal noise as a disturbance with a nonspecific frequency spectrum. The source of the disturbance may lie with the broadcaster, with the receiver, or in the channel. Screenshot, private archive of Birgit Schneider.

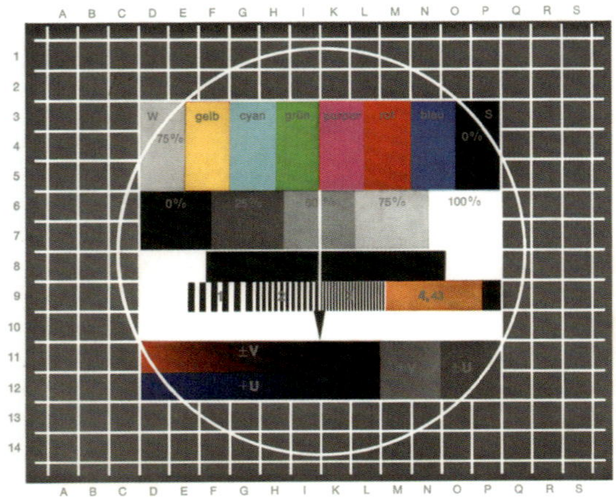

FIG. 4: Broadcast even today when interference occurs in television transmission, the test pattern, of which there are many versions, is synonymous with disrupted reception. The "electrical color test pattern" originally served to test antennas, television distribution cables, and broadcasters, because the test pattern's structure, consisting of a circular form, a grid, and red and gray areas, allows technicians to adjust the brightness, contrast, and picture position and to visualize interference from various sources. Informationsblatt Sender Freies Berlin: Elektrisches Farbtestbild, 1971.

This practice of eliminating inadequate recordings again shows that the history of technically generated images is usually written as a teleological story of progress, involving continually improving machines and increasingly precise resulting images—even as interference remains normal in the daily work of image producers. In attempts to expand the boundaries of what can be made visible using optical media in particular, such as in Joseph Gerlach's nineteenth-century microphotographs, the maximum enlargement is often indistinguishable from an image in which the photographic process presents itself in the picture, in the "visibility of silver precipitation" (Gerlach 1863, 11; Breidbach 1998). In his work on image interference, the art historian Peter Geimer has accordingly proposed that scholars remove the stigma of failure from such pictures: the boundary between image noise and the discovery of a new phenomenon by means of visualization techniques is often a fluid one (Geimer 2001, 2002). The production of "difference" (Rheinberger 2001) in the process of gaining scientific knowledge accordingly necessitates follow-up investigations that may provide information on the status of an initially confusing element of an image (fig. 5) as either signal or interference. —FB

LITERATURE

Benz, Arnold. "Das Bild als Bühne der Mustererkennung: Ein Beispiel aus der Astrophysik." In *Mit dem Auge denken: Strategien der Sichtbarmachung in wissenschaftlichen und virtuellen Welten*, ed. Bettina Heintz and Jörg Huber, 65–78. Zurich: Edition Voldemeer, 2001.

Breidbach, Olaf. "Der sichtbare Mikrokosmos: Zur Geschichte der Mikrofotografie im 19. Jahrhundert." In *Der Photopionier Hermann Krone: Photographie und Apparatur. Bildkultur und Phototechnik im 19. Jahrhundert*, ed. Wolfgang Hesse and Timm Starl, 131–58. Marburg: Jonas-Verlag, 1998.

Canguilhem, Georges. *The Normal and the Pathological* [1966], translated by Carolyn R. Fawcett in collaboration with Robert S. Cohen; with an introduction by Michel Foucault. New York: Zone Books, 1989.

Chéroux, Clément. *Fautographie: Petite histoire de l'erreur photographique.* Crisnée: Éditions Yellow Now, 2003.

Eder, Josef Maria. *Ausführliches Handbuch der Photographie.* Halle an der Saale: Knapp, 1930.

Elkins, James. "Marks, Traces, *Traits*, Contours, *Orli*, and *Splendores*: Nonsemiotic Elements in Pictures." *Critical Inquiry* 21, no. 4 (1995): 822–60.

Foucault, Michel. "Message ou bruit?" *Concours médical*, no. 88 (1966): 6285–6286.

Fritsche, Kurt. *Das große Fotofehler-Buch.* Leipzig: Fotokinoverlag, 1969.

FIG. 5: Franz Gießibl, Internal structures (individual orbitals) of an atom in an atomic force microscopic image. Gießibl and his colleagues in Augsburg published this image in 2000 as the first experimental evidence of internal atomic structures, leading to a controversy with scientists from Basel, who claimed that it showed interference caused by the instrument's electronics. In 2006, a working group in California was able to replicate Gießibl's results, indicating that they were not simply due to interference. The issue of whether the cloud-like form was noise or a trace of subatomic structures remained unresolved for seven years. Franz J. Gießibl et al., "Subatomic Features on the Silicon (111)-(7×7) Surface Observed by Atomic Force Microscopy," *Science* 289 (2000): 422–25, fig. 3. Franz J. Gießibl.

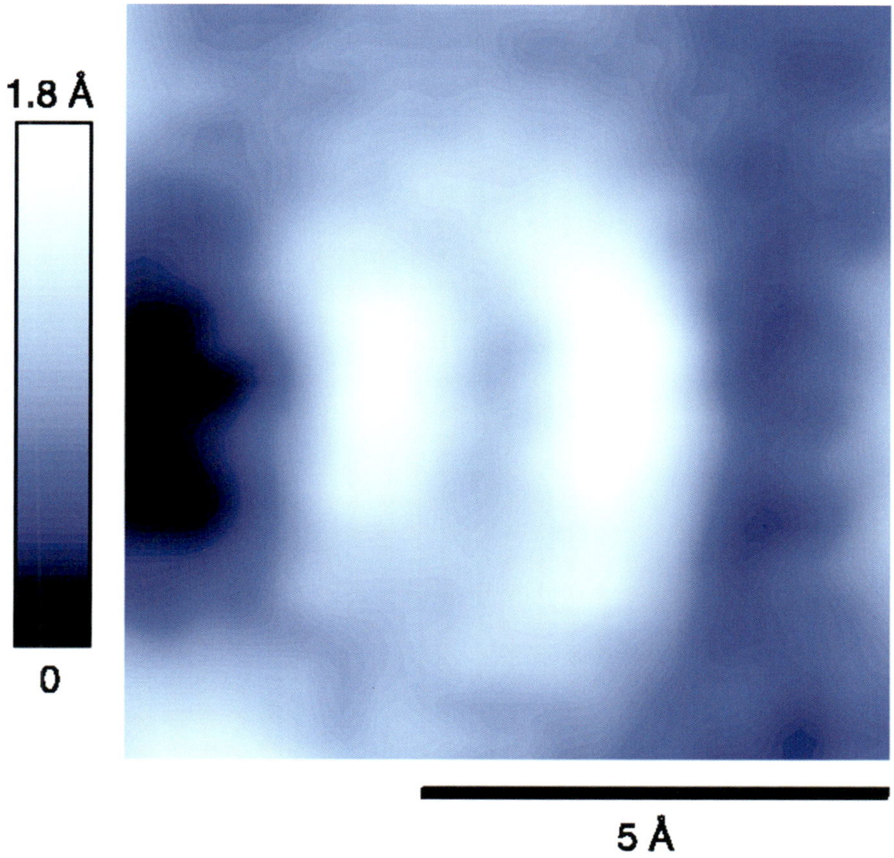

Geimer, Peter. *Bilder aus Versehen. Eine Geschichte fotografischer Erscheinungen*. Hamburg: Philo Fine Arts, 2010.

Geimer, Peter. "Was ist kein Bild?: Zur 'Störung der Verweisung.'" In *Ordnungen der Sichtbarkeit: Fotografie in Wissenschaft, Kunst und Technologie*, ed. Peter Geimer, 313–41. Frankfurt am Main: Suhrkamp, 2002.

Gerlach, Joseph. *Die Photographie als Hülfsmittel mikroskopischer Forschung*. Leipzig: Engelmann, 1863.

Hennig, Jochen. "The Instrument in the Image: Revealing and Concealing the Condition of the Probing Tip in Scanning Tunneling Microscopic Image Design." In *Instruments in Art and Science—On the Architectonics of Cultural Boundaries in the 17th Century*, ed. Jan Lazardzig, Helmar Schramm, and Ludger Schwarte, 348–61. Berlin: de Gruyter, 2008.

Hiepko, Andreas, and Katja Stopka, eds. *Rauschen: Seine Phänomenologie und Semantik zwischen Sinn und Störung*. Würzburg: Königshausen & Neumann, 2001.

Innis, Harold A. *The Bias of Communication*. Intro. by Marshall McLuhan. Toronto: University of Toronto Press, 1964.

Kittler, Friedrich A. *Discourse Networks 1800/1900*. Palo Alto, CA: Stanford University Press, 1990.

Kittler, Friedrich A. "Geschichte der Kommunikationsmedien." In *Raum und Verfahren. Interventionen*, ed. Jörg Huber and Alois Martin Müller, 169–88. Basel: Stroemfeld, 1993.

Menkman, Rosa. *The Glitch Moment(um)*. Amsterdam: Institute of Network Cultures, 2011.

Müller, Hugo. *Die Misserfolge in der Photographie und die Mittel zu ihrer Beseitigung: Ein Hilfsbuch für Liebhaber der Lichtbildkunst*. Halle an der Saale: Knapp, 1894.

Paech, Joachim. "Figurationen ikonischer n...Tropie: Vom Erscheinen des Verschwindens im Film." In *Konfigurationen zwischen Kunst und Medien*, ed. Sigrid Schade, 122–36. Munich: Fink, 1999.

Rheinberger, Hans-Jörg. *Toward a History of Epistemic Things: Synthesizing Proteins in the Test Tube*. Palo Alto, CA: Stanford University Press, 1997.

Schneider, Birgit. "Die kunstseidenen Mädchen: Test- und Leitbilder des frühen Fernsehens." In *1929. Beiträge zur Archäologie der Medien*, ed. Stefan Andriopoulos and Bernhard J. Dotzler, 54–79. Frankfurt am Main: Suhrkamp, 2002.

Serres, Michael. *The Parasite*. Baltimore: Johns Hopkins University Press, 1982.

Shannon, Claude E., and Warren Weaver. *The Mathematical Theory of Communication*. Urbana: University of Illinois Press, 1949.

Weltzien, Friedrich. "Defekt-Effekt: Realitätseffekte des Fotografischen." In *Realitätseffekte: Ästhetische Repräsentation des Alltäglichen im 20. Jahrhundert*, ed. Alexandra Kleihues, 69–93. Munich: Fink, 2008.

Programmed Images: Systems of Notation in Seventeenth- and Eighteenth-Century Weaving

Birgit Schneider

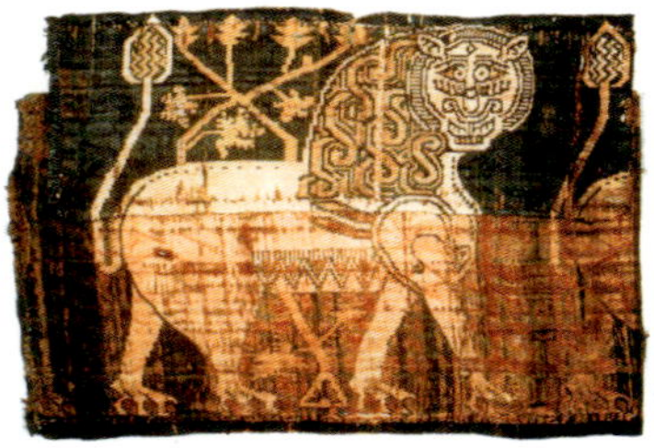

FIG. 1: Silk velvet with striding lions, reliquary sleeve, Persia, ninth or tenth century. The stepped outlines of the lion clearly reveal the fabric's textile structure, with lines (weft) and columns (warp). Leonie von Wilckens, *Die textilen Künste: Von der Spätantike bis um 1500* (Munich: Beck, 1991), 52. Maastricht, Treasury of the Basilica of Saint Servatius.

*Early notational systems used in seventeenth- and eighteenth-century pattern weaving represent a case of the interrelation between image and pictorial code—a more familiar example today is "digital imagery"—that far predates the history of the computer. Weaving exemplifies the underlying strata of the interplay between arts and media: it is defined by a particular and indissoluble union of art and technology. Ever since the invention of the loom, weavers have produced structurally "rastered" patterns and pictures (*fig. 1*); nowadays, the picture composed of pixels has become the standard form of the digitization of imagery in preparation for technological processing. Fabrics have always implemented the basic principles of technological image-generation, so they constitute one of the sources of digitization, from which connections may be drawn to a media history of technologically generated imagery.*

In the seventeenth century, weavers in Ulm, Germany, developed written profile notations in order to record geometric patterns as unambiguous designs to be implemented on looms. The forms of notation for fabric patterns created in Ulm are the earliest known attempts to encode images using a limited repertoire of unequivocal signs. Yet we have reason to assume that the practice is much older; the strict secrecy the guilds imposed on such trade knowledge prevented its earlier publication. Considered in the wider horizon of the genesis of technical and machine-generated imagery, the forms of notation developed in weaving are significant in several respects: they may be discussed, on the one hand, as the earliest medium used to store images in unambiguous form and, on the other hand, as a way of translating images into instructions to be fed into machines—that is to say, as notations and encoded images but also as a program for the loom.[1] A large body of scholarship has examined the development of notational systems in mathematics, music, and cryptology as well as the general significance of notational systems.[2] By contrast, the complexity of the historical notations of pictorial patterns in weaving has remained virtually unexplored; they have not been the subject of analysis in the perspectives of media history or the history of science.[3] We have every reason to ask, then, how these weaving notations worked and how they affected weaving practices.

The Ulm Fabric Notations of 1677

The structural relationship between weaving and encoding images becomes apparent where fabrics were drafted and recorded on paper.

1 The following considerations rely on Nelson Goodman's definition of *notation*. According to Goodman, a notation is defined by the following requirements: it must be syntactically and semantically disjoint, possess finite differentiation, and be unambiguous. Since the pictures discussed in the following meet the criteria for a notation, they also fulfill the requirements for a pictorial code. Scores are the particular implementations of notational systems—a score for piano, for example, is an application of the system of musical notation. See Nelson Goodman, *Languages of Art: An Approach to a Theory of Symbols* (New York: Hackett, 1968), 177–79, 127–57.

2 For exemplary discussions, see Sybille Krämer, Eva Kancik-Kirschbaum, and Rainer Totzke, eds., *Schriftbildlichkeit: Wahrnehmbarkeit, Materialität und Operativität von Notationen* (Berlin: Akademie, 2012); Gabriele Brandstetter, Franck Hofmann, and Kirsten Maar, eds., *Notationen und choreographisches Denken* (Freiburg im Breisgau: Rombach, 2010); Hubertus von Amelunxen, Peter Weibel, and Dieter Appelt, eds., *Notation: Kalkül und Form in den Künsten*

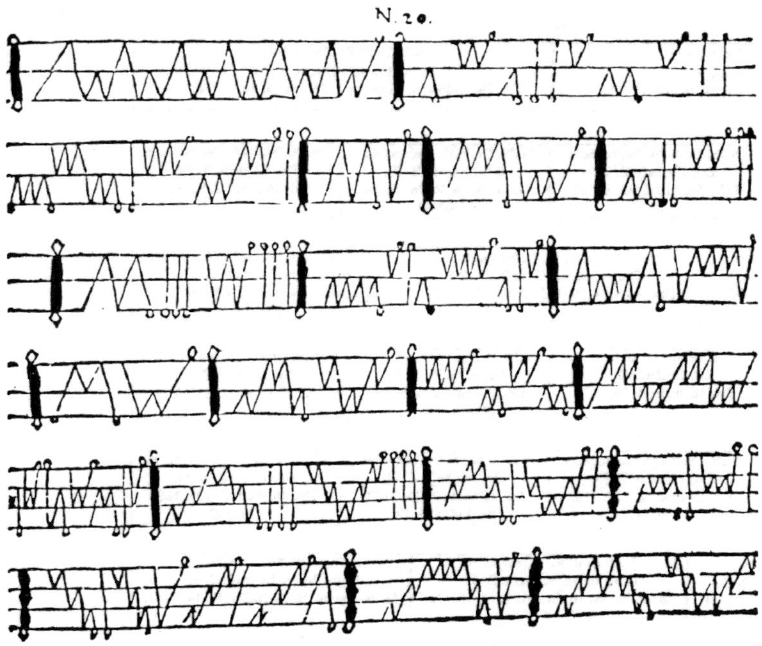

FIG. 2: Plate containing notations of twenty threadings for the setup of a loom, from Ziegler's *Weber Kunst und Bild Buch* (Ulm, 1677).

The early forms of such "plans" sometimes render the woven image in a manner that resembles the musical notation of a score (fig. 2). The picture, from a weavers' pattern book, shows moving lines constrained by the scaffold of a "stave": they zigzag up and down like the temperature curve of a fever patient; the beginning and end of each line are adorned by little curlicues; and black lines resembling bars intersect the parallel horizontal lines at irregular intervals—these, too, bear little curls on both ends. Leaping up and down in discrete bounds, the series of dashes coalesce into figures that recall the variation of a basic theme in the musical notation of a fugue, with repetition, mirroring, and inversion emerging as crucial operations that generate order. Another illustration contained in the same book exemplifies a different form of representation (fig. 3). It shows patterns of black squares within chessboard-like larger units; wide black margins divide the illustration into nine such fields, each subdivided into sixteen-by-sixteen squares. These images typify the two types of schemata the book contains, with a range of variations appearing in the plates.

These notations are taken from a printed publication that is also regarded as the first handbook of weaving: the *Weber Kunst und Bild Buch* published in 1677 by the Ulm-based master weaver Marx Ziegler.[4] The term *Bild* (image) continues to play a prominent role throughout the book that is of interest to scholarship on visual culture; it must be briefly explained in the present context because it refers to a matter of image technology. On the one hand, Ziegler used *Bild* in the early sense of the "figure" or "pattern" that appeared in the products of weaving; he accordingly called pattern-weaving "Bild-Arbeit," or "image-work." He distinguished *Bild* in this sense from "Boden," or "ground," to differentiate the notation of image-generating effects ("Bild") from that of the structural

(Berlin: Akademie der Künste, 2008); Sebastian Klotz, "Tonfolgen und die Syntax der Berauschung: Musikalische Zeichenpraktiken 1738–1788," in *Das Laokoon-Paradigma: Zeichenregime im 18. Jahrhundert*, ed. Inge Baxmann, Michael Franz, and Wolfgang Schäffner (Berlin: Akademie, 2000), 306–38; Ulrike Bergermann, *Ein Bild von einer Sprache: Konzepte von Bild und Schrift und das Hamburger Notationssystem für Gebärdensprachen* (Munich: Fink, 2001); Bernhard Siegert, *Passagen des Digitalen: Zeichenpraktiken der neuzeitlichen Wissenschaften 1500–1900* (Berlin: Brinkmann und Bose, 2003); Ursula Klein, *Experiments, Models, Paper Tools: Cultures of Organic Chemistry in the Nineteenth Century* (Palo Alto, CA: Stanford University Press, 2003).

3 For a fuller development of the analysis presented in the following, see Birgit Schneider, *Textiles Prozessieren: Eine Mediengeschichte der Lochkartenweberei* (Zurich: Diaphanes, 2007), 83–124.

4 See Patricia Hilts, "Roses and Snowballs: The Development of Block Patterns in the German Linen-Weaving Tradition," *Ars Textrina* 5 (1986): 167–248. Hilts, a historian of weaving, has published an annotated facsimile edition of Ziegler's book together with Nathaniel Lumscher's *Neu eingerichtetes Weber Kunst und Bild Buch* (Culmbach, 1708); see Marx Ziegler, "Weber Kunst und Bild Buch" (1677), ed. Patricia Hilts, *Ars Textrina* 13 (December 1990); Nathaniel Lumscher, "Neu eingerichtetes Weber Kunst und Bild Buch" (1708), ed. Patricia Hilts, *Ars Textrina* 14 (December 1990).

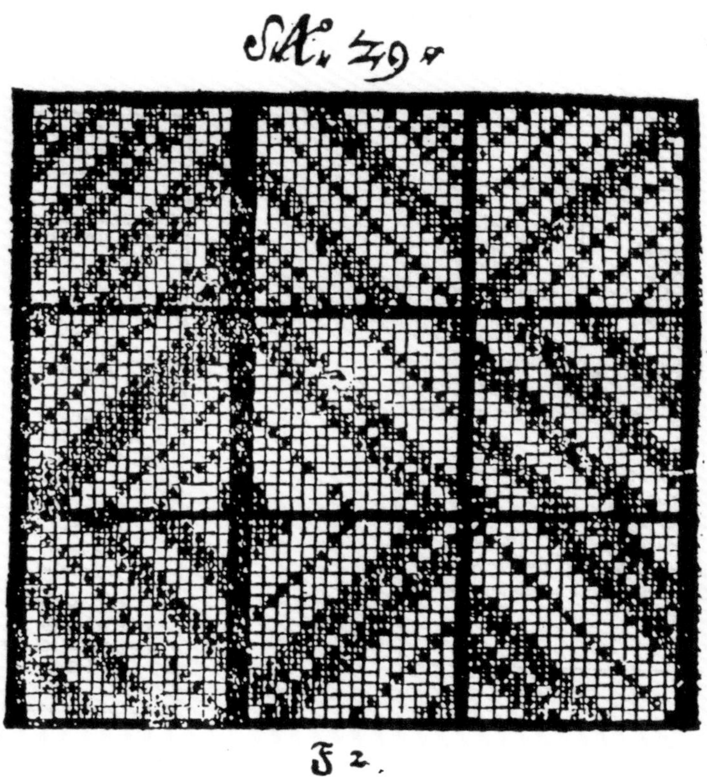

FIG. 3: Nine notations of possible tie-ups from Ziegler's *Weber Kunst und Bild Buch* (Ulm, 1677).

weave (the "ground" of the figure-ground distinction, as it were) in patterned fabrics. Yet *Bild* was in pattern-weaving also the general term for the schematic draft for a fabric made on paper; it is the historical term, that is to say, for pattern notation in weaving.[5]

The forms of patterns treated in Ziegler's publication represent a complicated special weaving technique practiced primarily in southern Germany.[6] These were geometrically patterned fabrics in which the weavers combined shapes such as squares, lines, and triangles to create abstract designs; the textiles were produced on so-called shaft looms. Aside from Ziegler's notation, there existed other conventions on how to notate fabric patterns, but these did not circulate publicly until the age of the encyclopedia in the eighteenth century.[7] Ziegler was thus one of the first to publish on the art and techniques of weaving, ending the guilds' exclusive control over what had heretofore been a strictly kept secret.[8] All of these forms of fabric pattern notation differed considerably from the well-known pattern drawdowns used in planning pictorial patterns (fig. 4), despite the sometimes close visual resemblance.[9] The latter patterns, which could be implemented in a variety of techniques, were widely disseminated in collections as far back as the early sixteenth century, but unlike the notations discussed here, they offered no directions regarding the technical realization on the loom. The following paragraphs will therefore address the technology of weaving and the question of what technical knowledge the notations connote.

5 For the historical use of the term *Bild* in weaving, see also Walther von Hahn, *Die Fachsprache der Textilindustrie im 17. und 18. Jahrhundert* (Düsseldorf: VDI, 1971).

6 This weaving technique produced patterns called "Spitzköper" (reverse twill) and "Schachwitz." The fabrics were also called "Bauerndamast" (country damask); today's standard term is *block damask.*

7 Schematic representations may be found in the French encyclopedic treatments of eighteenth-century weaving, where they appear under the label "translation"; see, e.g., Jean Paulet, *L'Art du fabricant d'étoffes de soie*, 7 vols. (Paris, 1773–1789). A third convention may be found in an Italian manuscript from Lucca that is dated to the 1680s, or only a few years before Ziegler's handbook of weaving. This notational format uses numerals. See Gino Arrighi, *Un manuale secentesco dei testori lucchesi* (Lucca: M. Pacini Fazzi, 1986).

8 Ziegler had various reasons to publish these materials, some of which related to his Protestant ideals about training and education. See Patricia Hilts, "Translator's Introduction," in Ziegler, "Weber Kunst und Bild Buch," 14.

9 On this practice, see Lesley Ellis Miller, "Representing Silk Design: Nicolas Joubert de l'Hiberderie and *Le Dessinateur pour les étoffes d'or, d'argent et de soie* (Paris, 1765)," *Journal of Design History* 17, no. 1 (2004): 29–53.

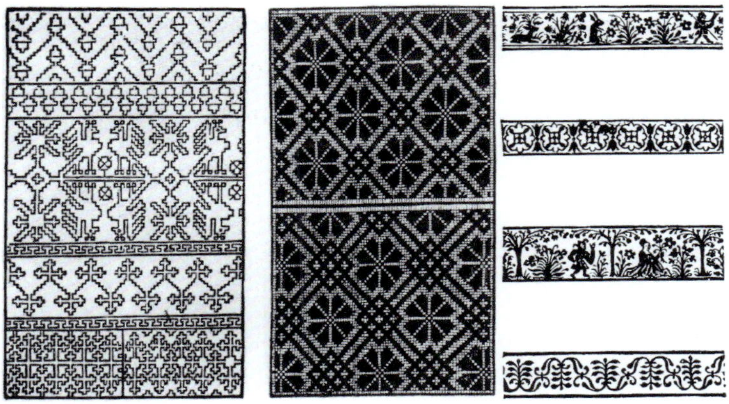

FIG. 4: Three plates from early pattern books: *Ein new Modelbuch* (probably Zwickau, 1526); *Ein new getruckt model Büchli* (Augsburg, 1529); *Furm oder modelbüchlein* (Augsburg, ca. 1523). Margarete Abegg, *Apropos Patterns for Embroidery, Lace and Woven Textiles* [1978] (Riggisberg: Abegg-Stiftung, 1998), 25, figs. 14, 15, and 4.

The Technical Procedure of Shaft-Weaving

Ziegler's notations cannot be appreciated without a basic familiarity with the loom for which they were set down, the shaft loom, which had come into general use in Europe in the thirteenth and fourteenth centuries; see figure 5 for a schematic illustration of a shaft loom. As the name indicates, the distinguishing feature of this weaving technology consisted of a system of shafts the weaver raised and lowered using pedals or treadles. To create a specific pattern using this process, the warp—the system of parallel threads held taut by the loom—is divided into groups and individually threaded through the eyelets (heddle eyes) of a system of threads running vertically between the shafts. Moving like the arms of a marionette, the shafts can raise the groups of warp threads, opening a "shed" through which the weaver passes the shuttle with the weft.[10] The operation of the shafts resembles that of an organ: by stepping on one of the pedals, the corresponding shafts raise a set of warp threads before the weaver picks the shuttle—that is, inserts a single weft thread—and drop back once the pick is complete. Operating another treadle reopens the shed for the

10 See Annemarie Seiler-Baldinger, *Systematik der textilen Techniken* [1973] (Basel: Wepf, 1991), 86; Eric Broudy, *The Book of Looms: A History of the Handloom from Ancient Times to the Present* (New York: Van Nostrand Reinhold, 1979), 102–23.

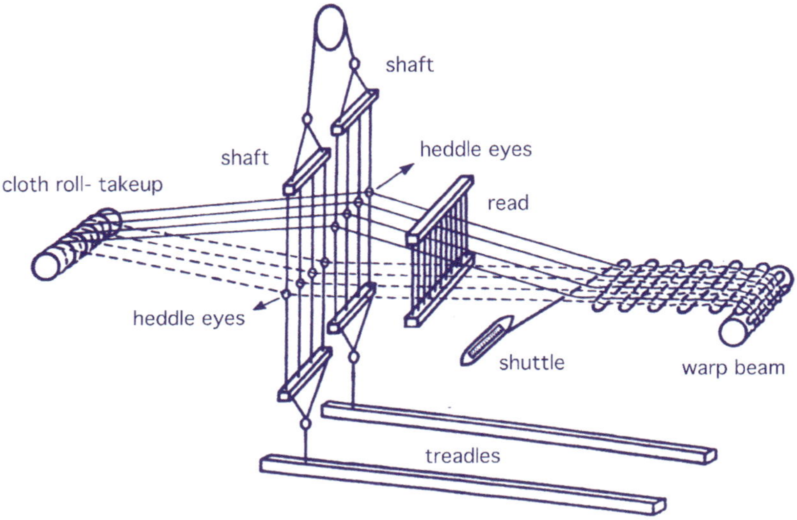

FIG. 5: Shaft loom with two shafts and two treadles (schematic illustration). Anna Döpfner, *Bindungen: Flechten und Weben* (Berlin: Museumspädagogischer Dienst, 1993), 11. Stiftung Deutsches Technikmuseum Berlin, Historisches Archiv.

subsequent row of the fabric, raising the next group of warp threads. The weaver throws the shuttle again, returning it to its initial position. The reed battens each row of the fabric to the completed fabric.

Depending on the weave to be created, different sets of warp threads are grouped together on several shafts. Two shafts (as shown in fig. 5) are sufficient for the plain or linen weave; all other weaves require more than two shafts. Threads may then be divided into a correspondingly larger number of groups and raised by operating the appropriate treadle. For instance, in order to produce a four-weave twill (the weave in denim, among other fabrics) with a "step" or offset between rows, the weaver counts off the warp ends from one to four and threads them into the first, second, third, and fourth shafts accordingly. Operating the treadles one after the other before beginning afresh creates the characteristic diagonal

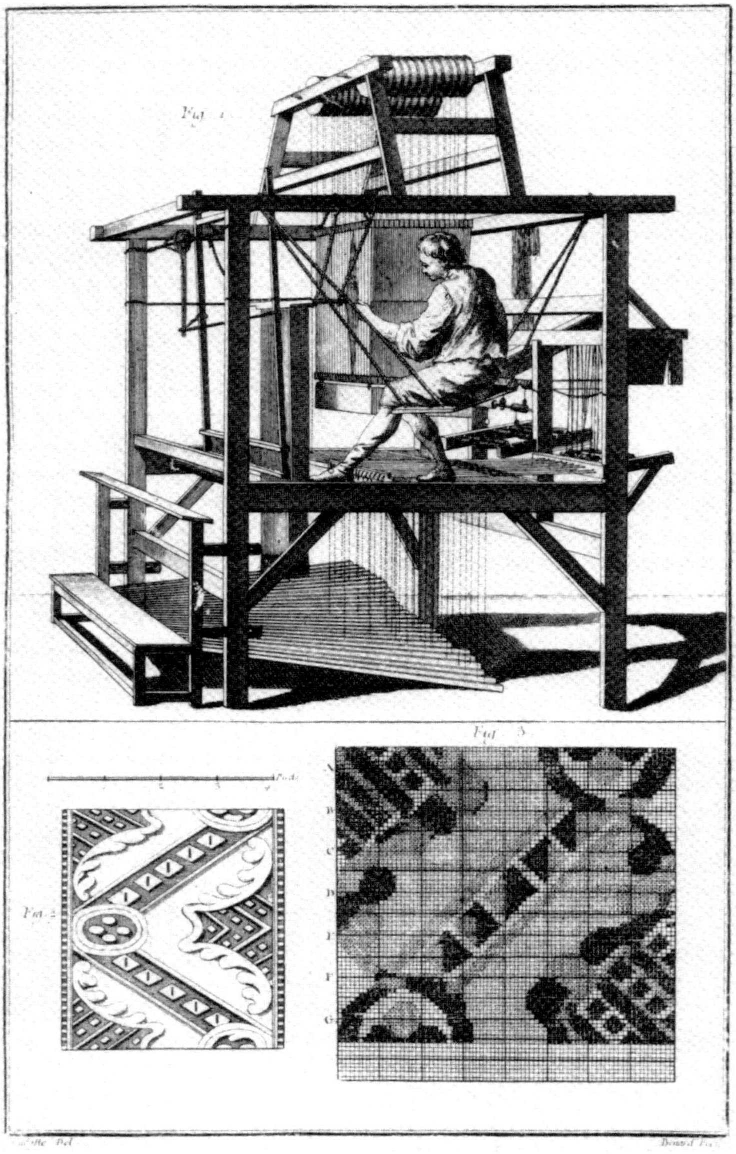

FIG. 6: Treadle loom for ribbon weaving from Diderot's *Encyclopédie* (Paris, 1751–1780). Denis Diderot and Jean-Baptiste le Rond d'Alembert, *Recueil de planches, sur les sciences, les arts libéraux, et les arts méchaniques, avec leur explication, L'art de la soie* (Paris, 1751–1780), *Passementerie*, plate vii.

ribbing. The possible combinations of a mere four shafts already allow for much longer patterns than groups of four picks; threading the warp into a larger number of shafts makes even more complex patterns possible. The pattern-weavers of Ulm used looms ranging from eight to more than thirty shafts similar to the shaft loom for *passementerie* weaving shown in an illustration in Diderot and d'Alembert's *Encylopédie* (fig. 6), taking the principle of shaft-weaving to the limits imposed by the need to arrange the shafts and treadles in the loom's legroom.

Marx Ziegler's System of Notation Compared to Contemporary Practice

The treadles, shafts, and warp threads of the loom return in abstract form in the notation on paper. The parallel arrangement of systems of threads and shafts in the device appears as orthogonal systems of lines forming rows and columns. By adapting the number of lines in the notations to the desired number of threads and shafts, the craftsman-artist can draft different patterns, not unlike the composer working on music paper, and plan their technical realization. To this end, the varying patterns are inscribed upon the scaffold of lines as series of strokes or dots, indicating to the weaver where he needs to tie knots and in what order he will have to operate the treadles.

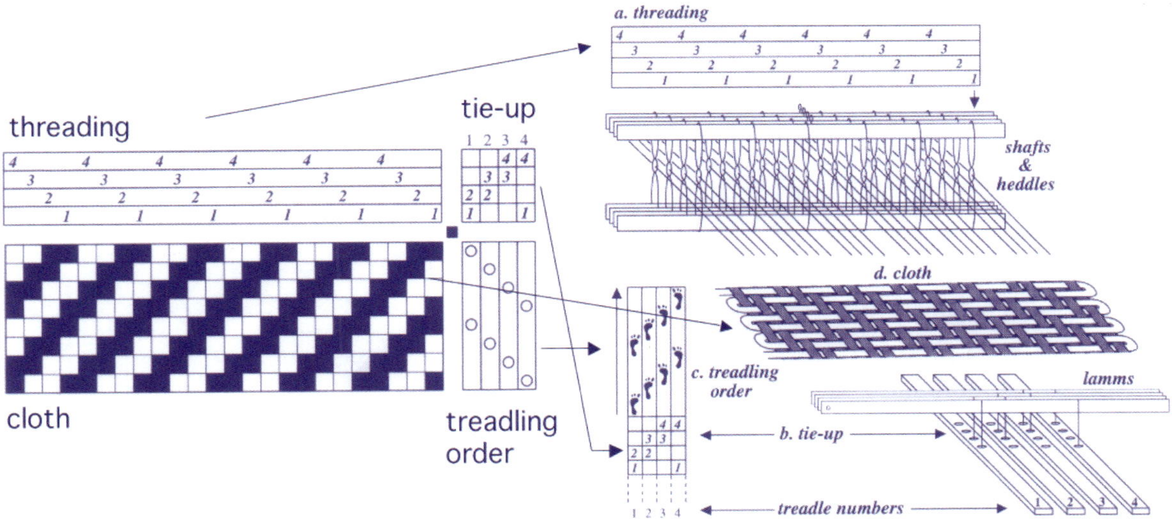

FIG. 7: Schematic illustration of the modern form of pattern notation in relation to the parts of the shaft loom. Madelyn van der Hoogt, *The Complete Book of Drafting for Handweavers* [1993] (Petaluma, CA: Unicorn, 2000), collage of ills., 4, 5.

The specific quality of the historical forms of notation is best understood by comparing them to today's prevailing convention, the draft.[11] Figure 7 illustrates the basic weave pattern for twill as it appears in a modern textbook for manual weaving on a shaft loom. The arrows connect the schemata to the components of the loom they control. The structure of staggered squares shows the resulting weave as a pattern of black and white boxes. The modern *drawdown* here represents the order in which the threads interlace in the fabric: a black square tells the weaver that the warp thread passes over the weft thread; a white square, that the warp passes beneath the weft.[12] Three additional schematic representations accompany the drawdown. Along its upper edge runs the *threading*, which specifies the order in which the warp ends are threaded through

11 See Madelyn van der Hoogt, *The Complete Book of Drafting for Handweavers* [1993] (Petaluma, CA: Unicorn, 2000).

12 For some fabrics, a black square indicates that the warp thread passes *beneath* the weft thread.

heddles on specific shafts. The bottom row represents the first shaft from the weaver's perspective. As mentioned above, twill requires four shafts, and the warp ends are threaded "straight through," or through the first, second, third, and fourth shafts in that order. Next to the right edge of the drawdown, moreover, a four-column diagram registers the *treadling order*. The curls indicate the order in which the treadles of the loom must be depressed. Read from above, this particular diagram denotes a sequence starting with the leftmost pedal. And finally, the rows of the threading cross the columns of the treadling order in the top right corner of the diagram, forming a structure of boxes called the *tie-up* that indicates which treadles must be tied to which shafts. Besides the drawdown, then, the draft notates three additional orders: the threading, the treadling sequence, and the tie-up.

The comparison to the modern notation reveals that the structure of little squares in Ziegler's book (fig. 3) does not show the resulting pattern as it will appear in the fabric; it is not a drawdown in the modern sense. What Ziegler notates here is the tie-up of treadles and shafts, which is to say, the order in which the treadles must be tied to the shafts for a specific pattern. A black square accordingly indicates that a particular shaft must be tied to a particular treadle. The plate shows altogether nine such tie-up schemes, delimited by the larger structure of strong black lines.

The pattern of dashes in a grid that recalls sheet music (fig. 2), meanwhile, turns out to denote a threading. Here, too, the horizontal lines represent the staggered shafts as the weaver sees them, front to back. Yet where numerals or black squares appear in today's notations, Ziegler's plates show zigzagging dashes. To understand how exactly their trajectories are to be read, we must first realize that they constitute a compressed form of notation: the spaces between the lines of the scaffold each stand not for a single shaft but for a group of shafts, so three lines may be taken to represent eight, ten, twelve, or even more shafts. The advantage of this notation based on series of dashes is that patterns may be adapted to the desired number of shafts, allowing for easy variation also of their size.[13] This abridged representation leads to the form of the ascending and descending slanted dashes, "/" and " \". Like the numerals or black squares of contemporary threading notations, the zigzags indicate which warp end must be threaded through which heddle eye on the shaft for the latter to raise the warp thread that generates the pattern. The little curls and black circles attached to the ends of each series of dashes make the diagram easier to read by marking the shaft with which the threading through the heddles begins.[14] As with the notation of the tie-up, a single plate contains the threading for several patterns. Each vertical bar marks the beginning of a new threading, or in other words, a different pattern. Plate 20 of Ziegler's book, that is to say, represents a series of twenty threadings, allowing the weaver to create a commensurate variety of patterns.

The experienced reader can tell the pattern symmetries that will be produced by weaving merely from looking at the notation. Just as sheet music conveys to the musician a sense of a melody's trajectory, this score gives the weaver an impression of the final fabric. Depending on whether the series of dashes appear to be composed of "*M*"s or "*W*"s or form rows of parallel lines, fundamentally different kinds of pattern symmetries result. The arrangement of the dashes also gave rise to the historical terms used

13 To weave the first pattern in figure 2 on an eight-shaft loom, the weaver would accordingly thread the warp ends through heddles attached to shafts 1...8...1...4...1..., etc. A threading noted on a scaffold of four horizontal lines such as those in the bottom system of figure 2, on the other hand, would be applied to a number of shafts divisible by three; for example, eighteen shafts would be grouped as 1–6, 7–12, 13–18.

14 In block-pattern weaving, which would be developed a little later, the compressed form of weave-pattern notation would become even more crucial for pattern composition. This method is explained in Nathaniel Lumscher's weaving handbook; see Lumscher, "Neu eingerichtetes Weber Kunst und Bild Buch."

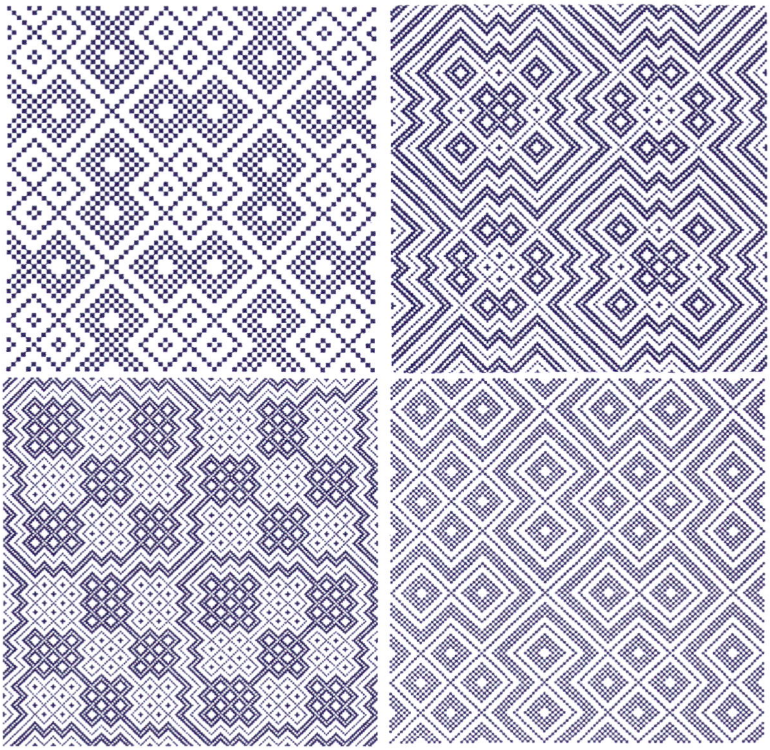

FIG. 8: Four possible combinations of the notations for threadings and tie-ups (see figs. 2 and 3) from Ziegler's pattern book (1677). Top to bottom: threading 1, tie-up 1; threading 2, tie-up 2; threading 3, tie-up 2; threading 4, tie-up 1 (computer-generated images). Marx Ziegler, *Weber Kunst und Bild Buch* (Ulm, 1677).

to designate these patterns: in the instance of pictorial star- and flower-shaped symmetries, weavers spoke of *Hin-und-Wieder-Arbeit*, or back-and-forth patterns ("*M*," "*W*"); the punctuated series of slanted dashes produced patterns called *gebrochen*, or broken.[15] The dash series discussed here represent broken patterns such as those that appear in figure 8.

At first glance, the historical formats used to record weave drafts seem to bear hardly any resemblance to the modern notation. Still, all weaving notations are recognizably based on the raster with prominent lines and/or columns as the central graphical method. The lines and columns may represent two structures: they refer either to the orthogonal order of the threads in the fabric or to the equally orthogonal order of the parts that control the loom, which is to say, to the system of treadles and shafts. The notations interlock the structures of the loom and the fabric. The transfer from the structure and function of the machine—the loom—to the structure of the notation is immediately comprehensible in the weaver's technical score. The graphical record of the pattern preparing its implementation on the loom illustrates how the score abstracts, layer by layer, from the loom and its operation.

15 The tie-ups specifying how the treadles and shafts are connected for *gebrochene* patterns consist exclusively of parallel diagonal lines running at a 45-degree angle; the diagrams are symmetrical to the central diagonal. The tie-ups for pictorial patterns, by contrast, are made up of intersecting diagonals or orthogonal lines. See Hilts, "Translator's Introduction," in Ziegler, "Weber Kunst und Bild Buch," 32–34.

Operationalizing the Graphical Recording for the Generation of New Patterns

In the weaver's practice, the notation fulfilled several functions that would have been inconceivable without graphical representations of the patterns. On the one hand, it met the requirements for a cursive script. The system was easy to learn and served as a sort of shorthand that compressed the process of planning a weave: putting pen to paper, the weaver could rapidly take down a pattern. Quickly written down, the resulting notations were also convenient to the reader, offering him a lucid representation of complex patterns and allowing him to understand the connections between the weave and the operation of the loom at a glance. The arrangement of series of dashes of unambiguous signification in a system of lines and columns in the Ulm notation combines two possible ways of reading: on the one hand, it renders the pattern intuitively comprehensible and allows the weaver to count its features as though on an arithmetic board. On the other hand, the unambiguous determinacy of the notations renders them usable as a storage medium. In this function, they provide the weaver with unequivocal instructions for the correct tie-up of treadles and shafts, guaranteeing his ability to reproduce a pattern once it has been notated—as a storage medium, the notations present a blueprint for the fabric by guiding the weaver as he implements the pattern on the machine.

Beyond this storage function, however, the notation offered another fundamental advantage that could not have been attained without a method of recording weave patterns. Ziegler's decision to give combined representations of multiple threadings and notate multiple tie-ups in compound diagrams illustrates that the intended use of the book went far beyond presenting the diagrams as a mere storage medium. By freely combining the notations of threadings and tie-ups, the weavers turned the plates into an instrument for the generation of new pattern variants. In our example, all twenty threadings may be combined with any of the nine tie-ups—the weaver merely needed to change the tie-up of the treadles to the shafts in accordance with figure 3 or thread the warp ends through the heddles of the sixteen shafts in a new order. Retying a few knots thus allowed the weaver to generate no less than 180 different weave patterns. Four possible patterns that result from this combinatorics are shown in figure 8.

The characteristic fabrics produced by so-called reverse-twill and block damask weaving as practiced in Ulm, which sometimes recall the look of op art, show how far the pattern weavers of southern Germany took the generative dimension of their notations. By operating with the compressed notations of geometric patterns, they were able to draft new designs in the abstract based purely on the notation, arriving at compositions that would have been out of reach if they had relied on drawdowns alone. Small changes to the tie-up and/or the treadling order produced an almost infinite number of variations on the level of geometric patterning. Because the various orders involved were each regular, their combination generated regular patterns. Reverse-twill and block damask weaving, that is to say, were weaving techniques in which patterns were drafted based on the possibility of combining mechanical orders; the capacity of the loom itself, in other words, was the primary source of creative invention. This combinatorics reveals the potential implicit in the notations: as written recordings of a fabric, they could serve as an instrument for the systematic creation of new patterns. The operative writing of the notation enabled weavers to use the structural linkage between loom and pattern, an implicit feature of loom weaving, as a systematic technique of pattern generation.

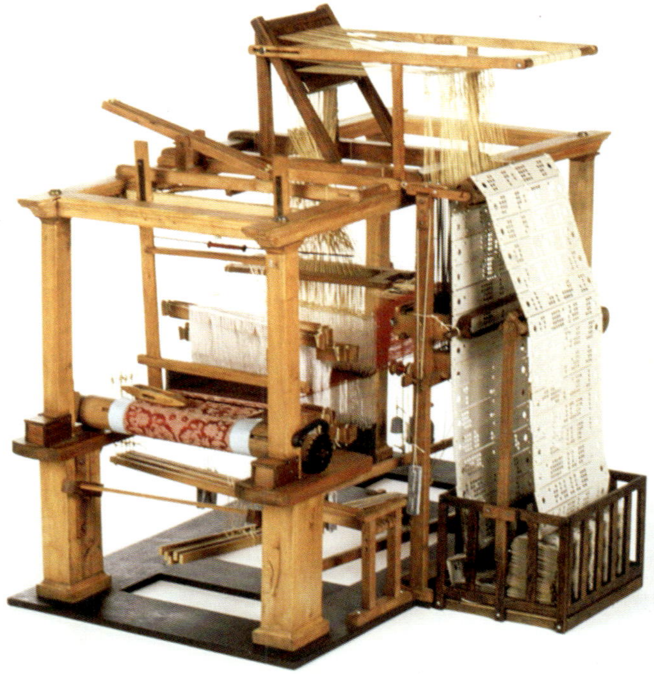

FIG. 9: Miniature model (1855) of Jean-Baptiste Falcon's punched-card loom, Lyons, ca. 1739. Musée des Tissus de Lyon, photographie D.R.

Pictorial Codes and Punched-Card Control

Considered in the horizon of a history of the technical and machine-generated image, weaving emerges as the field in which the earliest pictorial programs or codes were established. What rendered the execution of such a pictorial program unequivocal was the confluence of several factors that are specific to weaving: the use of "digital" notational systems, in Goodman's sense of the term, went hand in hand with the "digital" structure of weaving. No abstraction or reduction to a digital schema was required; the technical conditions for the reduction of semantic operations to purely syntactic ones were met thanks to the very structure of the fabric.

Punched-card controlled pattern looms such as those developed in France starting about 1725 amalgamated the notation on paper and the loom into a structural unit: the paper, bearing the score as a pattern of perforations, became the control module (fig. 9).[16] In combination with the scanning mechanism designed to read the card, the score no longer merely provided the code to be implemented by a weaver but also contained a program the machine itself was capable of interpreting.[17] Yet the punched card was no more than another form of pictorial notation, rendering the world of symbols compatible with that of the technical equipment. The early forms of pictorial coding in weaving are thus also an element in the historiography of symbolic machines.[18]

16 For a critical history of punched-card technology in eighteenth-century weaving leading up to Joseph-Marie Jacquard, see Schneider, *Textiles Prozessieren*, 125–66, 262–307.

17 In the 1830s, Charles Babbage emulated this principle as he conceived his analytical engine, the first mechanical calculator based on punched cards. See Bernhard Dotzler, "Technotation: Babbage und die Macht der Zeichen," *Weimarer Beiträge*, vol. 43 (1997): 99–109.

18 Sybille Krämer, *Symbolische Maschinen: Die Idee der Formalisierung in geschichtlichem Abriß* (Darmstadt: Wissenschaftliche Buchgesellschaft, 1988).

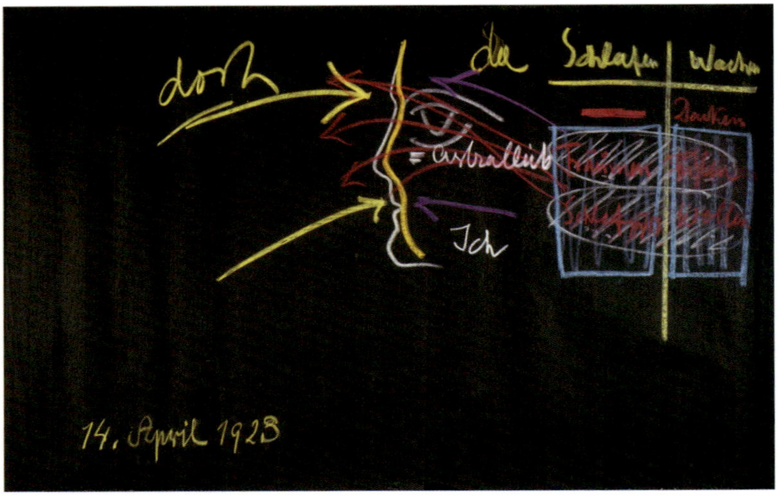

FIG. 1 (above): A blackboard drawing (1923) in which Rudolf Steiner sought to represent the states of waking and sleeping. The drawings have survived because Steiner subsequently covered the blackboard with black paper. These sketches combine letters and colors with linear forms such as loops, circles, arrows, and spirals. Steiner, a highly influential architect whose writings launched the anthroposophical movement, regarded his cosmological drawings as a gestural language designed to accompany speech. © Estate of Rudolf Steiner, Dornach, Switzerland.

FIG. 2 (below): TreeVisualizer, a program for visualizing genetic development trees. The different colors of the branches stand for Boolean operators (green), relational operators (blue), and numerical operators (red). http://www.geneffects.com/genetic-programming/ (accessed November 2013).

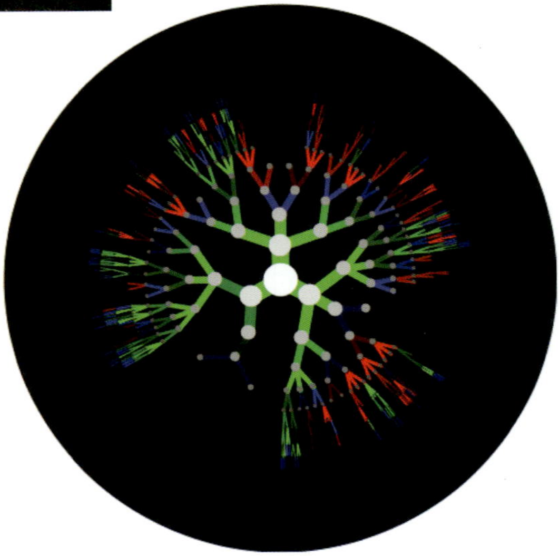

DIAGRAMMATICS

Diagrams have clear purposes: they explain and elucidate, illustrate and demonstrate, provide a concise overview, correlate information, and order content. They are often preceded by series of measurements or collections of data, which are visualized using graphical methods. Yet diagrams can also visualize abstract ideas, concepts, and thoughts on paper (fig. 1) or concrete things such as the structures of machines or architecture. The resulting knowledge provided in a diagram derives from a process of abstraction while at the same time rendering concrete some aspects of the object represented. Diagram types such as tree diagrams (fig. 2), network diagrams, and bar graphs are the result of a history of standardization and conventionalization and may fairly claim to be unambiguously legible.

The ancient core meaning of *diagram* alludes to a pragmatic context. The Greek word *diagramma* signified a "geometric figure" or "outline"; *graphein* meant "to record," "inscribe," or "carve." The word *diagram*, then, designates the characteristic intrinsic relation between writing and drawing. In the *Meno*, Plato elucidates the diagrammatic method, used, for example, to double the area of a square by geometrical means, which is to say, by drawing. The practical applications of this method lay in land surveying and astronomy, which produced land maps, architectural plans, and star charts.

Because of the high degree of conventionalization of diagrams and the way they interlaced writing and drawing, they were regarded as the rational part of the world of images and became the subject of semiotic studies. To the semiotic theorist Charles S. Peirce, diagrammatic practices were rooted in reasoning because, he believed, all reasoning was diagrammatic (Peirce 1906). He was interested in the power of diagrams to function as media

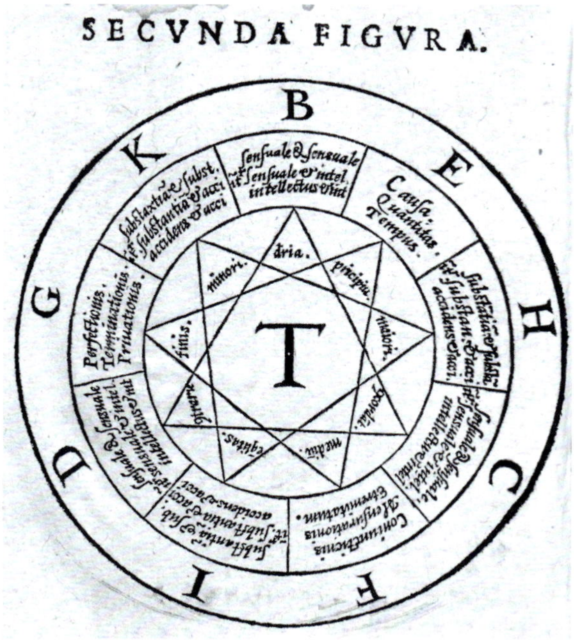

FIG. 3: Raimundus Lullus, circular diagram from *Ars Brevi* (1578). Lullus, a Catalan philosopher and theologian, was influenced by Arabic, Jewish, and Christian culture. *Ars Magna* was his term for the art of distinguishing truth from untruth in a logical way by means of a mechanical combination of terms. To this end, he used circular diagrams whose concentric discs could be rotated in opposing directions to resolve theological questions. Raimundus Lullus, *Ars Brevi*, ed. Bernardus de Lavinheta (Paris: Gilles Gourbin, 1578). Staatsbibliothek zu Berlin–Preußischer Kulturbesitz, Abteilung Historische Drucke, an: Nb 280 (R).

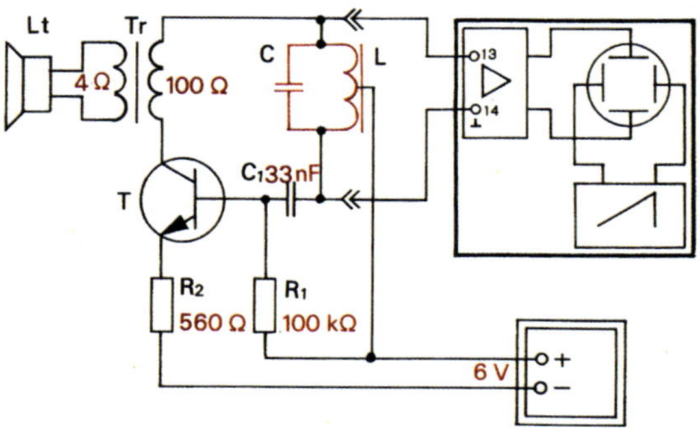

FIG. 4: Circuit diagram for a tone generator. Lothar König, *Rundfundtechnik selbst erlebt, Bauanleitungen und Experimente zum amplitudenmodulierten Hörfunk* (Leipzig: Urania-Verlag, 1988), 108.

of thought. Their main feature, according to Peirce, is to express relations. They resemble their objects due less to a representational relationship than to an abstract similarity to the object represented in the overall relationship of its parts. Peirce was therefore able to classify diagrams as iconic signs. In his observations, Peirce primarily focused on diagrams that may be described as "logical pictures," to use Wittgenstein's term: graphs, that is to say, geometrical and topological representations of relations between objects in which logical reasoning is represented. Syllogistic circular diagrams (fig. 3) as well as circuit diagrams (fig. 4) and the programming of diagrams in computers function in this way. Such forms of images can be used to quasi-mechanically produce "right" or "wrong" results.

The philosopher Nelson Goodman also discusses diagrams in his wider theory of symbols (Goodman 1968). He considers them examples of a theory of notation because they are structured in a way that ensures unambiguity. Goodman calls such diagrams "digital," so the bar graph shown in figure 5 and Playfair's import and export curves (fig. 6) would be digital in this sense, because both visualize a series of individual figures.

The cartographer Jacques Bertin likewise regarded diagrams as a rational domain of imagery. In his monumental *Semiology of Graphics* (1967), he analyzed maps, networks, and diagrams as "graphic representation systems" with the goal of providing a clear aid for the production of graphics that would be as "concise" as possible. Bertin measures concision as many other authors do, by the speed with which the eye comprehends a graphical statement. The optimum would be "comprehension at a glance."

The anthropologist and sociologist of science Bruno

FIG. 5: International comparison of national debts using the graphic method of bar chart diagrams. Left: per capita in German marks; center: total amounts in millions of German marks; right: annual debt service expenditures (principal and interest) in millions of German marks. Anton L. Hickmann, *Hickmanns Geographisch-Statistischer Universal-Taschen-atlas* (Wien/Leipzig: Freytag & Berndt, 1912), 55–56.

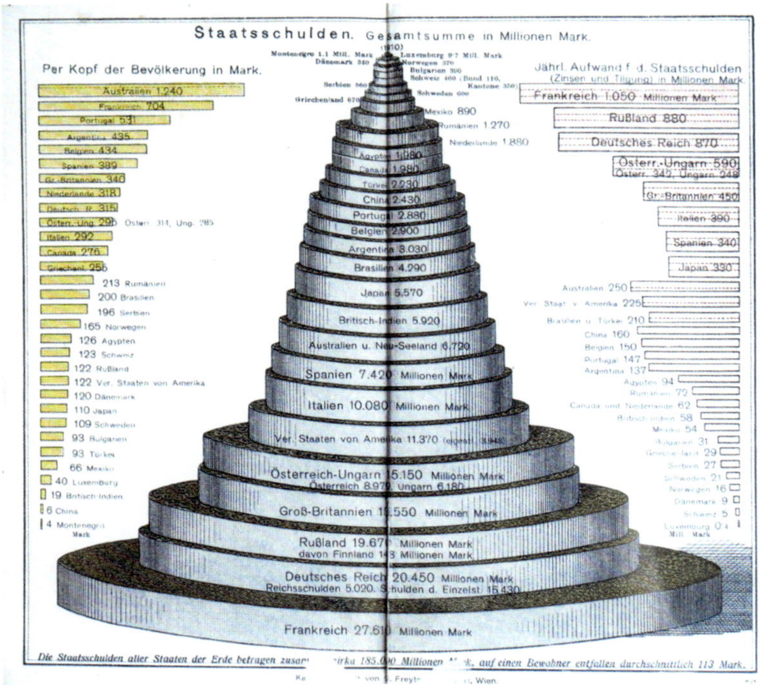

FIG. 6: Trade between England and Denmark/Norway between 1700 and 1780. William Playfair was the first to publish such charts on economic issues. In this correlation of two curves, he expresses an economic rule that exports should always exceed imports to ensure a country's prosperity. England met this condition from the 1750s on. William Playfair, *The Commercial and Political Atlas: Representing, by Means of Stained Copper-Plate Charts, the Exports, Imports, and General Trade of England* (London: Printed for J. Debrett, 1786).

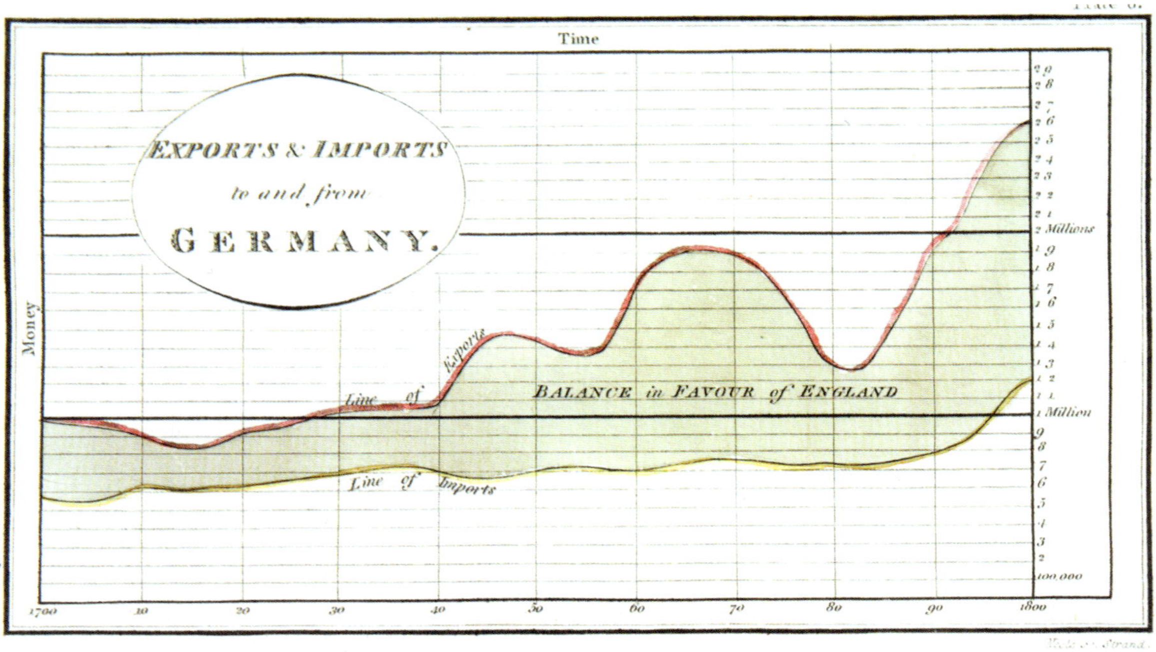

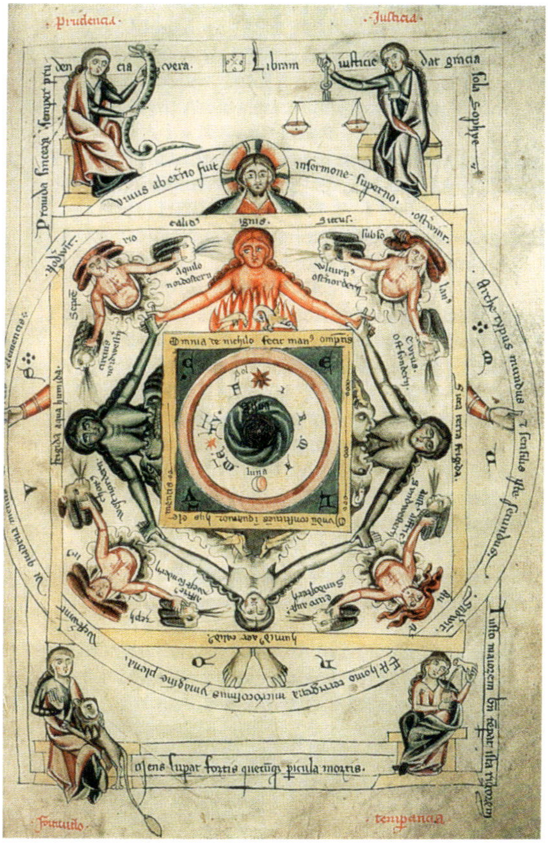

FIG. 7: Cosmological diagram showing the interaction of the elements, from a collective manuscript by Thomas of Cantimpré and William of Conches, late thirteenth century. The diagram shows as many of the relationships resulting from the combination of the basic characteristics as possible, with the diagrammatic structures—the nested squares and circles at the center—enriched by anthropomorphous figures and personifications, some of them Christian in origin. Alexander Patschovsky, *Die Bildwelt der Diagramme Joachims von Fiore: Zur Medialität religiös-politischer Programme im Mittelalter* (Ostfildern: Thorbecke, 2003), 216. Bayerische Staatsbibliothek München, Clm 2655, fol. 105r.

Latour takes an approach that diverges from a purely semiotic perspective. In his investigation of a field study in the Amazon jungle by an interdisciplinary team of researchers, he describes the development of a diagram between the already mentioned poles of abstraction and concretion as a highly artificial form of knowledge generation. The crucial issue is the concept of representation, which Latour dismantles layer by layer and replaces with the concept of "circulating reference." While the field study begins with the jungle's rough ground and wild vegetation, the diagram marks the end point of an incremental systematization and production of knowledge. The diagram appears as a result of ritualized practices of indexing, screening, and categorization, which are already inherent in operations such as the stretching of lines, the use of rulers, the creation of color tables, and the numbering of soil samples (Latour 1999).

Felix Thürlemann and Steffen Bogen's work continues the tradition of art-historical studies of diagrams; following Peirce, they propose the concept of "diagrammatics" (Bogen and Thürlemann 2003). They regard diagrams as a third genre whose properties exceed the text-image binomial and cannot adequately be defined as hybrid forms of image and text. They take a phenomenological approach in their analysis of diagrams. The characteristics specific to diagrams, they argue, include the pragmatic power to parallel categories of form and content by means of graphic relations, the dichotomic configuration of signs, and the structural significance of their topological arrangement. By proposing to investigate the general "diagrammatic features" of images rather than devising an ontology of the diagram, they are able to analyze not only definitive forms of diagrams such as infographics and medieval representations of the universe but also computer program interfaces or the arrangement of symbols on medieval paintings. Such an approach assumes that the diagrammatic parts of images are logically relevant structures of order that can be separated from the rest of the image like a skeletal framework. Dispensing with the strict categorization of a diagram as "image" or "writing" allows the field of diagrams to remain open at its margins. This is an approach that offers major advantages for an art-historical and empirical perspective on images, because it also includes elements of diagrams beyond the strict scope of semiotics such as allegories, ornamentation, or even pictorial motifs (fig. 7). —BS

LITERATURE

Auerbach, Felix. *Die graphische Darstellung.* Leipzig: Teubner, 1914.

Bauer, Matthias, and Christoph Ernst. *Diagrammatik: Einführung in ein kultur- und medienwissenschaftliches Forschungsfeld.* Bielefeld: Transcript, 2010.

Bender, John B., and Michael Marrinan. *The Culture of Diagram.* Palo Alto, CA: Stanford University Press, 2010.

Bertin, Jacques. *Semiology of Graphics: Diagrams, Networks, Maps.* Redlands, CA: Esri Press, 2011 [1967].

Blackwell, Alan F. *Thinking with Diagrams.* Dordrecht: Springer, 2001.

Bogen, Steffen, and Felix Thürlemann. "Jenseits der Opposition von Text und Bild: Überlegungen zu einer Theorie des Diagramms und des Diagrammatischen." In *Die Bildwelt der Diagramme Joachims von Fiore: Zur Medialität religiös-politischer Programme im Mittelalter,* ed. Alexander Patschovsky, 1–22. Ostfildern: Thorbecke, 2003.

Brasseur, Lee E. *Visualizing Technical Information: A Cultural Critique.* Amityville, NY: Baywood, 2003.

Brinton, Willard C. *Graphic Methods for Presenting Facts.* New York: The Engineering Magazine Company, 1914.

Evans, Michael. "The Geometry of the Mind." *Architectural Association Quarterly* 12, no. 4 (1980): 32–55.

Funkhouser, H. Gray. "Historical Development of the Graphical Representation of Statistical Data." *Osiris,* no. 3 (1937): 269–404.

Galison, Peter. "Feynman's War: Modelling Weapons, Modelling Nature." *Studies in History and Philosophy of Science Part B: Studies in History and Philosophy of Modern Physics* 29, no. 3 (1998): 391–434.

Goodman, Nelson. *Languages of Art: An Approach to a Theory of Symbols.* Indianapolis: Bobbs-Merrill, 1968.

Grafton, Anthony, and Daniel Rosenberg. *Cartographies of Time: A History of the Timeline.* New York: Princeton Architectural Press, 2009.

Gugerli, David, and Barbara Orland, eds. *Ganz normale Bilder: Historische Beiträge zur visuellen Herstellung von Selbstverständlichkeit.* Zurich: Chronos, 2002.

Hall, Bert S. "The Didactic and the Elegant: Some Thoughts on Scientific and Technological Illustration in the Middle Ages and the Renaissance." In *Picturing Knowledge: Historical and Philosophical Problems Concerning the Use of Art in Science,* edited by Brian S. Baigrie. Buffalo, NY : University of Toronto Press, 1996.

Ingold, Tim. *Lines: A Brief History.* London: Routledge, 2007.

Latour, Bruno. "Circulating Reference: Sampling the Soil in the Amazon Forest." In *Pandora's Hope: Essays on the Reality of Science Studies,* 24–79. Cambridge, MA: Harvard University Press, 1999.

Lenk, Krzysztof, and Paul Kahn. "To Show and Explain: The Information Graphics of Stevin and Comenius." *Visible Language* 26, nos. 3–4 (1992): 272–281.

Mitchell, W. J. T. "Spatial Form in Literature: Toward a General Theory." *Critical Inquiry* 6, no. 3 (1980): 539–67.

Peirce, Charles Sanders. "Prolegomena for an Apology to Pragmatism." In *New Elements of Mathematics: Vol. IV,* 313–30. The Hague: Mouton, 1976 [1906].

Peirce, Charles Sanders, and Patricia Ann Turrisi. *Pragmatism as a Principle and Method of Right Thinking: The 1903 Harvard Lectures on Pragmatism.* Albany: State University of New York Press, 1997.

Poggenpohl, Sharon H., and Dietmar R. Winkler, eds. "Special Issue: Diagrams as Tools for Worldmaking." *Visible Language* 26, nos. 3–4 (Summer–Fall 1992): 250–473.

Schmidt-Burkhardt, Astrit. *Maciunas' "Learning Machines": From Art History to a Chronology of Fluxus.* Berlin: Vice Versa, 2003.

Schmidt-Burkhardt, Astrit. *Stammbäume der Kunst: Zur Genealogie der Avantgarde.* Berlin: Akademie Verlag, 2005.

Schneider, Birgit, ed. *Diagramme und bildtextile Ordnungen. Bildwelten des Wissens: Kunsthistorisches Jahrbuch für Bildkritik,* ed. Horst Bredekamp and Gabriele Werner, vol. 3, no. 1. Berlin: Akademie Verlag, 2005.

Stjernfelt, Frederik. *Diagrammatology: An Investigation on the Borderlines of Phenomenology, Ontology, and Semiotics.* Dordrecht: Springer, 2007.

Tufte, Edward R. *Envisioning Information.* Cheshire, CT: Graphics Press, 1995.

Tufte, Edward R. *The Visual Display of Quantitative Information.* Cheshire, CT: Graphics Press, 1997.

Tufte, Edward R. *Visual Explanations: Images and Quantities, Evidence and Narrative.* Cheshire, CT: Graphics Press, 1997.

Early Modern Images of Musical Automata: On Athanasius Kircher's Trompe-l'Oreille Contemplations in the Quirinal Gardens in Rome

Angela Mayer-Deutsch

The following pages focus on several trompe-l'oreille pictures of musical automata in Athanasius Kircher's Musurgia Universalis *(1650), which hover between playful contemplation and the suggestion that what they depict may be realized. The spectrum ranges from the Aeolian harp—which Kircher was probably the first to design[1]—to an automated hydraulic organ. The discussion will examine the overall role images play as contemplative depictions of possible objects as well as representations of princely sovereignty: a "music for potentates," as we might say with Salomon de Caus, a constructor of automata who called pictures of machinery "mathematics for potentates."[2]*

The existing scholarship has largely focused on the question of whether the machines shown in these depictions might actually be built, blinding itself to crucial aspects of imaginative and contemplative seeing. Kircher, a Jesuit, quite consciously played with the ambivalent status of the (phonic) automaton, which goes back to antiquity, leaving the question of whether a device is real or fictional unanswered for many of the machines represented in the text and illustrations of his *Musurgia*; this strategy allowed him to emphasize the crucial contemplative aspect of his "Jesuitical" hypothetical and probabilistic science. Just as anamorphosis opened the door to the idea of different, non-Euclidean spaces, Kircher's images, many of which he simply copied or altered only slightly, may have opened up a space of possibility precisely thanks to their fictional elements, a space that resembled those of vision and contemplation. Given the fundamentally affirmative nature of images, this space is non-hierarchical and equivocal, yet as it straddles the boundary between telling and showing, it is also quite captivating. In the present context, *trompe-l'oreille* refers primarily to the idea, based on the principle of the *deus ex machina*, that unexpected sounds are attributed to a divine or miraculous agency. Time and again, images publicize this supernatural effect and at once reveal its being mechanically produced.

The trope of the organ[3] presents a particularly vivid example of the specific role that images played in the Jesuits' natural philosophy, which hewed to a metaphysical conception of the image.[4] The queen of instruments embodies the synthesis of sounds and worlds, the divine creation, and the principle of the *deus ex machina*. Initially a hydraulic and fully automated instrument, the organ was developed in the ancient, Byzantine, and Arab worlds; semi-automated organs gave way to today's fully

1 See the two woodcuts showing the instrument in Athanasius Kircher, *Musurgia Universalis, sive Ars Magna consoni et dissoni in X libros digesta [...]*, 2 vols. (Rome: Heirs of Francesco Corbelletti, 1650), repr. ed. Ulf Scharlau, *Musurgia Universalis*, vols. A and B (Hildesheim: Olms, 2004), vol. B, 352–53.

2 Salomon de Caus, preface to *Von Gewaltsamen bewegungen: Beschreibung etlicher, so wol nützlichen alß lustigen Machiner [...]*, 3 vols. (Frankfurt am Main: Abraham Pacquart, 1615), repr. *Kunst der Mechanik: Die phantastischen Erfindungen des Salomon de Caus*, ed. Stiftung Kloster Michaelstein (Halle an der Saale: Stekovics, 2003).

3 On the points discussed in the following, see also Reinhold Hammerstein, *Macht und Klang: Tönende Automaten als Realität und Fiktion in der alten und mittelalterlichen Welt* (Bern: Francke, 1986).

4 There is no Jesuitical natural philosophy as a clearly defined concept any more than there is a Jesuitical style, although scholars have repeatedly sought to define the latter. The author nonetheless concurs with Evonne Levy that there is something specifically Jesuitical about the emphasis on formation and the probabilistic aspect that informs the Jesuits' contributions to the various disciplines; see Evonne Levy, *Propaganda and the Jesuit Baroque* (Berkeley: University of California Press, 2004); Ilse von zur Mühlen, "Imaginibus honos—Ehre sei dem Bild: Die Jesuiten und die Bilderfrage," in *Rom in Bayern: Kunst und Spiritualität der ersten Jesuiten*, ed. Reinhold Baumstark (Munich: Hirmer, 1997), 161–70.

player-controlled instrument. It is little wonder, then, that the organ invited the aggregation of other sonic automata as well. Even today, the church organ, resounding as though of its own volition, its notes reaching the listener from on high, retains vestiges of the old automaton's semantics: it is heard as a likeness of cosmic harmonies or of the divine as such.

A School of Musical Vision

Kircher's *Musurgia* may be read as a picture book, a school of vision as Steffen Bogen has defined the term: the imagery conveys diagrammatic as well as eidetic content. According to Bogen, eidetic seeing, or the perception of pictures, associates a material form with a mental image, whereas the diagrammatic principle asks the beholder to read a graphical relation by understanding its construction and applying it to a relation between reference quantities.[5]

5 Steffen Bogen, "Algebraische Notation und Maschinenbild: Eine Rechenmaschine avant la lettre," in *Visuelle Argumentationen: Die Mysterien der Repräsentation und die Berechenbarkeit der Welt*, ed. Horst Bredekamp and Pablo Schneider (Munich: Fink, 2006), 189.

Printed in Rome in an edition of 1,500 copies, the lavishly produced two-volume book running to over a thousand pages united didactic-practical purposes with a theoretical exposition of the foundations of music in the widest sense of the term; its publication would have been impossible without the patronage of Emperor Ferdinand III and his brother, Archduke Leopold. Kircher himself made sure the work reached a virtually global audience the same year it was printed by having three hundred copies donated to the many members of his order who came to Rome from the provinces and missions to elect a new Superior General.

Book 9 of the second volume contains a section on musical automata; the concluding book 10 represents an essay on the world organ, a symbol of the principle of creation, the cosmic world-machine. Running to around sixty pages, the fifth section of the experimental book 9 treats

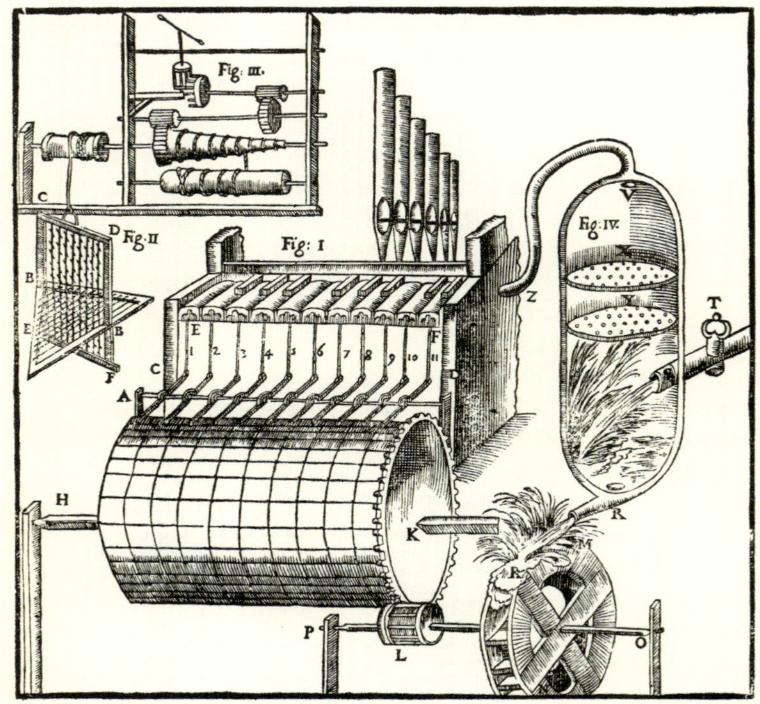

FIG. 1: A hydraulic organ and Robert Fludd's "barbiton." Woodcut from Athanasius Kircher, *Musurgia Universalis, sive Ars Magna consoni et dissoni in X libros digesta [. . .]*, vol. 2 (Rome: Corbelletti, 1650; reprint edited by Ulf Scharlau (Hildesheim: Olms, 1970), 334. © Gottfried Wilhelm Leibniz Bibliothek – Niedersächsische Landesbibliothek Hannover, 71/8521.

various organs and musical automata. Kircher sets out confidently with instructions of a sort on how to build a so-called Aeolian chamber,[6] a device used to generate compressed air, as well as a barrel to produce the sounds. He then introduces eleven different *machinamenta* based on this technology, including Vitruvius' hydraulic organ, Robert Fludd's "barbiton" (an automated stringed instrument; see fig. 1, ii and iii), carillons, the hydraulic organ in Rome on which the following discussion will focus (the two illustrations are accompanied by extensive sample scores), the Aeolian harp, and other pneumatic instruments. There are eight detailed engravings, most of which take up full pages, illustrating the organ—six show the hydraulic organ in particular—as well as numerous woodcuts and sample scores and five developed views of pinned barrels. The following pages will analyze several of these illustrations selected for their exemplary visual rhetoric and embed them in a chain of visualizations; particular attention will be devoted to the two engravings detailing the structure of the hydraulic organ in the Quirinal Gardens in Rome, which was reconstructed between 1984 and 1994.

6 Named after Aeolus, whom Zeus appointed keeper of the winds.

Hydraulic Organs

The illustration, a vertical-format engraving indexed with Latin and Greek letters, shows two vaulted grottoes, a wider one on the left and a narrower one on the right, each housing an automated organ with auxiliary registers and a non-mechanical figure representing a character from mythology (fig. 2). The grotto on the left—a circular opening in the ceiling suggests that the exterior constitutes an extended auditorium of sorts—contains the larger organ with two auxiliary registers, a cock and a cuckoo, as well as a flute-playing Pan seated atop the wind-chest. The coordinated movements of the cuckoo's bill, wings, and tail are regulated by a mechanical system actuated by the perforated barrel. Also associated with the bird are two pipes set in the register underneath it that produce the characteristic minor third of its call. As the "oar-blades" (*palmulae*) beneath the barrel snap in and out of the perforations, they operate the action opening and closing the wind supply to the transverse pipes; the chord rings out. The perforated barrel also controls the mechanism for the cock, where three "oar-blades" actuate three levers to produce the bird's crow. The barrel has been brought forward from inside the wall and, it seems, magnified to a gigantic scale in comparison to the pipe ensembles. The smaller organ appears in the right third of the sheet, with a little nymph standing motionless next to it. Beneath it, the wall has been broken up to reveal the Aeolian chamber. Pipes conduct the air upward, feeding the smaller instrument, by entering beneath its barrel as well as the wind-chest below the large organ's register. Water pipes supply water to the lower part of the chamber and hence to the wheel beneath, which drives the large perforated barrel via a set of cogwheels.

Reality and Imagination

An actual hydraulic organ existed in the Quirinal Gardens in Rome around 1650.[7] The Perugian organ maker Luca Blasi had built the original instrument, which was played by hand, between 1596 and 1598, during the pontificate of Clement VIII; in 1599, Heinrich Schickhardt described it in his diary. Between 1647 and 1648, Kircher and the Roman organ

7 See www.neac2.eu/water/english/arts/water_organ_quirinale.htm, as well as Simona Antellini Donelli, ed., *La Fontana dell' Organo nei Giardini del Quirinale: Nascita, storia e trasformazioni* (Rome: Palombi, 1995).

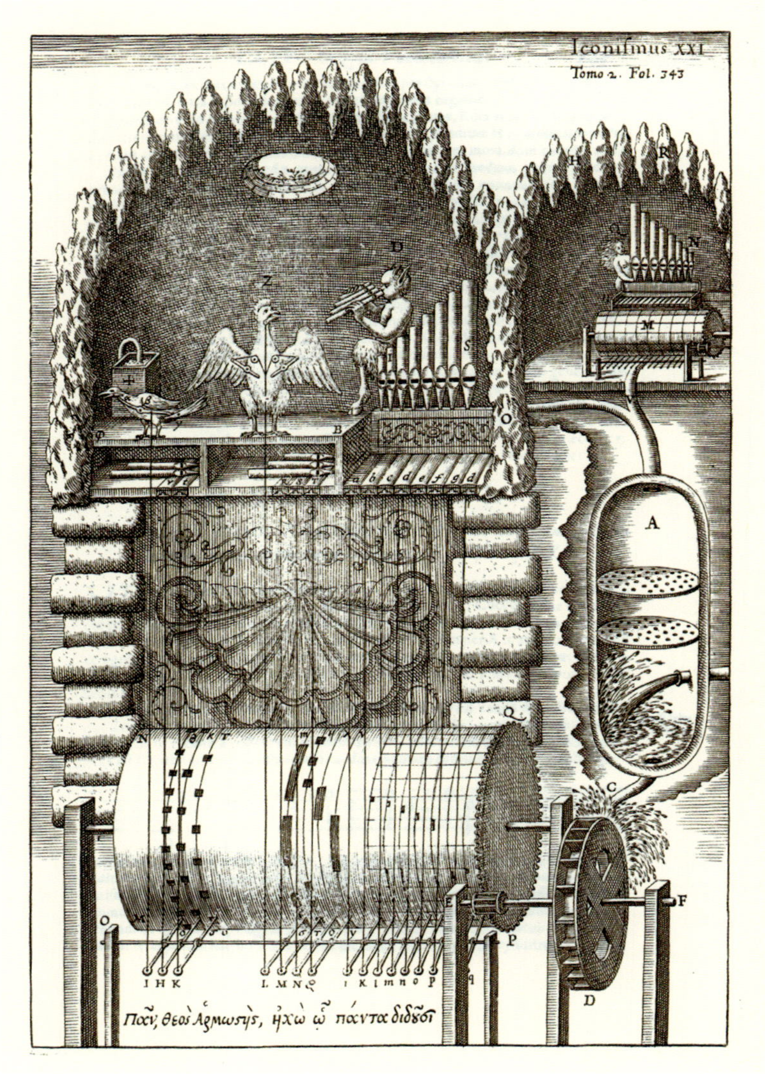

FIG. 2: Hydraulic organs in grottoes. Engraving from Athanasius Kircher, *Musurgia Universalis, sive Ars Magna consoni et dissoni in X libros digesta [. . .]*, vol. 2 (Rome: Corbelletti, 1650; reprint edited by Ulf Scharlau (Hildesheim: Olms, 1970), 342, plate xxi. © Gottfried Wilhelm Leibniz Bibliothek–Niedersächsische Landesbibliothek Hannover, 71/8521.

builder Matteo Marione replaced that organ, which had been destroyed, with a new automated instrument.

The aforementioned section on the construction of Aeolian chambers, however, begins with a description of this organ that calls it a work by Kircher alone: "Around the same time when I wrote this, I had been commissioned at the behest of Pope Innocent X to construct the hydraulic organ in the Quirinal Gardens. We had had the Aeolian chamber built to great success indeed, and so this instrument reasonably followed."[8] A good thirty pages further in, Kircher reasserts his authorship of the organ—"until it will be sufficiently attested to" (*satis testabitur*).[9]

The pictorial representation of the instrument is only loosely based on the real architecture, where the hydraulic organ occupied three recesses. The prominent role the illustration gives to the wind chamber and the barrel demonstrates Kircher's claim to have made an extraordinary contribution to technological progress. The barrel, in

8 Kircher, *Musurgia Universalis*, vol. B, 309.

9 Ibid., 345.

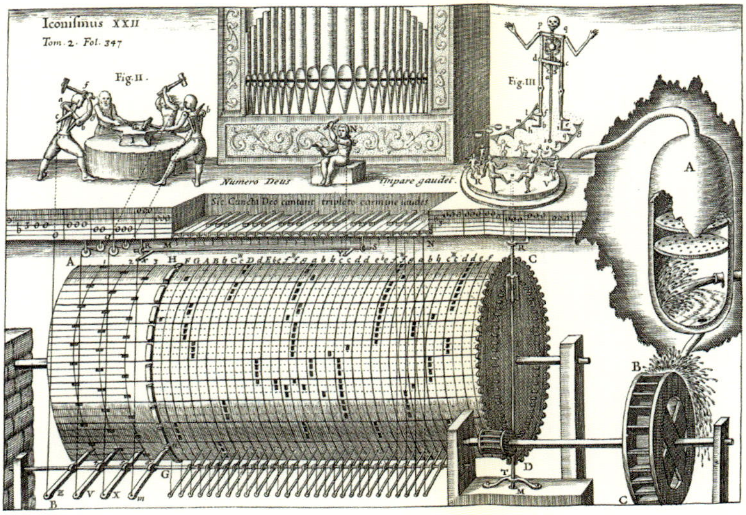

FIG. 3: An automated hydraulic organ. Engraving from Athanasius Kircher, *Musurgia Universalis, sive Ars Magna consoni et dissoni in X libros digesta [...]*, vol. 2 (Rome: Corbelletti, 1650; reprint edited by Ulf Scharlau (Hildesheim: Olms, 1970), 346, plate xxii. © Gottfried Wilhelm Leibniz Bibliothek–Niedersächsische Landesbibliothek Hannover, 71/8521.

particular, was without parallel in Italy, having been made entirely of durable metal rather than wood or stone.[10] The automata are embedded in the familiar world of myth and emblems. The illustration thus deliberately blends diagrammatic and eidetic elements of the school of vision.

The second engraving referring to the hydraulic organ is introduced as showing an automaton for the production of "Pythagorean music," a tuning based on simple intervals (fig. 3). The giant perforated and pinned barrel now also actuates three little automated smiths striking an anvil with their hammers (fig. 3, fig. ii); an additional cogwheel on one end of the barrel operates a group of dancers on a revolving platform (fig. 3, fig. iii). There are two sources of propulsion driving this ensemble, one fictional—the feet of an oversized jumping-jack skeleton, which hook into the pair of toothed racks attached to two dancers underneath; the other real—the transmission connecting it to the cogwheel. In the text, Kircher prints a fairly long mixed chorus in eight parts. Chorus 1 seems to be programmed onto the barrel shown, but chorus 2 is not, suggesting that the first chorus would be performed automatically, whereas the second was to be sung. The time is kept by a little wooden puppet holding a baton whose arm is likewise actuated by the pinned barrel. For variety's sake, the score includes alternating rests for the two choruses and three general rests, during which only the hammering noise of the smiths and the jumps of the dancing figures will be heard. The group of Vulcanian smiths, which probably imitates a similar ensemble in the grottoes at Pratolino, hints at the legend of how Pythagoras discovered intervals. It was not realized until about 1720, when a group of marble statues was installed in the neighboring loggia.[11] The "Pythagorean" character of this music is ultimately apparent only in the triad of the hammers, whose sequence is clearly legible on the barrel; the composition as printed in the accompanying text does not contain any chords. Once again, the depiction blends eidetic and diagrammatic, imaginary and operative elements; this time, the architecture is left out in favor of the visual representation of technology.

10 See Antonio Latanza, "Il ripristino dell'organo del Quirinale," in *La Fontana dell' Organo nei Giardini del Quirinale*, ed. Antellini Donelli, 83, 89–90. Perforating a cylinder measuring approx. 183 cm (approx. 72 inches) in diameter was not yet technically feasible, so the cylinder was encased in a perforated copper cladding.

11 See the photograph in Antellini Donelli, ed., *La Fontana dell' Organo nei Giardini del Quirinale*, 21.

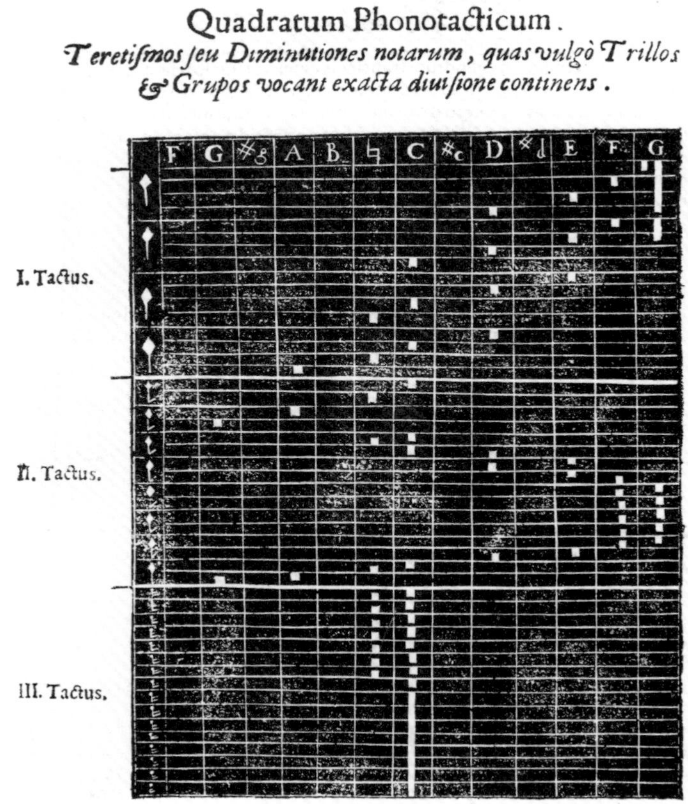

FIG. 4: "Quadratum Phonotacticum," developed view of a pinned-barrel surface. Engraving from Athanasius Kircher, *Musurgia Universalis, sive Ars Magna consoni et dissoni in X libros digesta [...]*, vol. 2 (Rome: Corbelletti, 1650; reprint edited by Ulf Scharlau (Hildesheim: Olms, 1970), 323. © Gottfried Wilhelm Leibniz Bibliothek–Niedersächsische Landesbibliothek Hannover, 71/8521.

The superimposition of the two pictorial principles is especially evident in another illustration of a pinned barrel, which is intended to help the reader understand how a composition is translated onto the cylindrical barrel, a process called pinning (fig. 4). The "phonotactic square," a section of a cylinder representing an octave that has been developed into the plane, shows the positioning and length of the pins. The columns correspond to the different strings or pipes of the instrument, the bold horizontal lines to the division into three bars. Ornaments such as the trill between F-sharp and G-sharp in the second bar (in 8/8 meter) and between B and C in the third bar (in 16/16 meter) are clearly visible.

Kircher also shows how up to eight pieces may be stored on a single barrel. The associated mechanism is illustrated by a woodcut presumably based on a sketch from Kircher's own hand (fig. 5). In this instance, visual abstraction has reduced the cylinder to a horizontal rectangle subdivided into three times eight sections marked with letters (*c* through *k*, skipping *j*). The cylinder rotates around an axis whose round shape is suggested by hatchings. The text describes how the barrel moves sideways after each full rotation: the lever *x* is actuated by the sequence of pins on the circle *c*, then *d*, *e*, and so on. The oblique depiction and shading of the three surfaces marked *x* is probably meant as another eidetic indicator of the device's three-dimensional structure. In the top right corner, a diagram proposes pins and "bridges" of different lengths corresponding to the note values of a whole, half, quarter, and eighth note.

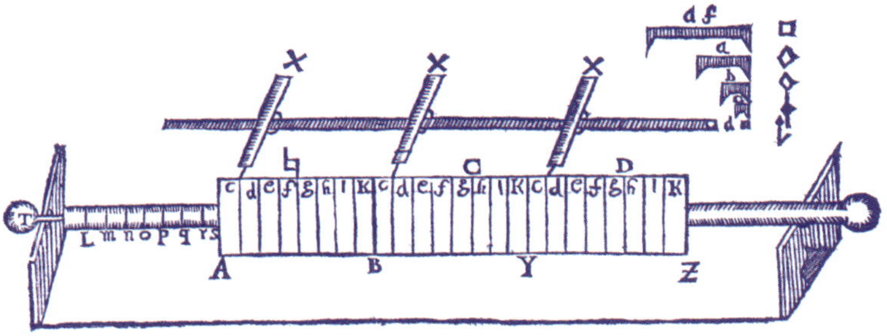

FIG. 5: A barrel prepared for multiple pieces. Woodcut from Athanasius Kircher, *Musurgia Universalis, siue Ars Magna consoni et dissoni in X libros digesta [...]*, vol. 2 (Rome: Corbelletti, 1650; reprint edited by Ulf Scharlau (Hildesheim: Olms, 1970), 320. © Gottfried Wilhelm Leibniz Bibliothek–Niedersächsische Landesbibliothek Hannover, 71/8521.

Contemplating the Hypothetical

Kircher's optical metamorphosis, or chain of visualizations, leads from the score to the three-dimensional image of the barrel, but one crucial link is missing: the depiction of the interaction between a section of the barrel and the keyboard. This missing link is provided, for example, by the boldly conceived detail view illustrating problem xxx in the book on mechanical engineering by the aforementioned Salomon de Caus (fig. 6). In this picture, the pinned surface of the barrel functions as a sort of mechanical score, as the next two engravings in the book make clear: they represent the exact same piece as a score. Rotating the image of the barrel by ninety degrees in our mind's eye, we may easily compare the first six bars in the picture and the notation.[12] Kircher, by contrast, doesn't offer an unambiguous visualization of the barrel as a mechanical score, nor do his pictures indicate unequivocally whether the pins shown are intended as a *pars pro toto* or the proposed mechanism actually features perforations as well as pins. The pertinent depictions (figs. 1–3) either elide the representation of pins or perforations on the barrel altogether or show perforations and merely suggest pins. The text, by contrast, mentions "cogwheels" and "mobile and immobile cogs"; the former refer to the Causian principle of variable pinning in pre-bored holes spread evenly over the entire surface of the barrel.

So image and text contradict each other on this matter, pointing to a specific pictorial strategy employed by Kircher that is even more evident in an engraving illustrating the Aeolian chamber (fig. 7). The Jesuit devotes four pages of text and a full-page engraving to an extensive discussion of the technology of wind supply even though the organ builders of his time have already solved the problem. He claims that air is present in water, so one merely needs to send water through a pipe wound in a spiral shape and then make it strike a smooth stone surface with great momentum inside the Aeolian chamber in order to effect the separation of air and water (compare fig. 7, ii). A marginal note in the text accompanying this illustration claims that "a bent pipe produces more air than a cylindrical one,"[13] which is not the case. Contorted pipes such as the one the illustration (fig. 7, ii) proposes as an air duct—similar pipes were very popular as ear trumpets—do not appear as water pipes anywhere outside this engraving. That Kircher nonetheless had this figure included in the book points up the immense and fundamental import of the rhetoric of images and the contemplation of the hypothetical in Kircher's work. The appearance or visualization is of quasi-ontological import. Kircher's

12 De Caus, *Von gewaltsamen Bewegungen*, problema xxx, 38. See Bogen, "Algebraische Notation und Maschinenbild," 201–2.

13 Kircher, *Musurgia Universalis*, vol. B, 309.

FIG. 6: Detail view of a stone barrel and keyboard. Engraving illustrating problem xxx in Salomon de Caus, *Von Gewaltsamen bewegungen* (1615). Salomon de Caus, *Von Gewaltsamen bewegungen: Beschreibung etlicher, so wol nützlichen alß lustigen Machiner [...]*, 3 vols. (Frankfurt am Main: Abraham Pacquart, 1615), vol. 1, plate 38. © Technische Informationsbibliothek, Universitätsbibliothek Hannover, 2 Haupt 224.

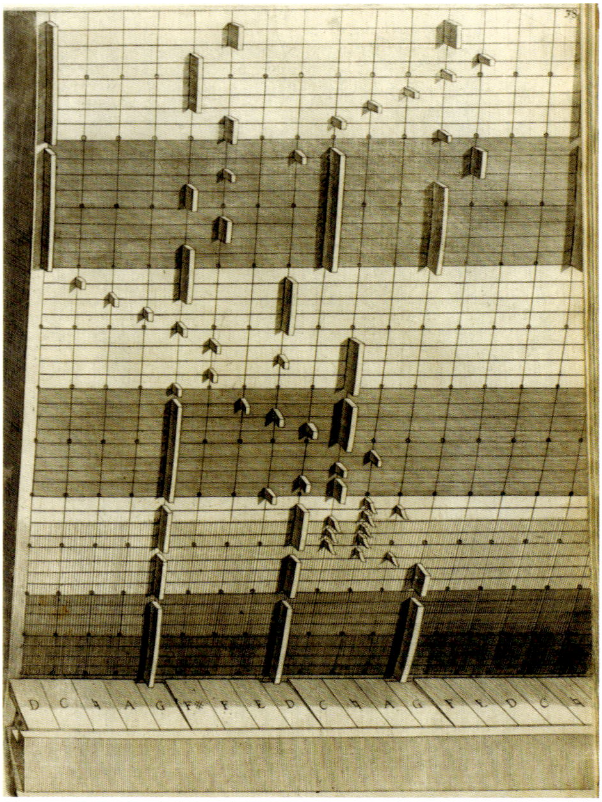

Spanish fellow Jesuit Baltasar Gracián (1601–1658) describes the connection between appearance and being in his *Art of Worldly Wisdom*: "130. Do and be seen doing. Things do not pass for what they are but for what they seem. To be of use and to know how to show it, is to be twice as useful. What is not seen is as if it was not [...] a good exterior is the best recommendation of the inner perfection."[14] Exaggerating slightly, we might say that Kircher has fictional spiral-shaped water pipes depicted in the engravings because he attaches greater value to the contemplation of the fictional than to the recording of what is feasible in reality. On the one hand, that is to say, Kircher places emphasis on great precision and realistic depiction, as in the images showing pinned and perforated barrels and gear ratios; on the other hand, he is not afraid to mix such depictions with fictions. This bewildering fluctuation between fiction and feasibility serves a strategic purpose, emphasizing the contemplative aspect; it is often associated with the encomia to Kircher's patrons, which again highlight the pretension to feasibility. That strategy is embodied by Kircher's popular encryption and composition chests made of wood and cardboard, several of which he sent, complete with instructions, to recipients including members of the house of Habsburg around the middle of the century (a few specimens have survived): a theoretically radical but impractical technique to heal the Babylonian loss of language and a purely mechanical method of composition provide the prince with a false sense of sovereign command in a wide range of fields. The crucial point, however, is the

14 Baltasar Gracián, *The Art of Worldly Wisdom*, trans. Joseph Jacobs (Boston: Shambhala, 1993), 73–74.

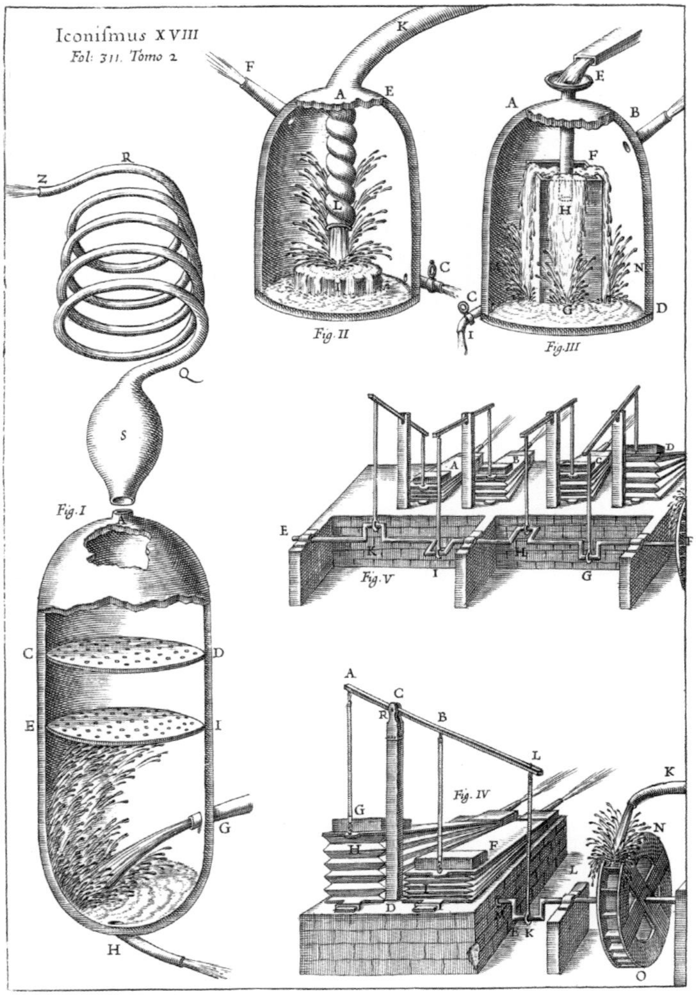

FIG. 7: Proposed wind supply equipment: hydraulic machinery and bellows. Engraving from Athanasius Kircher, *Musurgia Universalis, sive Ars Magna consoni et dissoni in X libros digesta [...]*, vol. 2 (Rome: Corbelletti, 1650; reprint edited by Ulf Scharlau (Hildesheim: Olms, 1970), 311, plate xviii. © Gottfried Wilhelm Leibniz Bibliothek–Niedersächsische Landesbibliothek Hannover, 71/8521.

celebration of his plenitude of power. Kircher received no more than a single letter in polygraphic writing; the correspondence about the devices, however, is immense.

Many critics have confused the pictures of organs with design drawings and dismissed them as technically unviable or flawed. Today, scholars recognize that Kircher was primarily interested in the potential impact of the contemplation of hypothetical objects; he found the suitable medium in the image, with its genuinely mediated relation to reality.

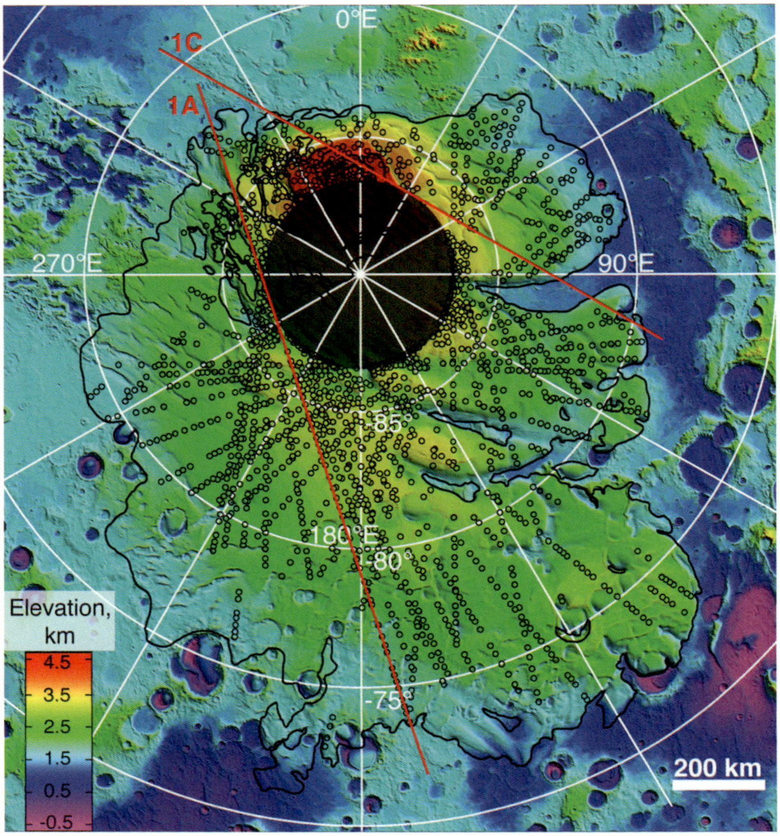

FIG. 1: Visualization of data from two Mars missions as maps showing the ice cap at the south pole of Mars, published in the specialist journal *Science*, 2007. Due to the orbits, only data from latitudes up to 87 degrees south is available, so the area around the pole is marked in black. The network of coordinates, indexing of latitudes and longitudes, and use of a color gradient comply with cartographic conventions. Individual measuring points are marked as circles; red lines mark areas shown in cross section in additional images. A distance scale and color scale make it possible to read quantities from the image. New knowledge about the composition, topography, thickness, and volume of the ice field was disseminated through the interplay between the images and the accompanying article. Jeffrey J. Plaut et al., "Subsurface Radar Sounding of the South Polar Layered Deposits of Mars," *Science* 316, no. 5821 (2007): 93. Reprinted with permission from AAAS.

POPULARIZING SCIENCE

The terms *popularization* and *popular science* were first used in the sociology of science and were coined in particular by the early-twentieth-century behaviorist Edward L. Thorndike. The hierarchical diffusion model of the "popularization of science," according to which elites produce simplifying trivializations for uneducated but interested audiences, dates further back, to the second half of the nineteenth century. Since the 1980s, however, researchers have assumed that the dissemination, transformation, and reconstitution of science and knowledge go hand in hand. The epistemologist Baudouin Jurdant took this view in the 1970s in analyzing the two spheres of "science" and "the public" based on a communication model (Jurdant 1973). According to this view, production and distribution are part of a continuum along which "scientists, popularizers, and the public no longer appear as separate poles in a linear process, but as participants in the dialogue between producers and recipients" (Kretschmann 2003, 9). Popularizers do not just disseminate knowledge; their activities also result in the creation of new and independent areas of knowledge.

Historians of science, especially in England, the United States, and France, have investigated the history of this popularization. With the "practical turn" in the 1980s (Lynch and Woolgar 1990), historians and sociologists of science increasingly focused on processes of visualization or "inscription" in techniques of writing, printing, demonstration, and visualization (Schaffer 1983; Latour 1986; Golinski 1992). They thus also established the (scientific) image as a legitimate object of research on science, overcoming the discipline's preference for the theoretical domain.

In their laboratory study on image production in

FIG. 2: Version of the image in figure 1 from the news magazine *Spiegel Online,* March 15, 2007. The process of popularizing these results tapped into the discourse about life on Mars. A relatively small step in the research into water deposits on Mars was incorporated into a "Grand Narrative" about extraterrestrial life. This re-contextualization also transforms the images. In this version, only the color gradient can be seen; the systems of coordinates, scaling, and indexing are no longer visible. It is therefore no longer possible to scientifically evaluate the image, which serves only as a placeholder for the scientific activity the article ties in with the discussion about life on Mars. *Spiegel Online,* March 15, 2007, http://www.spiegel.de/fotostrecke/radar-messungen-eis-am-mars-suedpol-fotostrecke-20051-3.html (accessed November 2013). Courtesy NASA/JPL-Caltech.

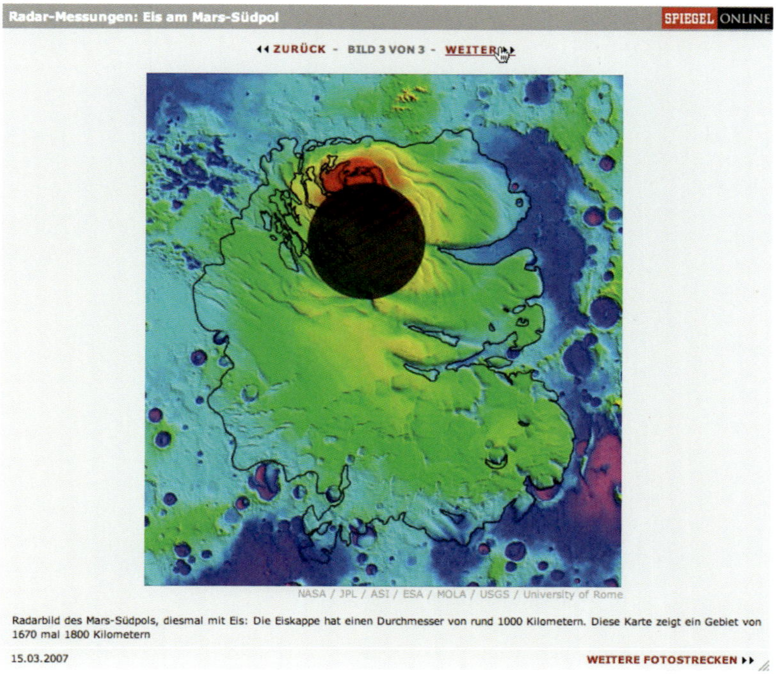

astrophysics, Edgerton and Lynch pointed out how the participating actors perceived themselves: they strictly separated the practice of science and knowledge production from the "beautification" of scientific images for public audiences. Their work involved aesthetic choices, but they denied that such interventions contributed in any way to the production of knowledge (Edgerton and Lynch 1988). Lynch observed and described the distinction between "lookers" and "users" of documents such as photographs. "Users" use them in their laboratory work, while "lookers" regard them as an illustration of a completed work or publication (Lynch 1985). Images prepared for the mass media, in which the information content is reduced, often allow only for the role of the "looker." The media scholar Ralf Adelmann has shown how the information content of an image is reduced over the course of the communication process in the example of measurements of the thickness of ice on Mars. In an article in the journal *Science,* a system of coordinates, a color legend, and a scale were included in the image (fig. 1), but these components had vanished in the version shown by the news magazine *Spiegel Online* (fig. 2), making it impossible to decipher the image (Adelmann et al. 2008). An image from the field of climate research popularized by Al Gore in his film *An Inconvenient Truth* (2004) similarly reveals the complex transformations that scientific images undergo in communication processes. This graphic gives the impression of an almost unavoidable climatic disaster, showing temperatures and CO_2 concentrations over the past 650,000 years as detected in samples from ice cores; today's CO_2 concentration is indicated by a yellow dot placed on the rapidly rising branch of the red graph (fig. 3). The glaciologist Eric Wolff adopted this kind of visual dramatization in

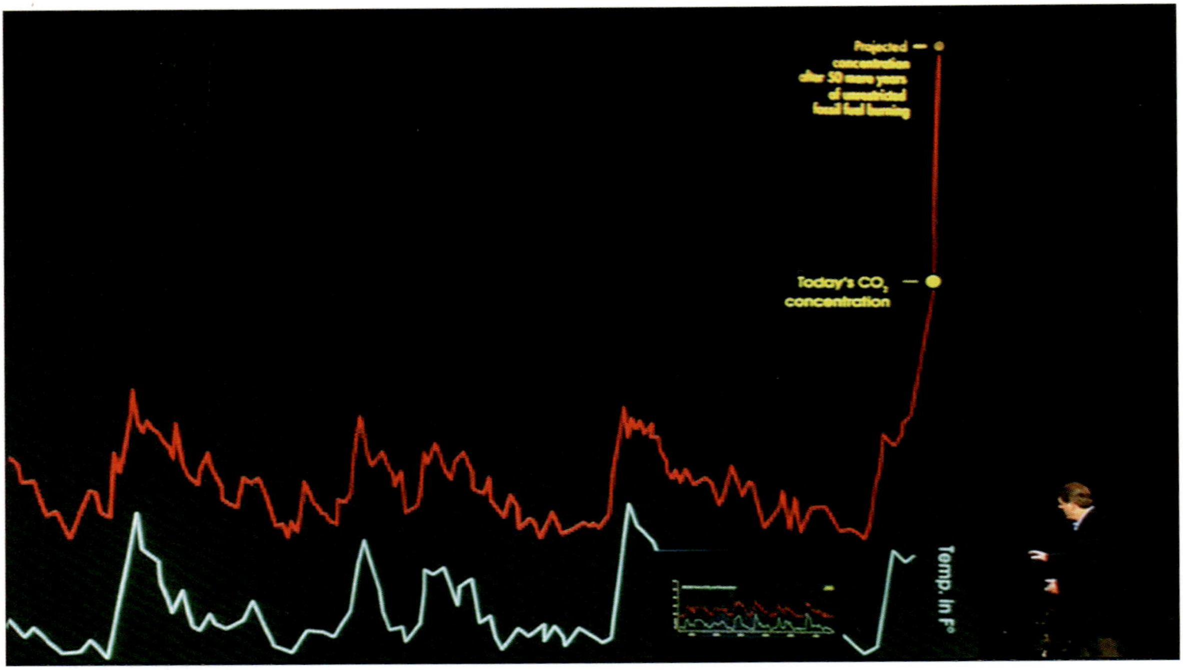

FIG. 3: Graph showing temperatures and CO_2 concentrations over the past 600,000 years, retrieved from Antarctic ice cores (detail). Still from Al Gore's film *An Inconvenient Truth,* 2006 (dir. Davis Guggenheim). © Paramount Pictures, 2006.

a PowerPoint slide for a public lecture, suggesting how popularization strategies may retroactively affect scientific contexts (fig. 4; Schneider 2011).

Further case studies are presented in volumes such as Pauwel 2006, Shteir and Lightman 2006, Nikolow and Schirrmacher 2007, and Hüppauf and Weingart 2008. While these authors have developed basic approaches toward linking the specificity of the image with the contextual interpretation of scientific communication, we still seem to lack overarching concepts that would do justice to the specificity of knowledge communication and iconicity, including in terms of media theory.

From the perspective of semiotic theory, Martina Hessler has noted that scientific images exist in a fundamental dilemma between their signification and their existence as objects (Hessler 2006). On the one hand, they have a status as signs within the scientific community; on the other hand, they often become "image objects" with an "artificial presence" over the course of scientific communication (Wiesing 2005). Experts can therefore easily decode them (as signs), but the uninitiated can only look at them (as objects). According to Hessler, context constitutes the epistemological status of images; being a sign is not intrinsic to them but a status the beholder must confer on them. This ambivalence between the sign and object qualities of scientific images may be rooted in the status of scientists themselves, who are primarily their own popularizers (Jurdant 1993, 2009). —AMD/JH

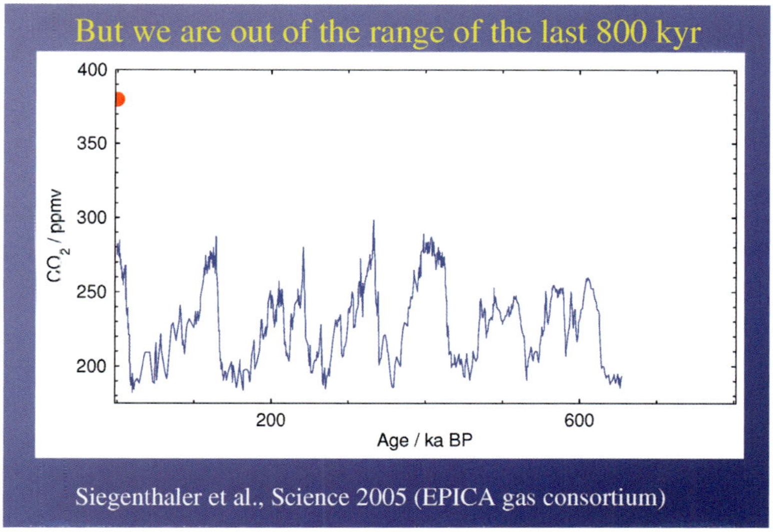

FIG. 4: PowerPoint slide by the glaciologist Eric Wolff, also showing a CO_2 curve based on measurements from an ice core. He similarly added a red dot for today's CO_2 concentrations as a reference value. Since glaciologists work with a reverse timeline, the red dot appears on the left edge of the graph. "Meltdown: Evidence of Climate Change from Polar Science," PowerPoint presentation by Eric Wolff, Environmental Protection Agency, Climate Change lecture series, January 2008, http://www.epa.ie/downloads/pubs/other/events/oclr/eric wolff lecture_reduced_13022008.pdf (accessed February 2013). Reproduced with permission of Eric Wolff.

LITERATURE

Adelmann, Ralf, Jochen Hennig, and Martina Heßler. "Visuelle Wissenskommunikation in Astronomie und Nanotechnologie: Zur epistemischen Produktivität von Bildern." In *Wissensproduktion und Wissenstransfer. Wissen im Spannungsfeld von Wissen, Politik und Öffentlichkeit*, ed. Friedhelm Neidhardt, Renate Mayntz, Peter Weingart, and Ulrich Wengenroth, 41–74. Bielefeld: Transcript, 2008.

Bluma, Lars, and Sybilla Nikolow. "Bilder zwischen Öffentlichkeit und wissenschaftlicher Praxis: Neue Perspektiven für die Geschichte der Medizin, Naturwissenschaften und Technik." *NTM: Internationale Zeitschrift für Geschichte und Ethik der Naturwissenschaften, Technik und Medizin* 10, no. 4 (2002): 201–8.

Edgerton, Samuel Y., and Michael Lynch. "Aesthetics and Digital Image Processing: Representational Craft in Contemporary Astronomy." In *Picturing Power: Visual Depiction and Social Relations*, ed. Gordon Fyfe and John Law, 184–220. London: Routledge, 1988.

Golinski, Jan. *Science as Public Culture: Chemistry and Enlightenment in Britain, 1760–1820*. Cambridge: Cambridge University Press, 1992.

Hentschel, Klaus. *Mapping the Spectrum: Techniques of Visual Representation in Research and Teaching*. Oxford: Oxford University Press, 2002.

Heßler, Martina. "Einleitung." In *Konstruierte Sichtbarkeiten: Wissenschafts- und Technikbilder seit der frühen Neuzeit*, ed. Marina Heßler, 11–37. Munich: Fink, 2006.

Hüppauf, Bernd, and Peter Weingart, eds. *Science Images and Popular Images of Sciences*. New York: Routledge, 2008.

Jasanoff, Sheila, Gerald E. Markle, James Petersen, and Trevor Pinch, eds. *Handbook of Science and Technology Studies*. Thousand Oaks, CA: Sage, 1995.

Jurdant, Baudouin. "Popularisation of Science as the Autobiography of Science." *Public Understanding of Science*, no. 2 (1993): 365–73.

Jurdant, Baudouin. *Les problèmes théoriques de la vulgarisation scientifique*. Paris: Éditions des Archives Contemporaines, 2009 (originally issued as doctoral thesis, Université Louis Pasteur de Strasbourg, 1973).

Kretschmann, Carsten, ed. *Wissenspopularisierung: Konzepte der Wissensverbreitung im Wandel*. Berlin: Akademie Verlag, 2003.

Latour, Bruno. "Visualization and Cognition: Thinking with Eyes and Hands." *Knowledge and Society: Studies in the Sociology of Culture Past and Present*, no. 6 (1986): 1–40.

Latour, Bruno, and Peter Weibel. *Making Things Public: Atmospheres of Democracy*. Cambridge, MA: MIT Press, 2005.

Lynch, Michael. *Art and Artifact in Laboratory Science: A Study of Shop Work and Shop Talk in a Research Laboratory*. London: Routledge & Kegan Paul, 1985.

Lynch, Michael, and Steve Woolgar, eds. *Representations in Scientific Practice*. Cambridge, MA: MIT Press, 1990.

Nikolow, Sybilla, and Arne Schirrmacher, eds. *Wissenschaft und Öffentlichkeit als Ressourcen füreinander: Studien zur Wissenschaftsgeschichte im 20. Jahrhundert*. Frankfurt am Main: Campus, 2007.

Nowotny, Helga, Peter Scott, and Michael Gibbons. *Re-Thinking Science. Knowledge and the Public in an Age of Uncertainty*. Cambridge: Polity, 2001.

Pauwels, Luc, ed. *Visual Cultures of Science: Rethinking Representational Practices in Knowledge Building and Science Communication*. Hanover, NH: Dartmouth College Press, 2006.

Remmert, Volker. *Widmung, Welterklärung und Wissenschaftslegitimierung: Titelbilder und ihre Funktion in der wissenschaftlichen Revolution*. Wiesbaden: Harrassowitz, 2005.

Schaffer, Simon. "Natural Philosophy and Public Spectacle in the 18th Century." *History of Science* 21 (1983): 1–43.

Schneider, Birgit. "Image Politics: Picturing Uncertainty: The Role of Images in Climatology and Climate Policy." In *Climate Change and Policy: The Calculability of Climate Change and the Challenge of Uncertainty*, ed. Johann Feichter and Gabriele Gramelsberger, 191–209. Berlin: Springer, 2011.

Shinn, Terry, and Richard P. Whitley, eds. *Expository Science: Forms and Functions of Popularisation*, Sociology of Sciences Yearbook, no. 9. Dordrecht: Reidel, 1985.

Shteir, Ann B., and Bernard Lightman, eds. *Figuring It Out: Science, Gender, and Visual Culture*. Hanover, NH: Dartmouth College Press, 2006.

Stafford, Barbara Maria. *Artful Science: Enlightenment, Entertainment, and the Eclipse of Visual Education*. Cambridge, MA: MIT Press, 1994.

Wiesing, Lambert. *Artifizielle Präsenz: Studien zur Philosophie des Bildes*. Frankfurt am Main: Suhrkamp, 2005.

Drawing and the Contemplation of Nature—Natural History around 1600: The Case of Aldrovandi's Images

Angela Fischel

Images have helped scholars gain knowledge of nature since the sixteenth century. The cabinets of the natural philosophers saw the compilation of large collections of drawings, documentary depictions of the natural world that recorded the forms of the animal and vegetal kingdoms.[1] *The collection built by the Bologna-based natural philosopher Ulisse Aldrovandi (1522–1605)*[2] *is among the most formidable of its kind. It has survived almost in its entirety, allowing us to illuminate how natural philosophy around 1600 worked with images. Aldrovandi saw himself as a pioneer of the modern natural sciences and championed the visual study of nature as the most important source of knowledge about it.*[3] *Closer inspection of his image collection, however, reveals that his drawings by no means derive directly from nature; many of the depictions of animals may be traced back to earlier printed sources. Other images stage their objects in suggestive compositions that exceed the purpose of objective documentation. This leads us to ask, then, what Aldrovandi meant by empirical study and, further, what specific functions images may have served in the context of early modern natural science.*[4]

One of the most spectacular drawings from Aldrovandi's collection shows two vipers (fig. 1).[5] The animals raise their heads in an aggressive posture; their bodies are intertwined to form a slightly asymmetrical ornament. The trompe-l'oeil drawing presents the vipers in a pictorial space that is visually continuous with the beholder's own environment, an effect underlined by the shadows and the use of perspective. The writhing snakes seem to come dangerously close to the beholder. The animals originally belonged to Francesco de' Medici, who sent the living specimens to Bologna to have them studied. At the same time, Aldrovandi had also asked for a drawing of the vipers—a prescient request, as it soon turned out: one of the two animals died during transport to Bologna, the other shortly after its arrival.[6] The death of the rare specimens left the collector with nothing but their likenesses. One of the functions of drawings in Aldrovandi's research practice was evidently to provide a vivid documentation of the forms of nature that would survive their physical demise. It remains remarkable, however, that a picture as dramatic and suggestive as the draftsman Jacopo Ligozzi's portrayal of the snakes would be used in this context. Ligozzi gave a very precise depiction of the animals, but what he shows is more than the phenotype of a rare species of viper: the use of trompe l'oeil also conveys a vigorous impression of the danger they pose, vividly illustrating an aspect of their nature as well.

1 The following discussion will be limited to drawings of animals in Aldrovandi's collection. For Aldrovandi's botanical drawings, see Enzo Crea, ed., *Hortus pictus: Dalla raccolta di Ulisse Aldrovandi* (Rome: Edizioni dell'Elefante, 1993). For Aldrovandi's image collection in general, see Giuseppe Olmi, *L'inventario del mondo: Catalogazione della natura e luoghi del sapere nella prima età moderna* (Bologna: Il Mulino, 1992); Paula Findlen, *Possessing Nature: Museums, Collecting, and Scientific Culture in Early Modern Italy* (Berkeley: University of California Press, 1996); and Brian Ogilvie, The Science of Describing (Chicago: University of Chicago Press, 2006). For recent literature on images in early modern zoology, see Sachiko Kusukawa, "Patron's Review: The Role of Images in the Development of Renaissance Natural History," *Archives of Natural History* 38, no. 2 (2011): 189–213. The zoological drawings from Conrad Gessner's collection are currently receiving a great deal of scholarly interest: Sachiko Kusukawa, "The Sources of Gessner's Pictures for the *Historia Animalium*," *Annals of Science* 67, no. 3 (2010): 303–28; Angela Fischel, "The 'Verae Icones' of Natural Philosophy: New Concepts of Cognition and the Construction of Visual Reality in Conrad Gessner's *Historia animalium*," *Yearbook for European Culture of Science* 6 (2011): 129–40; Florike Egmond, "A Collection within a Collection: Rediscovered Animal Drawings from the Collections of Conrad Gessner and Felix Platter," *Journal of the History of Collections* 25, no. 2 (2013): 149–70.

2 Ulisse Aldrovandi, who did most of his work at the University of Bologna, has always been considered a leading Italian natural philosopher of the sixteenth century. His most extensive project, a definitive natural history, remained incomplete. See Sandra Tugnoli Pàttaro, *Metodo e sistema delle scienze nel pensiero di Ulisse Aldrovandi* (Bologna: Clueb, 1981); Olmi, *L'inventario del mondo*, 22–157.

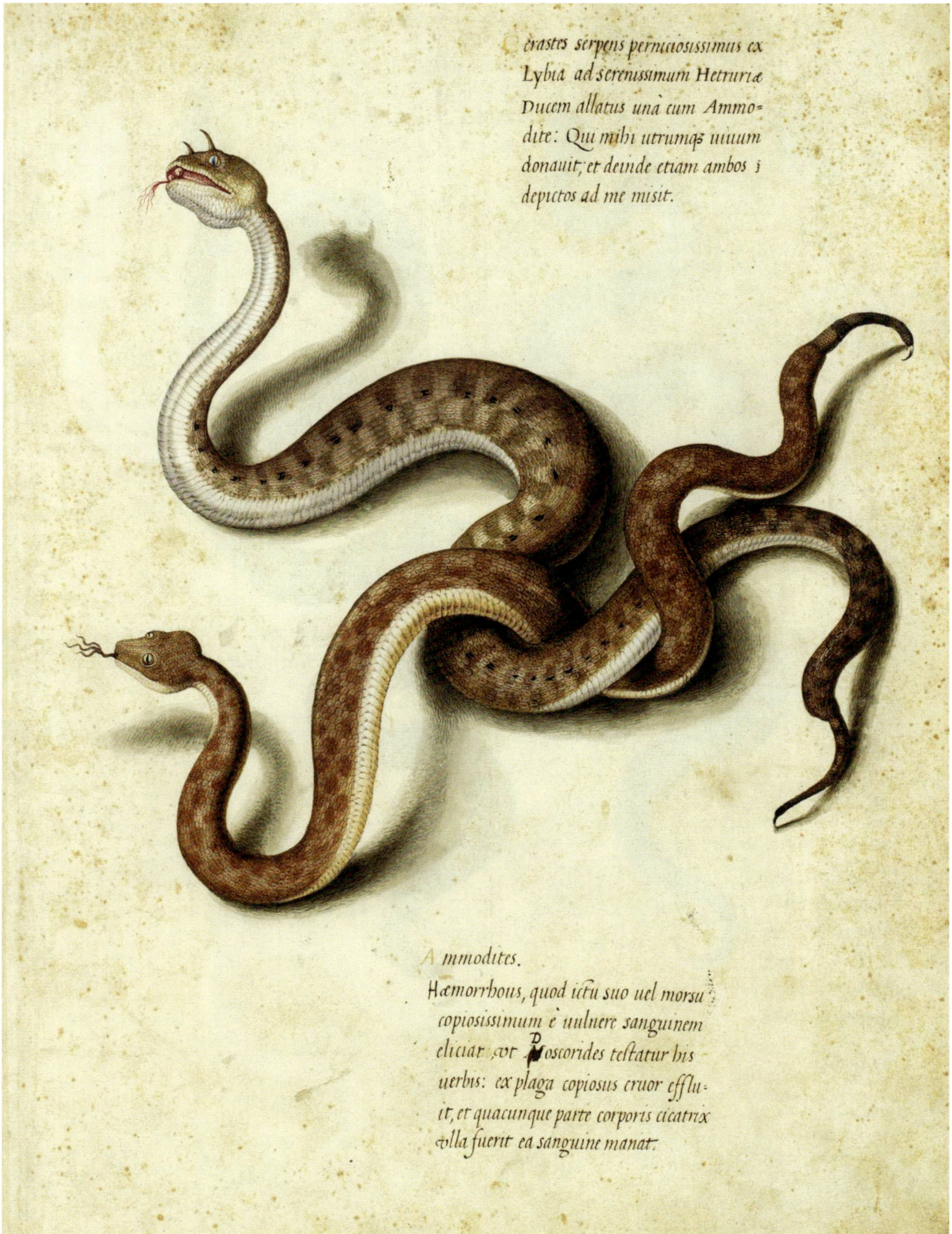

FIG. 1: Jacopo Ligozzi, Vipers, from Aldrovandi's collection of drawings, ca. 47.5 × 36 cm, 1577. Aldrovandi's collection of drawings, archive of the University Library of Bologna, Italy. *Tavoli di Animali* IV, c. 132, with permission of the University Library of Bologna.

Drawings constituted one part of Aldrovandi's natural history collection, which also included natural objects and preserved specimens;[7] herbariums containing dried vegetal specimens and collections of drawings of plants, numerous wood engravings made after the drawings, and a library.[8] The collection of animal drawings, which is now in the library of the University of Bologna, consists of six large leather-bound tomes.[9] The order in which the folios appear today obeys no recognizable system, instead laying out a vast and confusing mosaic of all sorts of conceivable—probable as well as less probable—forms. There are depictions of reptiles (fig. 1), birds (fig. 2), mammals, fishes (fig. 3), insects (fig. 4), prodigious births and monsters, seashells and snail shells, and fossils (fig. 5) but also empty sheets and others containing unfinished drafts (fig. 6), indicating the collection's incompleteness and openness to further expansion.

As shown in the following pages, Aldrovandi always devoted particular attention to his drawing collection. The importance he ascribed to it is also suggested by his writings, where he frequently refers to his drawing collection with particular pride. In addition, he expounded on the significance of images for the study of nature in numerous image-theoretical treatises.

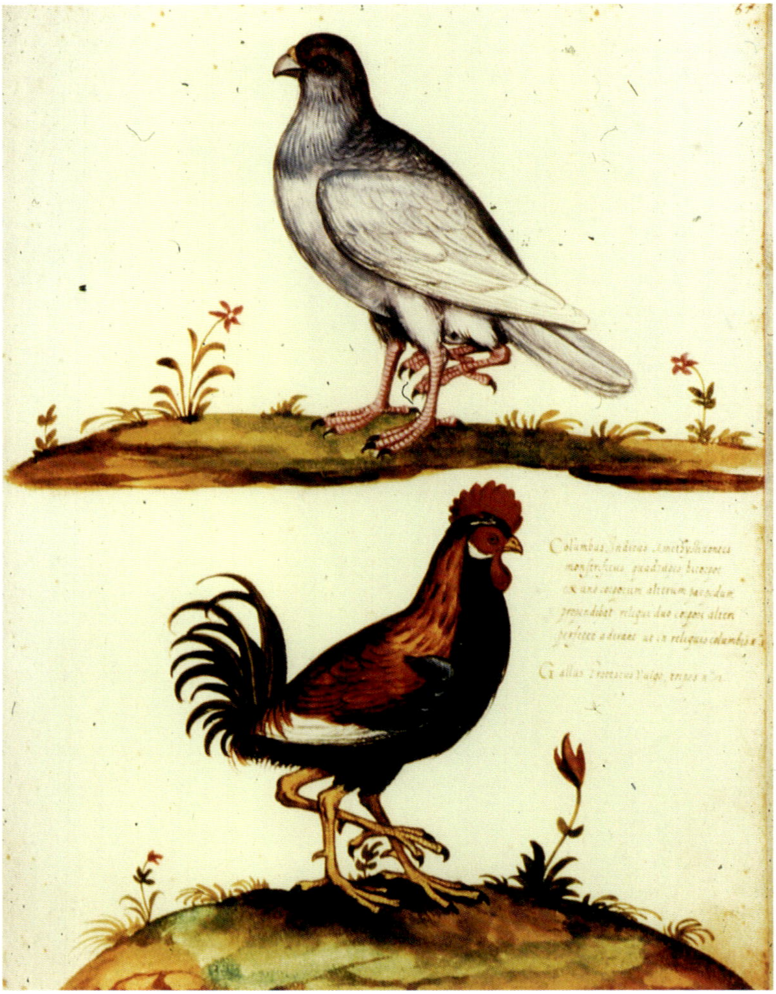

FIG. 2: Birds, from Aldrovandi's collection of drawings, ca. 47 × 35 cm, second half of the sixteenth century. Aldrovandi's collection of drawings, archive of the University Library of Bologna, Italy. *Tavoli di Animali* I, c. 67, with permission of the University Library of Bologna.

3 It is illuminating to note that method, to Aldrovandi, meant more than merely a research procedure. As Tugnoli Pàttaro suggests, Aldrovandi's use of the term also comprised techniques of teaching and learning. Visual inspection and excursions were accordingly part of his tuition. See Tugnoli Pàttaro, *Metodo e sistema delle scienze*, 65–73.

4 Claudia Swan has also made important contributions to the study of drawing and concepts of truth in this context. See Claudia Swan, "*Ad vivum, naer het leven*, from the Life: Considerations on a Mode of Representation," *Word and Image* 11 (1995): 353–72; Claudia Swan, *Art, Science, and Witchcraft in Early Modern Holland: Jacques de Gheyn II (1565–1629)* (Cambridge: Cambridge University Press, 2005).

5 Aldrovandi had received this drawing from Francesco I de' Medici in Florence. A detailed account of its history can be found in Findlen, *Possessing Nature*, 241–48.

6 Aldrovandi experimented on the cadavers of the vipers, trying to find a recipe for theriac. The latter, a concoction already known to Galen that was regarded as the "antidote of antidotes," had featured importantly in the pharmacopoeia since the Middle Ages. Almost all early modern natural scientists sought to find a way to prepare this legendary compound. For extensive references, see Findlen, *Possessing Nature*, 241–43.

7 Raffaella Simili, ed., *Il Teatro della natura di Ulisse Aldrovandi* (Bologna: Compositori, 2001); Walter Tega, ed., *Guide to Palazzo Poggi Museum: Science and Art* (Bologna: Compositori, 2002).

8 Irene Ventura Folli, "La natura scritta: la libraria di Ulisse Aldrovandi (1522–1605)," *L'Archiginnasio* (Florence) 49 (1993): 495–506. For Aldrovandi's work in written formats, see Christa Riedl-Dorn, *Wissenschaft und Fabelwesen: Ein kritischer Versuch über Konrad Gessner und Ulisse Aldrovandi* (Vienna: Böhlau, 1989); Fabrian Krämer, "Aldrovandi's Pandechion," in "Paper Technology: Wissenstechniken in der frühneuzeitlichen Wissenschaft," ed. Volker Hess and Andrew Mendelsohn, special issue,

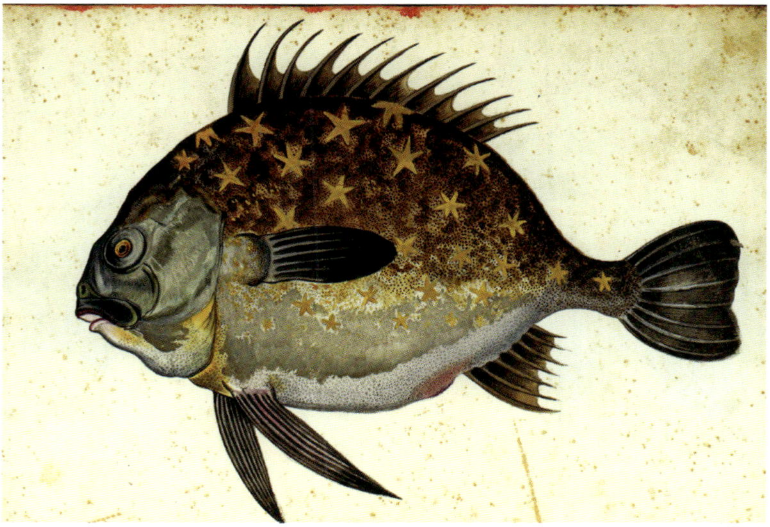

FIG. 3: Fish, from Aldrovandi's collection of drawings, ca. 47 × 35 cm, second half of the sixteenth century. Aldrovandi's collection of drawings, archive of the University Library of Bologna, Italy. *Tavoli di Animali* VI, c. 11, with permission of the University Library of Bologna.

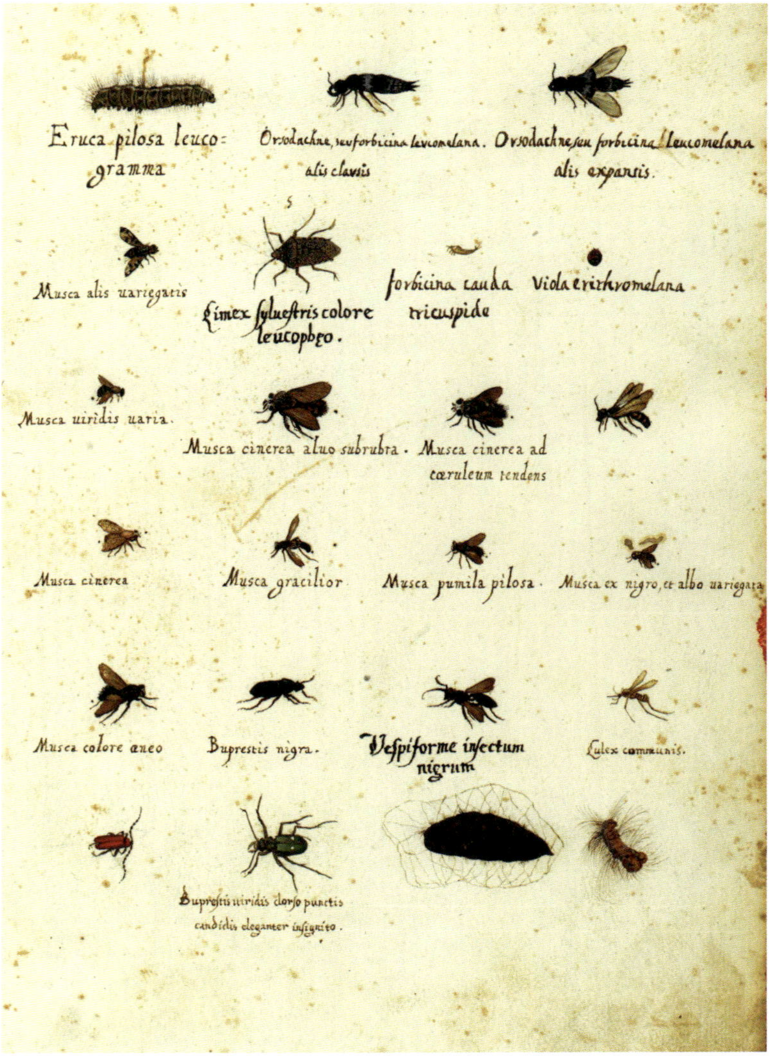

FIG. 4: Insects, from Aldrovandi's collection of drawings, ca. 47 × 35 cm, second half of the sixteenth century. Aldrovandi's collection of drawings, archive of the University Library of Bologna, Italy. *Tavoli di Animali* VII, c. 15, with permission of the University Library of Bologna.

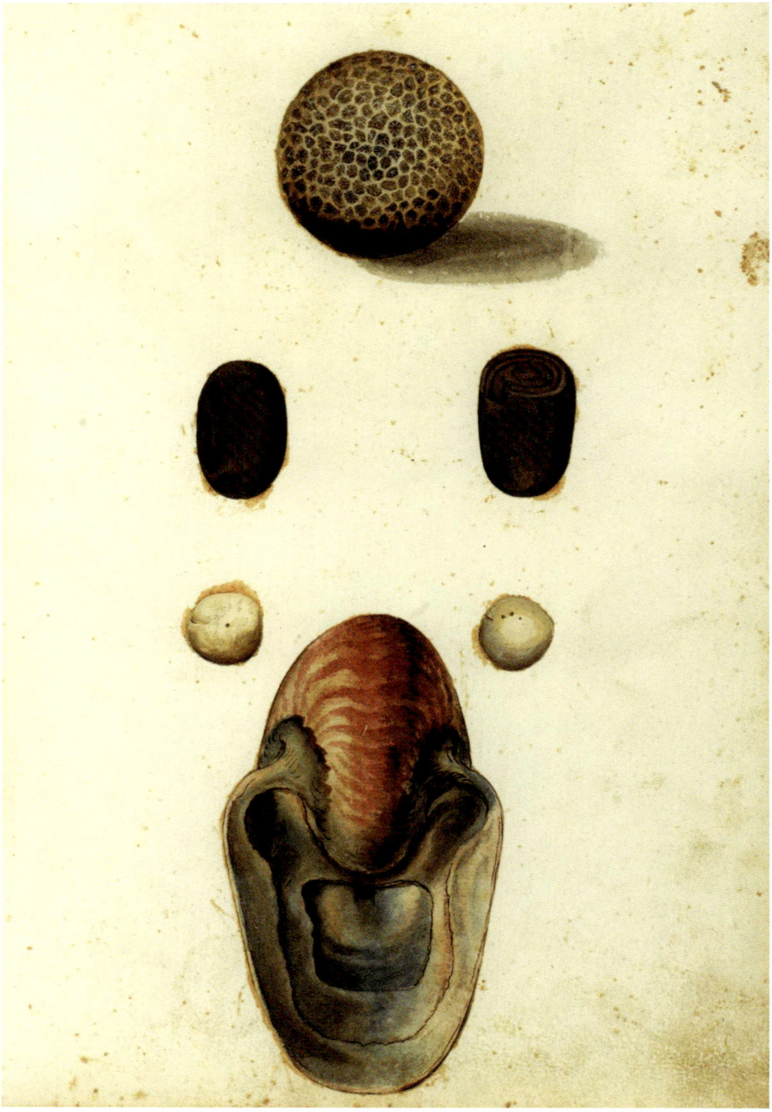

FIG. 5: Fossils and a nautilus shell, from Aldrovandi's collection of drawings, ca. 47 × 35 cm, second half of the sixteenth century. Aldrovandi's collection of drawings, archive of the University Library of Bologna, Italy. *Tavoli di Animali* VI, c. 69, with permission of the University Library of Bologna.

Empirical Science and the Politics of the Image

Aldrovandi was one of the few early modern zoologists to examine the matter of the image as such at length; with Conrad Gessner, he was among the first to address the particular significance imagery had for the philosophy of nature.[10] Aldrovandi's writings advocate the employment of images and give a prominent role to visual perception and sensory experience in connection with his call for a transformation of natural history into a science founded on empirical data. Tactile and visual perception and experience, he argues, must form the basis for any profound study of nature; only the outward senses provide the access to the world that enables the human understanding to know it. In this context, Aldrovandi developed an interesting reading of Aristotle, who, he writes, characterized the experience of individual objects of physical nature as the first step toward knowledge of the substance of the world.[11]

NMT: Zeitschrift für Geschichte der Wissenschaften, Technik und Medizin 21 (2013): 11–36.

9 For an overview of the collection, see Antonino Biancastello, ed., *Animali e creature mostruose di Ulisse Aldrovandi*, exhibition catalogue (Milan: Federico Motta, 2004). A complete digital copy of the collection of drawings may be found at www.filosofia.unibo.it/aldrovandi/ (accessed May 3, 2012).

10 See Sachiko Kusukawa and Ian Maclean, eds., *Transmitting Knowledge: Words, Images, and Instruments in Early Modern Europe* (Oxford: Oxford University Press, 2006); Angela Fischel, *Natur im Bild: Zeichnung und Naturerkenntnis bei Conrad Gessner und Ulisse Aldrovandi* (Berlin: Mann, 2009).

11 This reference is all the more remarkable since the term *substance* usually describes ideal characteristics but not individual physical properties. Aldrovandi, however, tries to connect his new, empirical ideal of natural history to this classical Aristotelian term. This new view of Aristotle differs strongly from earlier (for example, late medieval) references to Aristotle's philosophy. Prior to Aldrovandi and his precursor, Conrad Gessner, knowledge of nature was by no means based on perception of the outer appearance of an animal or plant. For an early modern interpretation of Aristotle in biology, see James G. Lennox, *Aristotle's Philosophy of Biology: Studies in the Origins of Life Science* (Cambridge: Cambridge University Press, 2001). For Aldrovandi's reference to Aristotle, see also Tugnoli Pàttaro, *Metodo e sistema delle scienze*, 69–73; Holger Steinemann, *Eine Bildtheorie zwischen Repräsentation und Wirkung: Kardinal Gabriele Paleotti's "Discorso intorno alle imagini sacre e profane"* (1582) (Hildesheim: Olms 2006). Cf. below, p. 179 and n. 27.

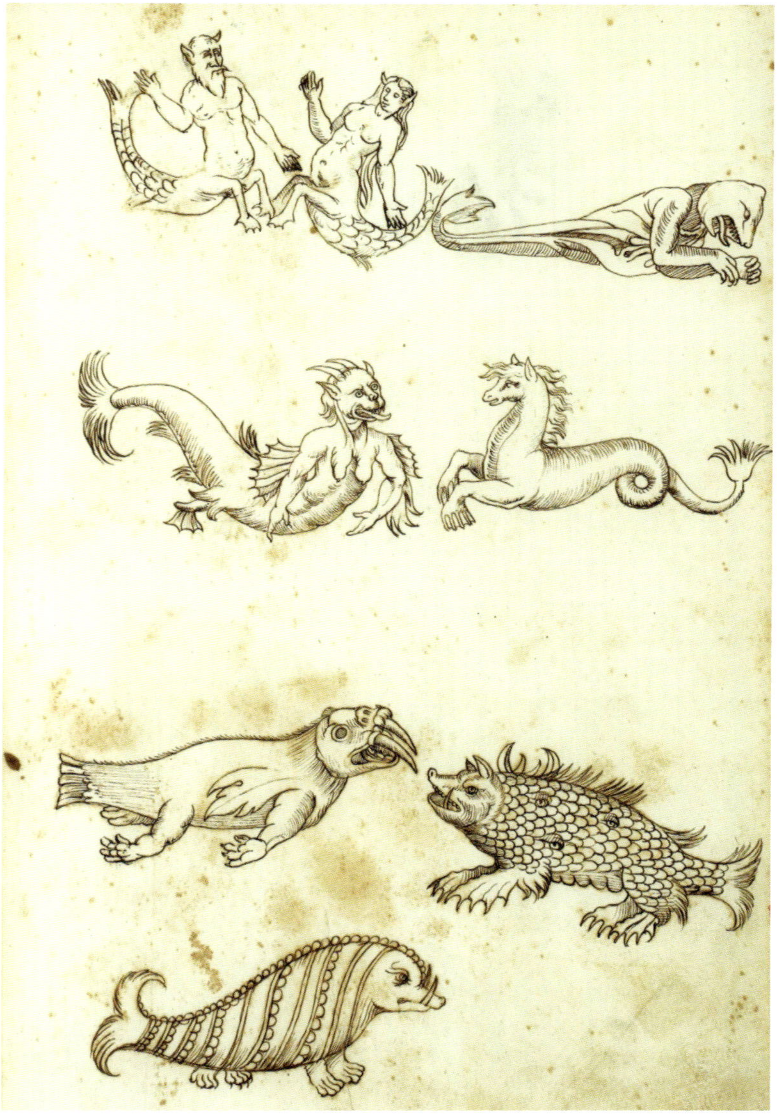

FIG. 6: Sea monster, unfinished drawing, from Aldrovandi's collection of drawings, ca. 47 × 35 cm, second half of the sixteenth century. Aldrovandi's collection of drawings, archive of the University Library of Bologna, Italy. *Tavoli di Animali* VI, c. 72, with permission of the University Library of Bologna.

His description of his own work clearly expresses the same ideal: in "my natural history [. . .] I have described not a single object I did not see with my own eyes, touch with my own hands, and dissect into its external and internal parts. [. . .] Over time, I have collected these objects in my small natural world, where anyone can come—and they do come all day long—to see and contemplate them, preserved in likenesses drawn from life, in our museum."[12]

The empirical study of nature, this account indicates, did not simply mean perception of natural objects or their immediate study *in situ*. On the contrary, turning perceptions of nature into data that would be generalizable and communicable in scientific terms required technical mediation. In the museum, the following discussion aims to show, the drawing archive is a prerequisite for the scholar's study of nature's forms and appearances.[13]

12 Ulisse Aldrovandi, "Discorso naturale," in Tugnoli Pàttaro, *Metodo e sistema delle scienze*, 180.

13 In the "Discorso" and even more extensively in his last will, Aldrovandi also offers concrete information about the scope and status of his picture collection. Ulisse Aldrovandi, "Discorso naturale," 183; Giovanni Fantuzzi, *Memorie della vita di Ulisse Aldrovandi, Medico e Filosofo Bolognese* (Bologna: Lelio dalla Volpe, 1774), 67–85.

The ambition to found natural history on tactile and visual experience is closely associated with the early history of the documentary image as a scholarly instrument. It is accordingly little wonder that Aldrovandi's description of the ideal image sketches a kind of second nature: "On the whole, painting must be the truthful imitation of nature. Kinship and purposiveness must govern the relation between nature and art so that if nature were compelled to accomplish the works of art, they would not be made any other way."[14] So Aldrovandi's ideal of art goes far beyond the notion of art as mimetic depiction, aiming at the idea of a nature-like art. In his words, art as a whole even becomes an "image," a trace, of nature.[15] This close and indeed almost symbiotic affinity between art and nature, in which the former is positively assimilated to the latter, is characteristic of Aldrovandi's conception of art in the context of science. Art is here more than a mere portrayal of nature: it is seen as a mode of articulation that is adequate to nature itself.[16]

The case of the artist Jacopo Ligozzi, the creator of the picture of the two snakes described above, renders this ideal explicit. To Aldrovandi, Ligozzi represented the ideal natural-history draftsman. In his art-theoretical writings, he repeatedly compared him to the ancient artist Parrhasius.[17] Ligozzi's works were virtually perfect implementations of the standards Aldrovandi set for scientific images. In particular, he praised the artist's ability to deceive the beholder, lending such animation to his pictures of animals that the latter seemed to be physically present. According to Aldrovandi's ideal, the painting and the living creature became virtually indistinguishable in the eye of the beholder, blurring the difference between art and nature on the visual plane. This perspective allows us to arrive at a more precise understanding of the striking composition of Ligozzi's picture of the snakes. The suggestive pictorial mise-en-scène served a specific purpose of natural philosophy, also indicating what qualities Aldrovandi thought of as the crucial criteria of nature, as worthy of depiction. Aldrovandi's primary object was what he called the "essence" of nature: its individuality and its animation. "We must first and foremost demand pictures," Aldrovandi wrote, "that render the life and spirit, however different and variegated [the objects depicted] may be."[18] Aldrovandi's employment of imagery, in other words, was geared directly to human perception; his interests did not yet have anything to do with the demands of the classification of living beings (a problem of eighteenth-century zoology). Images, to his mind, were instruments of animation. Their task was to record the essence of nature for the purposes of scientific contemplation.

Recording and Conveying Perceptions of Nature

Yet most of the drawings, by far, in Aldrovandi's collection were not created from live models. The majority of the folios are commissioned works, most of which were produced on the basis of printed sources or preserved specimens from the natural history collection, with only a few derived from the observation of live fauna. On several occasions, Aldrovandi had motifs that already existed in print transferred into his drawing archive. That is the case, for example, of the comparative study of a human and an avian skeleton, which Aldrovandi copied from Pierre Belon's book on birds.[19] More significantly, the monsters and deformations that Aldrovan-

14 Ulisse Aldrovandi, "Avvertimenti del Dottore Aldrovandi sopra le pitture mostrifiche et prodigiose" [1581], in Olmi, *L'inventario del mondo*, 113. A precise description of the collection is also contained in Ulisse Aldrovandi's last will, published in Fantuzzi, *Memorie della vita di Ulisse Aldrovandi*, 67–85.

15 Ulisse Aldrovandi, "Avvertimenti sopra le pitture," 113.

16 Aldrovandi also repeatedly argued that the drawing was an important tool of science and reminded the reader that the hand was the Aristotelian "instrument of instruments" also for the scholar: "la maestra mano, la quale, come testifica Aristotele, è l'istromento degli istromenti." Ulisse Aldrovandi, "Modo di esprimere per la pittura tutte le cose dell'universo mondo: Enarrazione di tutti i generi principali delle cose naturali et artificiali che ponno cadere sotto la pittura," in *Scritti d'arte del cinquecento*, vol. 1, ed. Paola Barocchi (Milan: Ricciardi, 1981), 930.

17 Parrhasius emerged victorious from the painter's contest that pitted him against Zeuxis, as described in volume 15 of Pliny's *Natural History*. Aldrovandi himself makes reference to this competition in a text about "monstrous pictures"; see Ulisse Aldrovandi, "Avvertimenti sopra le pitture," 112.

18 Ulisse Aldrovandi, "Modo di esprimere per la pittura," 925.

19 Library of the University of Bologna, Ulisse Aldrovandi, *Tavoli di animali*, vol. 1, ch. 1. Such use of sources may be traced in other pictures as well; see Fischel, *Natur im Bild*, 55–67, 103–17.

di's collection documents—at the time an issue of great interest to natural scientists[20]—constituted a firmly established part of public visual culture. Consider, for example, the motifs on an unfinished folio (fig. 6) showing the outlines of hybrid creatures that blend anatomical properties of humans, mammals, and fish. The uppermost section of the sheet contains a depiction of a pair of nereids and, to their right and prone, a monstrous fish with the head and upper extremities of a dog. At the center appear a horned devilish monster and a little nereid's horse. Near the bottom of the sheet, we find an animal resembling a walrus and two other sea monsters. The figures are drawn in black ink with firmly delineated contours; there are no exploratory or tentative strokes of the pen. These motifs may be traced back to the illustrations in Ambroise Paré's book *Des monstres et prodiges*, published in 1573.[21] This author, for his part, had copied the visual materials from other sources, most importantly Conrad Gessner's *De piscium et aquatilium animantium natura* (1558).[22] Gessner (1516–1565), a physician and the most important natural historian of the early modern era, was the author of numerous works of natural history. The illustrations printed in his books were copied by many authors and then reproduced from one publication to the next. The pair of nereids, for example, was adopted in Paré's book.[23] It later reappears in Aldrovandi's collection, where several details of the printed image, primarily interior structures and hatchings, have been omitted (figs. 6 and 7). Aldrovandi's collection of drawings, that is to say, not only documented immediate perceptions of nature but also incorporated forms derived from earlier pictorial traditions. So the technique of drawing was not an instrument used to record immediate visual impressions of animate nature.

The relationship between drawing and scientific contemplation merits closer study. It becomes clearer when we consider another folio

20 Lorraine Daston and Katharine Park, *Wonders and the Order of Nature, 1150–1750* (New York: Zone Books, 1998); see also Fischel, *Natur im Bild*, 118–36, with further references.

21 Included in Ambroise Paré, *Deux livres de Chirurgie* (Paris: Wechel, 1573). Yet Gessner, too, used second- and third-hand images, a practice that is as popular now as it was then.

22 Conrad Gessner, *Historia animalium*, vol. 4: *De piscium et aquatilium animantium natura* (Zurich: Froschauer, 1558). On the natural history of fish of this time, cf. Philippe Glardon, *L'histoire naturelle au XVIe siècle: Introduction, étude et édition critique de "La nature et diversité des poissons" de Pierre Belon* (1555) (Geneva: Droz, 2011).

23 Ambroise Paré, *Des monstres et prodiges* [1573], ed. Jean Céard (Geneva: Droz, 1971), 102; cf. Ambroise Paré, *On Monsters and Marvels*, trans. Janis L. Pallister (Chicago: University of Chicago Press, 1982), 12.

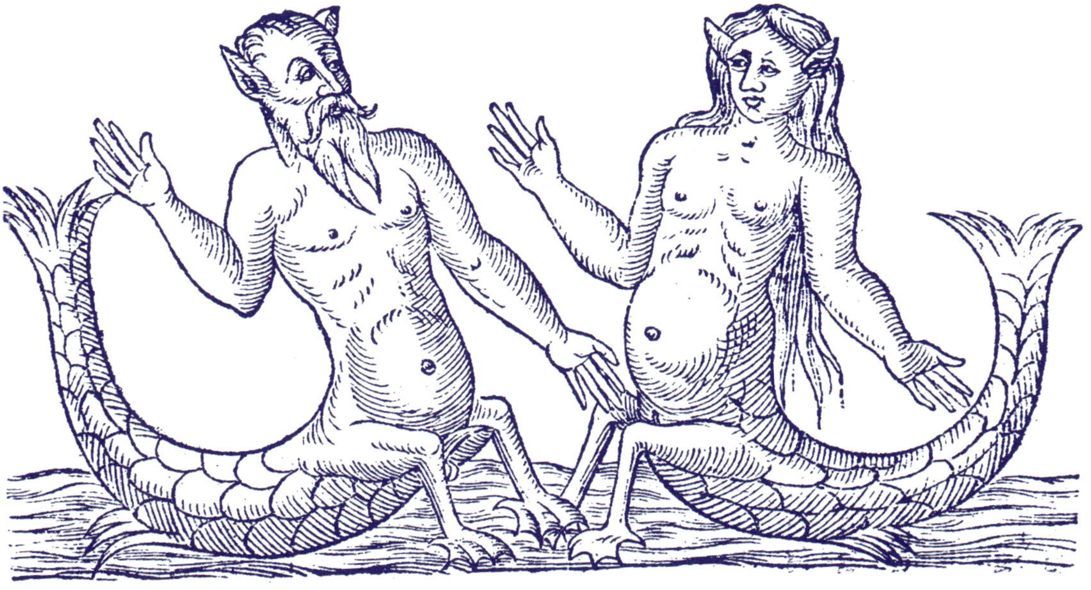

FIG. 7: Ambroise Paré, pair of nereids, from *Des monstres et prodiges* (1573), fig. 40. Ambroise Paré, *On Monsters and Marvels* [1573], trans. and ed. Janis L. Pallister (Chicago: University of Chicago Press, 1982), 102, fig. 40. © BIU Santé (Paris).

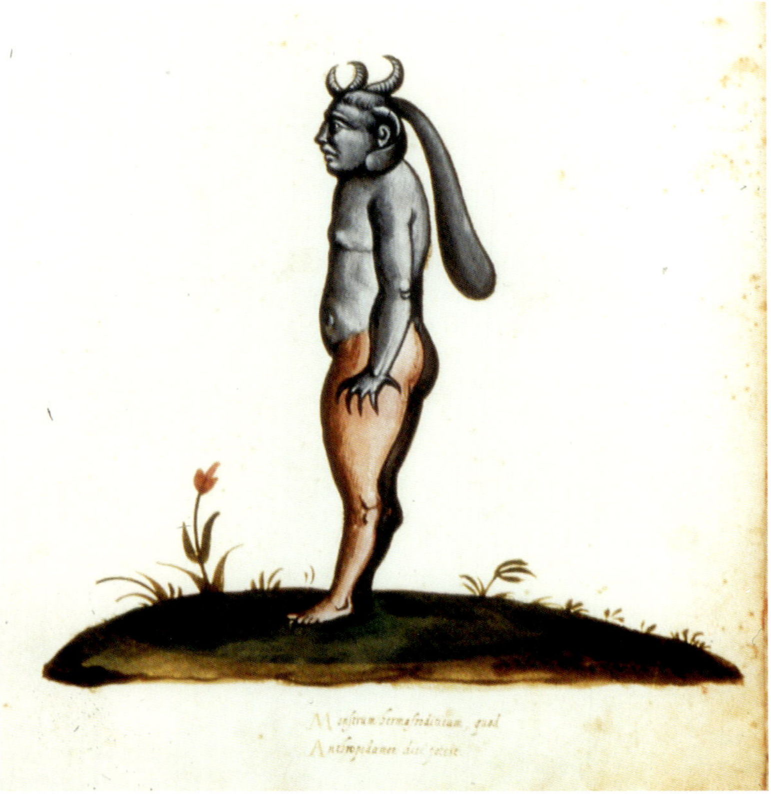

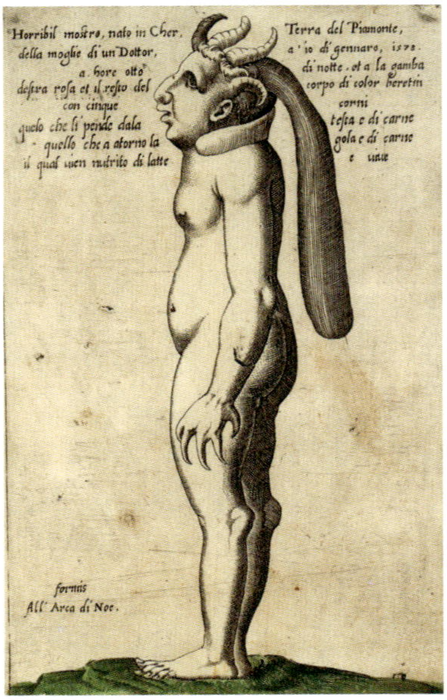

FIG. 8 (above): Hermaphroditic monster, from Aldrovandi's collection of drawings, ca. 47 × 35 cm, second half of the sixteenth century. Aldrovandi's collection of drawings, archive of the University Library of Bologna, Italy. *Tavoli di Animali* I, c. 115, with permission of the University Library of Bologna.

FIG. 9 (right): "Horribil mostro," pamphlet, 19.1 × 12.2 cm (1578). Irene Ewinkel, *De monstris: Deutung und Funktion von Wundergeburten auf Flugblättern im Deutschland des 16. Jahrhunderts* (Tübingen: Niemeyer, 1995), 331, fig. 42. Zentralbibliothek Zürich, Graphische Sammlung und Fotoarchiv, Signatur: PAS II 15/8.

24 Paré, *On Monsters and Marvels*, 11.

25 Fabian Krämer, "The Persistent Image of an Unusual Centaur: A Biography of Aldrovandi's Two-Legged Centaur Woodcut," *Nuncius* 24, no. 2 (2009): 313–40.

26 See Paré, *On Monsters and Marvels*, 11.

from Aldrovandi's collection, which may, moreover, be set in relation to his theory of the image (fig. 8). The drawing presents a motif that circulated on numerous sixteenth-century pamphlets and in books of the period: a monster that was found in Piedmont (fig. 9). It was also included in the above-mentioned book by Ambroise Paré, where it appears, facing left, in a black-and-white print, without any backdrop or ground to stand on. Paré gives a detailed account of the Piedmontese monster, supplying information about the circumstances of its appearance, the place and time, and historical interpretations; he also describes the coloration of its body in detail: "The foot and the right leg were an intense red color. The remainder of the body was a smoky gray color."[24] In Aldrovandi's collection of drawings, it figures as a *Monstrum hermaphroditicum, quod Anthropodamon dici potest.*[25] Everything that would indicate the historical context of its appearance—which had heretofore been a central aspect of all reproductions of this figure—has been stripped away. The new name conferred on it by Aldrovandi instead reduces it to its physical features, the ambiguous sexual anatomy and the rotated knee joints, that mark the creature as an antipode. By contrast, all other historical information, such as the place, time, and circumstances of its birth, has been omitted. On the other hand, visual features such as the coloration, which Paré described in words, have been incorporated with great precision in the drawing in Aldrovandi's collection.[26] Moreover, the draftsman has set the figure on a stylized field of flowers. The transfer from print to drawing has led to several modifications of the motif. It has been isolated from its historical and geographical context, renamed, colorized, and embedded in a schematic natural environment. Such transfers, in other words, were not exact copies but rather translations. This adaptation of images is guided by a particular interest in those qualities of nature that may be perceived by the eye. As Aldrovandi's own writings indicate, these modifications are deliberate shifts that are essential to the study of nature. Painting, he argues, plays a prominent role in the representation of nature because it can give a colorful image of the visible world: "And I say, in affirmation [of Gabriele Paleotti's *Discorso intorno alle immagini sacre e profane*, 1582], that there is nothing in the world that could represent all things created by the great God better than painting, by means of the variety of colors that come into existence in all mixed bodies by virtue of the mixture of the four elements; for color is the true object of vision, as the philosopher [i.e., Aristotle] says, since we cannot see anything that is devoid of color. Color is indeed an accident of things that is of great use when it comes to understanding the substances. Aristotle said, *quod accidentia maxime conferunt ad quod quid erat esse*. [. . .] It is clear, then, how necessary an accident color is, since there is nothing in this humble world that would not need to be described and depicted by means of colors before we can attain an understanding of it."[27] The drawing, that is to say, conveys a visual surplus over the printed source on which it is based because it directly represents the coloration of its object. Its purpose was not so much to document immediate perception as to generate impressions adequate to nature.

Visual Common Sense

The visual documentation of the forms of nature, it turns out upon closer examination, was not a stand-in for absent animals or perishable natural objects. The drawing represents the mise-en-scène of what the natural historians of the sixteenth century thus defined as nature's essence, aiming primarily at the perception of its future beholders, whom it served as a cognitive and experiential basis. In addition to the visual and technical naturalization of printed images by means of their translation into a technique of representation adequate to nature itself, the third function that drawings took on in Aldrovandi's scientific practice was that of communicating visual data. Not all folios were created in the scholar's immediate milieu; many were drawn by naturalists at distant locations and sent to Bologna. Aldrovandi's treatises frequently mention pictures that had been mailed to him or whose arrival he anticipated.[28] Friends and sources all over Europe sent him drawings.[29] A lively exchange also connected various image collections. For example, information traveled between Bologna and Prague, where Rudolf II built his *Kunstkammer*, and lists of animal depictions in Rudolf's collections were communicated to Aldrovandi.[30]

Aldrovandi's correspondence shows, moreover, that he likewise sent pictures and natural objects to other collectors. In particular, substantial shipments went to Francesco de' Medici in Florence.[31] So the drawings not only served as visual aids in the individual scholar's work, they also provided a medium of exchange and the communication of visual records between natural scientists, transposing individual perceptions into the sphere of comparison and critique. They formed the foundation for the establishment of what we may literally call a "visual common sense"

27 Ulisse Aldrovandi, "Avvertimenti del dottore Aldrovandi all'Ill.mo e R.mo Cardinale Paleotti sopra alcuni capitoli della pittura," in *Trattati d'arte del Cinquecento: Fra manierismo e Controriforma*, vol. 2 (Bari: Laterza, 1961), 512–13. For the debate over images in religion and science, see Steinemann, *Eine Bildtheorie zwischen Repräsentation und Wirkung*, 126.

28 One such shipment is mentioned in one of Aldrovandi's image-critical treatises: a year earlier, the author mentions in passing, his nephew Giuliano Griffoni had drawn the likeness of a monstrous dog for him, and he was expecting the packet from Rome, including the picture and other painted documents, to arrive any day. Aldrovandi, "Avvertimenti sopra le pitture," 116.

29 Selections from his correspondence have been published in Ludovico Frati, ed., *Catalogo dei manoscritti di Ulisse Aldrovandi* (Bologna: Zanichelli, 1907).

30 According to these letters, Giuseppe Arcimboldo, who was court portraitist to Emperor Rudolf II, made two drawings in Aldrovandi's collection. See Eva Irblich, ed., *Thesaurus Austriacus: Europas Glanz im Spiegel der Buchkunst. Handschriften und Kunstalben von 800 bis 1600* (Vienna: Österreichische Nationalbibliothek, 1996), 233.

31 Registers of entire collections and drawings changed owners in this manner. See, for example, the letter to Grand Duke Francesco I of Tuscany, dated September 27, 1577, which includes a catalogue of natural objects ("Catalogus Rerum ad Magnum Hetruriae Ducem missarum ab Aldrovando") as well as additional letters to the grand duke containing information about natural-historical objects. Fantuzzi lists several letters containing entire catalogues of objects that Aldrovandi shipped to the grand duke, which frequently also mention pictures and wood engravings. See Fantuzzi, *Memorie della vita di Ulisse Aldrovandi*, esp. 126–27.

among natural scientists: an authoritative visual-pictorial memory of natural history.

It is with a view to these far-ranging functions that Aldrovandi's critical reflection on the image begins. The critique of the image, that is to say, is as old as its use as an instrument of science. Even the translation of printed visual sources into drawings, Aldrovandi recognized, was a procedure that posed risks: the draftsman's hand, his lack of knowledge or his creative imagination, might allow faulty information to slip into the drawing, where it would become manifest and part of the subsequent tradition. Aware of this danger, Aldrovandi developed several criteria that natural-history drawing was supposed to meet; these criteria also clearly marked the distinction between documentary and artistic drawing.[32] He emphatically warned against the hazards of distortion, denouncing records that were not adequate to nature. Pictures by painters who allowed their imagination to influence the representation of nature, he argued, were monstrous and grotesque.[33]

Elsewhere, Aldrovandi brands the productions of imprecise draftsmen depicting natural objects as artificial chimeras. A painter who sets out to render nature, he writes, needs to know it intimately, or else he will commit mistakes and, as it were, bring monsters into the world. For not only the imagination produces false likenesses; so does ignorance: "In instances in which the painter depicts parts of the bodies of humans or other animals [. . .], he must regard them with a very keen eye and imitate them closely so as to avoid painting some chimera."[34] Aldrovandi thus drew a clear distinction between scientific images and other kinds of imagery, laying the foundations for a communication of visual impressions as scientific information. Eliminating the imagination became a fundamental technique of scientific draftsmanship even before the establishment of the modern natural sciences.[35]

The copying of images from printed sources was indispensable to natural history from the early modern era on. This practice would seem incompatible with contemporary ideals of scientific documentation, but in view of the historical conditions in which research operated, especially with regard to the use of sources in the natural sciences around 1600, the problem takes on a different aspect.

Aldrovandi pioneered observation as a technique of knowledge and persistently championed visual perception as the basis of knowledge in natural history. The collection of drawings was a crucial auxiliary in the pursuit of this ideal. One of the most important roles that drawings played for Aldrovandi was to convey the visual appearance of an animal in a manner adequate to its character. The drawing became the basis for the contemplation of nature in the natural history collection but only infrequently documented an immediate perception of nature.[36] It seemed the instrument most suited to the task of integrating the characteristics of animals into the scholarly apparatus and transforming them into knowledge that could be communicated in scientific terms.

Although Aldrovandi, like other scientists of the late sixteenth century,[37] celebrated visual perception as an innovative technique of knowledge, it has much in common with the "traditional" techniques of managing knowledge: the techniques of reading, copying, citing, and systematizing texts that Aldrovandi himself practiced to perfection.[38] In

32 See Aldrovandi, "Avvertimenti sopra le pitture," 112–17; Aldrovandi, "Modo di esprimere per la pittura," 925; see also Fischel, *Natur im Bild*, 111–17.

33 Aldrovandi, "Modo di esprimere per la pittura," 930.

34 Ibid., 929.

35 The latter is usually dated to the routine employment of optical and scientific instruments and the establishment of scientific academies. Alternatively, some authorities identify it with the publication of Bacon's *Novum Organon* in 1611.

36 The most extensive discussion of this ideal in Aldrovandi appears in the "Discorso naturale."

37 Consider, for example, the pictorial argument presented slightly later by Galileo Galilei; see Horst Bredekamp, "Gazing Hands and Blind Spots: Galileo as Draftsman," *Science in Context* 13, nos. 3–4 (2000): 424–62.

38 See note 8.

the field of empirical study, too, Aldrovandi relied on the intercession of media, and his work remained rooted in traditions. The case of Aldrovandi's images illustrates how pictorial traditions can inform and even constitute people's conception of nature. Still, empirical study as practiced in this scenario amounted to more than the preservation of the pictorial tradition; it set new focuses in the study of nature, as the shift toward images in natural history was associated with a methodological sea change that established, in conjunction with the techniques for a visual analysis of nature, a new conception of nature itself.

The visual documentation of animals in early modern natural history can be described as a historically informed cultural technique that is by no means self-explanatory. Moreover, the case of the sixteenth-century collections of natural-history drawings discussed in these pages confronts us with a fundamental dispute over the use of documentary images in science. The rising ideal of zoology as an empirical science founded on visual perception seems to contradict the constitutive influence of a pictorial memory on that perception. But that impression arises from a misunderstanding of the function of documentary images in Aldrovandi's time. In early modern natural history, drawings do not first and foremost serve to record the individual scholar's perceptions; rather, they communicate the visual appearance of animals and thus create the basis for the contemplation of nature. Documentary depictions of natural objects play an instrumental role in the scholar's communication of visual perception: they are a matrix for his future studies.[39]

39 I would like to thank Gerrit Jackson for his helpful critical remarks and Gerald Nixon for proofreading parts of this essay.

BIBLIOGRAPHY

The following bibliography lists essential contributions to the scholarship on scientific imagery, with an emphasis on literature in German and English. It is divided into sections according to subject areas; within each section, titles are listed alphabetically. A few titles are listed in more than one section. The bibliography presents a survey of the existing scholarship and makes no claim to completeness. For particular issues, please also consult the references listed at the end of the pertinent glossary entries.

ART-HISTORICAL FOUNDATIONS AND METHODS IN VISUAL ANALYSIS

Foundations of Art-Historical Methodology

Bader, Lena, Martin Gaier, and Falk Wolf, eds. *Vergleichendes Sehen*. Munich: Fink, 2010.

Baxandall, Michael. *Patterns of Intention: On the Historical Explanation of Pictures*. New Haven, CT: Yale University Press, 1985.

Belting, Hans. *Likeness and Presence: A History of the Image before the Era of Art* [1990]. Trans. Edmund Jephcott. Chicago: University of Chicago Press, 1994.

Belting, Hans, Heinrich Dilly, Wolfgang Kemp, Willibald Sauerländer, and Martin Warnke, eds. *Kunstgeschichte. Eine Einführung*. 6th ed. Berlin: Reimer, 2003.

Bredekamp, Horst. "A Neglected Tradition? Art History as 'Bildwissenschaft.'" *Critical Inquiry* 29, no. 3 (2003): 418–28.

Bredekamp, Horst, Michael Diers, Kurt W. Forster, Nicolas Mann, Salvatore Settis, and Martin Warnke, eds. Aby Warburg, *Der Bilderatlas Mnemosyne; Gesammelte Schriften*, Zweite Abteilung, no. II.1. Berlin: Akademie Verlag, 2000.

Bruhn, Matthias. *Das Bild: Theorie, Geschichte, Praxis*. Berlin: Akademie Verlag, 2009.

Bruhn, Matthias, and Vera Dünkel. "The Image as Cultural Technology." In *Visual Literacy*, edited by James Elkins, 165–78. New York: Routledge, 2008.

Damisch, Hubert. *A Theory of /Cloud/: Toward a History of Painting* [1972]. Translated by Janet Lloyd. Stanford, CA: Stanford University Press, 2002.

Freedberg, David. *The Power of Images: Studies in the History and Theory of Response*. Chicago: University of Chicago Press, 1989.

Hatt, Michael, and Charlotte Klonk. *Art History: A Critical Introduction to Its Methods*. Manchester: Manchester University Press, 2006.

Kubler, George. *The Shape of Time: Remarks on the History of Things*. New Haven, CT: Yale University Press, 1962.

Matyssek, Angela. *Kunstgeschichte als fotografische Praxis: Richard Hamann und Foto Marburg*. Berlin: Mann, 2009.

Meyer, Leonard B., and Berel Lang, eds. *The Concept of Style*. Philadelphia: University of Pennsylvania Press, 1979.

Mitchell, Frank B., and Daniel Adler, eds. *German Art History and Scientific Thought*. Farnham: Ashgate, 2012.

Panofsky, Erwin, "The Concept of Artistic Volition" [1920], trans. Kenneth J. Northcott, and Joel Snyder. *Critical Inquiry* 8, no. 1 (1981): 17–33.

Panofsky, Erwin. "On the Problem of Describing and Interpreting Works of the Visual Arts." *Critical Inquiry* 38, no. 3 (Spring 2012): 467–82.

Panofsky, Erwin. *Sinn und Deutung in der bildenden Kunst*. Cologne: DuMont Schauberg, 1955.

Riegl, Alois. *Problems of Style: Foundations for a History of Ornament*. 1893; repr. Princeton, NJ: Princeton University Press, 1992.

Wittkower, Rudolf. *Architectural Principles in the Age of Humanism*. London: Warburg Institute, University of London, 1949.

Wölfflin, Heinrich. *Principles of Art History: The Problem of the Development of Style in Later Art*. 7th ed. 1915; repr. New York: Dover, 1932.

Wood, Christopher S., ed. *The Vienna School Reader: Politics and Art Historical Method in the 1930s*. New York: Zone Books, 2000.

Bildwissenschaft and Visual Culture Studies

Although Visual Culture Studies as conducted in the English-speaking world and the German *Bildwissenschaft* have different disciplinary histories, their approaches often blend into each other, and both frequently address the same issues and materials. Painting in very broad strokes, we may say that *Bildwissenschaft* examines the picture as such, whereas Visual Culture Studies focus on the relationship between pictures and their beholders while also analyzing pictorial techniques. Still, a rigorous distinction between the two is virtually impossible to make and in any case not helpful, and so the following lists contributions to both fields.

Bal, Mieke. *A Mieke Bal Reader*. Chicago: University of Chicago Press, 2006.

Belting, Hans. *Bild-Anthropologie: Entwürfe für eine Bildwissenschaft*. Munich: Fink, 2001.

Belting, Hans. *Bildfragen: Die Bildwissenschaften im Aufbruch*. Munich: Beck, 2007.

Berger, John. *Ways of Seeing*. London: British Broadcasting Corporation and Penguin Books, 1972.

Boehm, Gottfried, ed. *Was ist ein Bild?* Munich: Fink, 1994.

Bryson, Norman, Michael A. Holly, and Keith P. F. Moxey, eds. *Visual Culture: Images and Interpretations*. Hanover, NH: University Press of New England [for] Wesleyan University Press, 1994.

Davis, Whitney. *A General Theory of Visual Culture*. Princeton, NJ: Princeton University Press, 2011.

Dikovitskaya, Margaret. *Visual Culture: The Study of the Visual after the Cultural Turn*. 1st ed. Cambridge, MA: MIT Press, 2005.

Elkins, James. *The Domain of Images*. Ithaca, NY: Cornell University Press, 1999.

Elkins, James, ed. *Visual Literacy*. New York: Routledge, 2008.

Elkins, James. *Visual Studies: A Skeptical Introduction*. New York: Routledge, 2003.

Freedberg, David. "Iconography between the History of Art and the History of Science: Art, Science, and the Case of the Urban Bee." In *Picturing Science, Producing Art*, edited by Caroline A. Jones and Peter Galison, 272–96. New York: Routledge, 1998.

Hall, Stuart, ed. *Representation: Cultural Representations and Signifying Practices*. London: Sage, 1997.

Heßler, Martina, and Dieter Mersch, eds. *Logik des Bildlichen: Zur Kritik der ikonischen Vernunft*. Bielefeld: Transcript, 2009.

Holly, Michael A. *Past Looking: Historical Imagination and the Rhetoric of the Image*. Ithaca, NY: Cornell University Press, 1996.

Jay, Martin. "That Visual Turn: The Advent of Visual Culture." Interview, *Journal of Visual Culture*, no. 1 (2002): 87–92.

Maar, Christa, and Hubert Burda, eds. *Iconic Turn: Die neue Macht der*

Bilder. Cologne: Dumont, 2004.
Mirzoeff, Nicholas, ed. *The Visual Culture Reader*. London: Routledge, 1999.
Mitchell, William J. T. *Iconology*. Chicago: University of Chicago Press, 1994.
Mitchell, William J. T. "The Pictorial Turn." *Artforum*, March 1992, 89–95.
Mitchell, William J. T. *Picture Theory: Essays on Verbal and Visual Representation*. Chicago: University of Chicago Press, 1995.
Mitchell, William J. T. *What Do Pictures Want?: The Lives and Loves of Images*. Chicago: University of Chicago Press, 2005.
Paul, Gerhard, ed. *Visual History: Ein Studienbuch*. Göttingen: Vandenhoeck & Ruprecht, 2006.
Sachs-Hombach, Klaus. *Bildwissenschaft: Disziplinen, Themen, Methoden*. Frankfurt a.M.: Suhrkamp, 2005.
Schulz, Martin. *Ordnungen der Bilder: Eine Einführung in die Bildwissenschaft*. Munich: Fink, 2005.
Stafford, Barbara. *Good Looking: Essays on the Virtue of Images*. Cambridge, MA: MIT Press, 1996.

Visual Thinking and Theories of Perception

Alpers, Svetlana, and Michael Baxandall. *Tiepolo and the Pictorial Intelligence*. New Haven, CT: Yale University Press, 1994.
Antonelli, Paola. *Design and the Elastic Mind*. New York: Museum of Modern Art, 2008.
Arnheim, Rudolf. *Visual Thinking*. Berkeley: University of California Press, 1969.
Bexte, Peter. *Blinde Seher: Die Wahrnehmung von Wahrnehmung in der Kunst des 17. Jahrhunderts*. Dresden: Verlag der Kunst, 1999.
Crary, Jonathan. *Suspensions of Perception: Attention, Spectacle, and Modern Culture*. Cambridge, MA: MIT Press, 1999.
Crary, Jonathan. *Techniques of the Observer: On Vision and Modernity in the Nineteenth Century*. Cambridge, MA: MIT Press, 1990.
Bundy, Murray W. *The Theory of Imagination in Classical and Mediaeval Thought*. Urbana: University of Illinois, 1927.
Clark, Stuart. *Vanities of the Eye: Vision in Early Modern European Culture*. Oxford: Oxford University Press, 2007.
Crombie, A. C. *Styles of Scientific Thinking in the European Tradition: The History of Argument and Explanation Especially in the Mathematical and Biomedical Sciences and Arts*. London: Duckworth, 1994.
Crow, Thomas E. *The Intelligence of Art*. Chapel Hill: University of North Carolina Press, 1999.
Damisch, Hubert. *The Origin of Perspective*. Cambridge, MA: MIT Press, 1995.
Didi-Huberman, Georges. *Was wir sehen, blickt uns an*. Munich: Fink, 1999.
Ehrenzweig, Anton. *The Hidden Order of Art: A Study in the Psychology of Artistic Imagination*. London: Weidenfeld & Nicolson, 1967.
Elkins, James. *The Object Stares Back*. San Diego: Harcourt Brace, 1997.
Foster, Hal, ed. *Vision and Visuality*. Seattle: Bay Press, 1988.
Gombrich, Ernst H. *Art and Illusion: A Study in the Psychology of Pictorial Representation*. New York: Pantheon Books, 1960.
Hall, Edward T. *The Hidden Dimension*. Garden City, NY: Doubleday, 1966.
Heyden, Anne van der, and Claudia Blümle. *Blickzähmung und Augentäuschung: Jacques Lacans Bildtheorie*. Zurich and Berlin: Diaphanes, 2006.
Hunter, Matthew C., and Francesco Lucchini, eds. *The Clever Object*. Hoboken, NJ: John Wiley & Sons, 2013.
Jay, Martin. *Downcast Eyes: The Denigration of Vision in Twentieth-Century French Thought*. Berkeley: University of California Press, 1993.
Kemp, Martin. *The Science of Art: Optical Themes in Western Art from Brunelleschi to Seurat*. New Haven, CT: Yale University Press, 1990.
Kemp, Martin. "Seeing and Picturing: Visual Representation in Twentieth-Century Science." In *Science in the Twentieth Century*, edited by John Krige and Dominique Pestre, 361–90. Amsterdam: Harwood Academic Publishers, 1997.
Kemp, Martin. *Seen/Unseen: Art, Science, and Intuition from Leonardo to the Hubble Telescope*. Oxford: Oxford University Press, 2006.
Kemp, Wolfgang, ed. *Der Betrachter ist im Bild: Kunstwissenschaft und Rezeptionsästhetik*. Cologne: DuMont, 1985.
Kepes, György. *Language of Vision*. Chicago: Paul Theobald, 1944; reissued New York: Dover Publications, 1995.
Lindberg, David C. *Theories of Vision from Al-kindi to Kepler*. Chicago: University of Chicago Press, 1976.
Nöe, Alva. *Action in Perception*. Cambridge, MA: MIT Press, 2006.
Panofsky, Erwin. *Perspective as Symbolic Form*. New York: Zone Books, 1997.
Pick, Daniel. "Stories of the Eye." In *Rewriting the Self: Histories from the Renaissance to Present*, edited by Roy Porter, 186–99. London: Routledge, 1997.
Schneider, Birgit, and Vera Dünkel. "Rundbild und Augenblick." In *Der Ball ist rund: Kreis Kugel Kosmos*, exhibition catalogue, Pergamonmuseum Berlin, June 8–August 27, 2006, edited by Bernd Ebert and Moritz Wullen, 108–15. Berlin: Pergamonmuseum Berlin, 2006.
Swan, Claudia. "Eyes Wide Shut: Early Modern Imagination, Demonology, and the Visual Arts." *Zeitsprünge: Forschungen zur Frühen Neuzeit* 7 (2003): 156–81.
Yates, Frances A. *The Art of Memory*. London: Routledge and Kegan Paul, 1966.
Zeki, Semir. *Inner Vision: An Exploration of Art and the Brain*. Oxford: Oxford University Press, 1999.

Visual Aspects of Writing and Issues in Semiotics

Bal, Mieke. *A Mieke Bal Reader*. Chicago: University of Chicago Press, 2006.
Brandt, Reinhard. *Die Wirklichkeit des Bildes: Sehen und Erkennen; Vom Spiegel zum Kunstbild*. Munich and Vienna: Hanser, 2002.
Bredekamp, Horst, and Sybille Krämer, eds. *Bild, Schrift, Zahl*. Munich: Fink, 2003.
Elkins, James. *The Domain of Images*. Ithaca, NY: Cornell University Press, 1999.
Foucault, Michel. *The Order of Things: An Archaeology of the Human Sciences* [1966]. New York: Vintage Books, 1970.
Goodman, Nelson. *Languages of Art: An Approach to a Theory of Symbols*. London: Oxford University Press, 1969.
Halliwell, Stephen. *The Aesthetics of Mimesis: Ancient Texts and Modern Problems*. Princeton, NJ: Princeton University Press, 2002.

Krämer, Sybille. "'The Mind's Eye': Visualizing the Non-Visual and the 'Epistemology of the Line.'" In *Image and Imaging in Philosophy, Science, and the Arts.* Preproceedings of the 33rd International Wittgenstein Symposium, Vol. 2, edited by Elisabeth Nemeth, Richard Heinrich, and Wolfram Pichler, 275–95. Kirchberg am Wechsel: Austrian Ludwig Wittgenstein Society, 2010.

Krämer, Sybille, ed. *Schriftbildlichkeit: Wahrnehmbarkeit, Materialität und Operativität von Notationen.* Berlin: Akademie Verlag, 2012.

Krämer, Sybille. *Symbolische Maschinen: Die Idee der Formalisierung in geschichtlichem Abriß.* Darmstadt: Wissenschaftliche Buchgesellschaft, 1988.

Kress, Gunther, and Theo Van Leeuwen. *Reading Images: The Grammar of Visual Design.* London: Routledge, 2006.

Mitchell, W. J. T. *Iconology: Image, Text, Ideology.* Chicago: University of Chicago Press, 1986.

Mitchell, W. J. T. *Picture Theory: Essays on Verbal and Visual Representation.* Chicago: University of Chicago Press, 1994.

Marin, Louis. *On Representation* [1994]. Translated by Catherine Porter. Stanford, CA: Stanford University Press, 2001.

Peirce, Charles Sanders. *Peirce on Signs: Writings on Semiotic by Charles Sanders Peirce.* Edited by James Hoopes. Chapel Hill: University of North Carolina Press, 1991.

Schneider, Pablo, and Moritz Wedell, eds. *Grenzfälle: Transformationen von Bild, Schrift und Zahl.* Weimar: VDG, Verlag und Datenbank für Geisteswissenschaften, 2004.

Scholz, Oliver Robert. *Bild, Darstellung, Zeichen: Philosophische Theorien bildhafter Darstellung.* Freiburg: Alber, 1991.

Wiesing, Lambert. *Artifizielle Präsenz: Studien zur Philosophie des Bildes.* Frankfurt am Main: Suhrkamp, 2005.

IMAGES BETWEEN ART AND SCIENCE / AESTHETICS IN SCIENCE

Ackerman, James S. *Distance Points: Essays in Theory and Renaissance Art and Architecture.* Cambridge, MA: MIT Press, 1991.

Ackerman, James S. *Origins, Imitation, Conventions: Representation in the Visual Arts.* Cambridge, MA: MIT Press, 2002.

Alpers, Svetlana. *The Art of Describing: Dutch Art in the Seventeenth Century.* Chicago: University of Chicago Press, 1983.

Baigrie, Brian S., ed. *Picturing Knowledge: Historical and Philosophical Problems Concerning the Use of Art in Science,* 2nd revised edition. Toronto: University of Toronto Press, 1996.

Bätzner, Nike, Werner Nekes, and Eva Schmidt, eds. *Blickmaschinen, oder wie die Bilder entstehen: Die Zeitgenössische Kunst schaut auf die Sammlung Werner Nekes.* Exhibition catalogue. Cologne: Dumont, 2008.

Belting, Hans. *Das echte Bild: Bildfragen als Glaubensfragen.* Munich: C. H. Beck Verlag, 2005.

Blume, Eugen, Thomas Schnalke, Annemarie Hürlimann, and Daniel Tyradellis, eds. *Schmerz: Kunst + Wissenschaft.* Exhibition catalogue, Hamburger Bahnhof, Museum für Gegenwart, Berlin und Medizinhistorisches Museum der Charité, Berlin, April 5–August 5, 2007. Cologne: Dumont, 2007.

Blümle, Claudia, and Armin Schäfer, eds. *Struktur, Figur, Kontur: Abstraktion in Kunst- und Lebenswissenschaften.* Berlin and Zurich: Diaphanes, 2007.

Brain, Robert Michael. "The Pulse of Modernism: Experimental Physiology and Aesthetic Avant-Gardes circa 1900." *Studies in History and Philosophy of Science* 39 (2008): 393–417.

Bredekamp, Horst, Jochen Brüning, and Cornelia Weber, eds. *Theatrum Naturae et Artis: Theater der Natur und Kunst; Wunderkammern des Wissens.* Exhibition catalogue, Martin-Gropius-Bau, Berlin, December 10, 2000–March 4, 2001. Berlin: Henschel, 2000.

Clair, Jean, ed. *L'âme au corps: Arts et Sciences 1793–1993.* Exhibition catalogue. Grand Palais, Paris. Paris: Gallimard, 1993.

Clarke, Bruce, and Linda Dalrymple Henderson, eds. *From Energy to Information: Representation in Science and Technology, Art, and Literature.* Palo Alto, CA: Stanford University Press, 2002.

DaCosta Kaufmann, Thomas. *The Mastery of Nature: Aspects of Art, Science, and Humanism in the Renaissance.* Princeton, NJ: Princeton University Press, 1993.

Daston, Lorraine, ed. *Things That Talk: Object Lessons from Art and Science.* New York: Zone Books, 2004.

Daston, Lorraine, and Katharine Park. *Wonders and the Order of Nature, 1150–1750.* New York: Zone Books, 1998.

Didi-Huberman, Georges. *Atlas: How to Carry the World on One's Back?* Madrid: Museo Nacional Centro de Arte Reina Sofía, 2010.

Dürbeck, Gabriele, ed. *Wahrnehmung der Natur. Natur der Wahrnehmung: Studien zur Geschichte visueller Kultur um 1800.* Dresden: Verlag der Kunst, 2001.

Elkins, James. *How to Use Your Eyes.* New York, 2000.

Feyerabend, Paul. *Wissenschaft als Kunst.* Frankfurt am Main: Suhrkamp, 1984.

Flach, Sabine, and Margarete Vöhringer, eds. *Ultravision: Zum Wissenschaftsverständnis der Avantgarde.* Munich: Fink 2010.

Fox, Celina. *The Arts of Industry in the Age of Enlightenment.* New Haven, CT: Yale University Press for the Paul Mellon Centre for Studies in British Art, 2009.

Fox, Robert, and Agustí Nieto-Galan. *Natural Dyestuffs and Industrial Culture in Europe, 1750–1880.* Canton, MA: Science History Publications, 1999.

Frankel, Felice. *Envisioning Science: The Design and Craft of the Science Image.* Cambridge, MA: MIT Press, 2002.

Friedrich, Markus, Karin Leonhard, and Gabriele Wimböck, eds. *Evidentia: Reichweiten visueller Wahrnehmung in der Frühen Neuzeit.* Münster: LIT, 2007.

Gerbino, Anthony, and Stephen Johnston. *Compass and Rule: Architecture as Mathematical Practice in England, 1500–1750.* New Haven, CT: Yale University Press, 2009.

Gombrich, Ernst H. *Das forschende Auge: Kunstbetrachtung und Naturwahrnehmung.* Frankfurt am Main: Campus, 1994.

Gördüren, Petra, and Dirk Luckow, eds. *Dopplereffekt. Bilder in Kunst und Wissenschaft.* Exhibition catalogue, Kunsthalle Kiel. Cologne: Dumont, 2010.

Grau, Oliver. *Virtuelle Kunst in Geschichte und Gegenwart: Visuelle Strategien.* Berlin: Reimer, 2001.

Hauser, Jens, ed. *Sk-interfaces: Exploding Borders: Creating Membranes in Art, Technology and Society.* Liverpool, UK: FACT, 2008.

Henderson, Linda Dalrymple. *The Fourth Dimension and Non-Euclidean Geometry in Modern Art.* Princeton, NJ: Princeton University Press, 1983.

Henderson, Linda Dalrymple, and Bruce Clark, eds. *From Energy to Information: Representation in Science and Technology, Art, and Literature.* Palo Alto, CA: Stanford University Press, 2002.

Hyman, John. "Art and Neuroscience." In *Beyond Mimesis and Convention: Representation in Art and Science,* ed. Roman Frigg and Matthew Hunter, 245–261. Dordrecht: Springer, 2010.

Jones, Caroline A. "The Modernist Paradigm: The Artworld and Thomas Kuhn." *Critical Inquiry* 26, no. 3 (2000): 488–528.

Jones, Caroline A., and Peter Galison, eds. *Picturing Science, Producing Art*. London: Routledge, 1998.

Kemp, Martin. *Seen/Unseen: Art, Science, and Intuition from Leonardo to the Hubble Telescope*. Oxford: Oxford University Press, 2006.

Kemp, Martin, and Marina Wallace, eds. *Spectacular Bodies: The Art and Science of the Human Body from Leonardo to Now*. Exhibition catalogue, Hayward Gallery. Berkeley: University of California Press, 2000.

Latour, Bruno, and Peter Weibel, eds. *Iconoclash: Beyond the Image Wars in Science, Religion, and Art*. Exhibition catalogue, ZKM Karlsruhe, May 4–August 4, 2002. Karlsruhe: Center for New Art and Media (ZKM) /MIT Press, 2002.

Lynch, Michael, and Samuel Y. Edgerton. "Aesthetics and Digital Image Processing: Representational Craft in Contemporary Astronomy." In *Picturing Power: Visual Depictions and Social Relations*, edited by Gordon Fyfe and John Law, 184–220. London: Routledge, 1988.

Mersch, Dieter, and Michaela Ott, eds. *Kunst und Wissenschaft*. Munich: Fink, 2007.

Métraux, Alexandre, and Andreas Mayer, eds. *Kunstmaschinen: Spielräume des Sehens zwischen Wissenschaft und Ästhetik*. Frankfurt am Main: Fischer, 2005.

Panofsky, Erwin. "Artist, Scientist, Genius: Notes on the 'Renaissance-Dämmerung.'" In *The Renaissance: Six Essays*, ed. W. Ferguson, 121–82. New York: Harper & Row, 1962.

Panofsky, Erwin. *Galileo as a Critic of the Arts*. The Hague: M. Nijhoff, 1954.

Parkinson, Gavin. *Surrealism, Art, and Modern Science: Relativity, Quantum Mechanics, Epistemology*. New Haven, CT: Yale University Press, 2008.

Pichler, Cathrin, and Susanne Neuburger, eds. *The Moderns: Wie sich das 20. Jahrhundert in Kunst und Wissenschaft erfunden hat*. Exhibition catalogue, Museum Moderner Kunst. Vienna: Springer, 2012.

Reeves, Eileen. *Painting the Heavens: Art and Science in the Age of Galileo*. Princeton, NJ: Princeton University Press, 1997.

Reichle, Ingeborg. *Kunst aus dem Labor: Zum Verhältnis von Kunst und Wissenschaft im Zeitalter der Technoscience*. Vienna: Springer, 2005.

Rifkin, Benjamin A., and Michael J. Ackermann, eds. *Human Anatomy: Five Centuries of Art and Science; From the Renaissance to the Digital Age*. New York: Harry N. Abrams, 2006.

Scherer, Ralf, and Burkhard Leismann, eds. *Diagnose Kunst: Die Medizin im Spiegel der zeitgenössischen Kunst*. Exhibition catalogue. Kunstmuseum Ahlen, October 22, 2006–February 11, 2007. Cologne: Wienand, 2006.

Schmidt-Burkhardt, Astrit. *Diagrammatik und Kunst*. Bielefeld: Transcript, 2012.

Schramm, Helmar, Ludger Schwarte, and Jan Lazardzig, eds. *Kunstkammer, Laboratorium, Bühne: Schauplätze des Wissens im 17. Jahrhundert*. Berlin: Walter de Gruyter, 2003.

Shapiro, Alan E. "Artists' Colors and Newton's Colors." *Isis* 85, no. 4 (1994): 600–630.

Shirley, John W., and F. David Hoeniger, eds. *Science and Arts in the Renaissance*. Washington, DC: Folger Shakespeare Library, 1985.

Smith, Pamela H. "Art, Science, and Visual Culture in Early Modern Europe." *Isis* 97 (2006): 83–100.

Smith, Pamela H. *The Body of the Artisan: Art and Experience in the Scientific Revolution*. Chicago: University of Chicago Press, 2004.

Stafford, Barbara Maria. *Artful Science: Enlightenment, Entertainment, and the Eclipse of Visual Education*. Cambridge, MA: MIT Press, 1994.

Stafford, Barbara Maria. *Body Criticism: Imaging the Unseen in Enlightenment Art and Medicine*. Cambridge, MA: MIT Press, 1991.

Stafford, Barbara Maria. *Visual Analogy: Consciousness as the Art of Connecting*. Cambridge, MA: MIT Press, 2001.

Swan, Claudia. *Art, Science, and Witchcraft in Early Modern Holland: Jacques de Gheyn II (1565–1629)*. Cambridge: Cambridge University Press, 2005.

Thomas, Ann, ed. *Beauty of Another Order. Photography in Science*. Exhibition catalogue, National Gallery of Canada, Ottawa 1997. New Haven, CT: Yale University Press, 1997.

Werner, Gabriele. "Das technische Bild—aus ästhetischer Sicht betrachtet." In *Mit den Augen denken: Strategien der Sichtbarmachung in wissenschaftlichen und virtuellen Welten*, edited by Bettina Heintz and Jörg Huber, 367–82. Vienna: Springer, 2001.

Werner, Gabriele. *Mathematik im Surrealismus*. Marburg: Jonas, 2002.

Wessely, Anna. "Transporting 'Style' from the History of Art to the History of Science." *Science in Context* 4, no. 2 (1991): 265–78.

Wright, Richard. "Art and Science in Chaos: Contested Readings of Scientific Visualizations." In *FutureNatural: Nature, Science, Culture*, edited by George Robertson, 218–36. London: Routledge, 1996.

REPRESENTATION IN THE SCIENCES

Overarching Questions, Anthologies, and Methodological Approaches to the Study of Images in the Sciences

Bensaude-Vincent, Bernadette, and William R. Newman, eds. *The Artificial and the Natural: An Evolving Polarity*. Cambridge, MA: MIT Press, 2007.

Beyer, Andreas, and Markus Lohoff, eds. *Bild und Erkenntnis. Formen und Funktionen des Bildes in Wissenschaft und Technik*. Berlin: Dt. Kunstverlag, 2006.

Bredekamp, Horst, and Gabriele Werner, eds. *Bilder in Prozessen*. In *Bildwelten des Wissens. Kunsthistorisches Jahrbuch für Bildkritik*, vol. 1, no. 1. Edited by Horst Bredekamp, Matthias Bruhn, and Gabriele Werner. Berlin: Akademie Verlag, 2003.

Breidbach, Olaf. *Bilder des Wissens: Zur Kulturgeschichte der wissenschaftlichen Wahrnehmung*. Munich: Fink, 2005.

Buschhaus, Markus, and Inge Hinterwaldner, eds. *The Picture's Image: Wissenschaftliche Visualisierungen als Komposit*. Munich: Fink, 2006.

Chadarevian, Soraya , and Nick Hopwood, eds. *Models: The Third Dimension of Science*. Stanford, CA: Stanford University Press, 2004.

Daston, Lorraine, and Peter Galison. *Objectivity*. New York: Zone Books, 2007.

Dommann, Monika, and Marietta Meier, eds. *Wissenschaft, die Bilder schafft / Science en Image*. Traverse. Zurich: Chronos-Verlag, 1999 (3).

Donald, Diana, and Jane Munro, eds. *Endless Forms: Charles Darwin, Natural Science and the Visual Arts*. Cambridge: Fitzwilliam Museum, 2009.

Dupré, Sven, and Christoph H. Lüthy, eds. *Silent Messengers: The Circulation of Material Objects of Knowledge in the Early Modern Low Countries*. Berlin: Lit, 2011.

Elkins, James, ed. *Visual Practices across the University*. Munich: Fink, 2007.

Fleck, Ludwik. *Genesis and Development of a Scientific Fact*. New edition. Chicago: University of Chicago Press, 1981.

Fleck, Ludwik. "Schauen, sehen, wissen." In *Erfahrung und Tatsache: Gesammelte Aufsätze*, edited by Lothar Schäfer, and Thomas Schnelle, 147–74. Frankfurt am Main: Suhrkamp, 1983.

Ford, Brian. *Images of Science: A History of Scientific Illustration*. London: British Library, 1992.

Freedberg, David. *The Eye of the Lynx: Galileo, His Friends, and the Beginnings of Modern Natural History*. Chicago: University of Chicago Press, 2002.

Frigg, Roman. "Models and Fiction." *Synthese* 172, no. 2 (2010): 251–68.

Frigg, Roman, and Matthew Hunter, eds. *Beyond Mimesis and Convention: Representation in Art and Science*. Dordrecht: Springer, 2010.

Fyfe, Gordon, and John Law, eds. *Picturing Power: Visual Depictions and Social Relations*. New York: Routledge, 1988.

Galison, Peter. *Image and Logic: A Material Culture of Microphysics*. Chicago: University of Chicago Press, 1997.

Gall, Alexander, ed. *Konstruieren, kommunizieren, präsentieren: Bilder von Wissenschaft und Technik*. Göttingen: Wallstein-Verlag, 2007.

Geimer, Peter. "Image as Trace: Speculations about an Undead Paradigm." *differences* 18, no. 1 (2007): 7–28.

Gugerli, David, and Barbara Orland, eds. *Ganz normale Bilder: Historische Beiträge zur visuellen Herstellung von Selbstverständlichkeit*. Zurich: Chronos, 2002.

Hacking, Ian. *Representing and Intervening*. Cambridge: Cambridge University Press, 1983.

Heintz, Bettina, and Jörg Huber, eds. *Mit dem Auge denken. Strategien der Sichtbarmachung in wissenschaftlichen und virtuellen Welten*. Zurich: Voldemeer; Vienna: Springer, 2001.

Heßler, Martina, ed. *Konstruierte Sichtbarkeiten: Wissenschafts- und Technikbilder seit der Frühen Neuzeit*. Munich: Fink, 2006.

Hodges, Elaine R. S. *The Guild Handbook of Scientific Illustration*, Hoboken, NJ: Wiley, 2003.

Holländer, Hans, ed. *Erkenntnis, Erfindung, Konstruktion: Studien zur Bildgeschichte von Naturwissenschaften und Technik vom 16. bis zum 19. Jahrhundert*. Berlin: Mann, 2000.

Huber, Jörg, and Martin Heller, eds. *Konstruktionen Sichtbarkeiten*. Vienna: Springer, 1999.

Kemp, Martin. *Visualizations: The Nature Book of Art and Science*. Oxford: Oxford University Press, 2000.

Knorr Cetina, Karin. *Epistemic Cultures: How the Sciences Make Knowledge*. Cambridge, MA: Harvard University Press, 1999.

Kostelnick, Charles, and Michael Hassett. *Shaping Information: The Rhetoric of Visual Conventions*. Carbondale: Southern Illinois University Press, 2003.

Krohn, Wolfgang, ed. *Ästhetik in der Wissenschaft: Interdisziplinärer Diskurs über das Gestalten und Darstellen von Wissen*. Hamburg: Meiner, 2006.

Latour, Bruno. "Circulating Reference: Sampling the Soil in the Amazon Forest." In *Pandora's Hope: Essays on the Reality of Science Studies*, 24–79. Cambridge, MA: Harvard University Press, 1999.

Latour, Bruno. "Drawing Things Together." In *Representation in Scientific Practice*, edited by M. Lynch and S. Woolgar, 19–68. Cambridge, MA: MIT Press 1990.

Latour, Bruno, and Steve Woolgar. *Laboratory Life: The Social Contruction of Scientific Facts*. Princeton, NJ: Princeton University Press, 1986.

Lee, Jennifer B., and Miriam Mandelbaum, eds. *Seeing Is Believing: 700 Years of Scientific and Medical Illustration*. Exhibition catalogue. New York: New York Public Library, 1999.

Lefèvre, Wolfgang, Jürgen Renn, and Urs Schoepflin, eds. *The Power of Images in Early Modern Science*. Basel: Birkhäuser, 2003.

Lenoir, Timothy, ed. *Inscribing Science: Scientific Texts and the Materiality of Communication*. Stanford, CA: Stanford University Press, 1998.

Lynch, Michael. *Art and Artifact in Laboratory Science. AStudy of Shop Work and Shop Talk in a Research Laboratory*. London: Routledge & Kegan Paul, 1985.

Lynch, Michael. "Discipline and the Material Form of Images: An Analysis of Scientific Visibility." *Social Studies of Science* 15 (1985): 37–66.

Lynch, Michael. "The Externalized Retina: Selection and Mathematization in the Visual Documentation of Objects in the Life Sciences." *Human Studies* 11 (1988): 201–34.

Lynch, Michael, and Steve Woolgar. *Representation in Scientific Practice*. Cambridge, MA: MIT Press, 1990.

Morgan, Mary S, and Margaret Morrison, eds. *Models As Mediators: Perspectives on Natural and Social Sciences*. Cambridge: Cambridge University Press, 1999.

Nowotny, Helga, and Martina Weiss, eds. *Shifting Boundaries of the Real: Making the Invisible Visible*. Zurich: vdf Hochschulverlag, 2000.

Pickering, Andrew, ed. *Science as Practice and Culture*. Chicago: University of Chicago Press, 1992.

Rheinberger, Hans-Jörg. *Toward a History of Epistemic Things: Synthesizing Proteins in the Test Tube*. Stanford, CA: Stanford University Press, 1997.

Rheinberger, Hans-Jörg, Michael Hagner, and Bettina Wahrig-Schmidt, eds. *Räume des Wissens: Repräsentation, Codierung, Spur.* Berlin: Akademie Verlag, 1997.

Roberts, Lissa, Simon Schaffer, and Peter Dear, eds. *The Mindful Hand: Inquiry and Invention from the Late Renaissance to Early Industrialisation*. Amsterdam: Koninkliijke Nederlandse Akademie van Wetenschappen, 2007.

Robin, Harry. *The Scientific Image: From Cave to Computer*. New York: Harry N. Abrams, 1992.

Rocke, Alan J. *Image and Reality: Kékulé, Kopp, and the Scientific Imagination*. Chicago: University of Chicago Press, 2010.

Schiebinger, Londa L., and Claudia Swan, eds. *Colonial Botany: Science, Commerce, and Politics in the Early Modern World*. Philadelphia: University of Pennsylvania Press, 2005.

Suárez, Mauricio, ed. *Fictions in Science: Philosophical Essays on Modeling and Idealization*. New York: Routledge, 2009.

Thicke, Mike. "Visual Representation and Science." *Spontaneous Generations: A Journal for the History and Philosophy of Science*, vol. 6, no. 1. (2012).

Topper, David. "Towards an Epistemology of Scientific Illustration." In *Picturing Knowledge: Historical and Philosophical Problems Concerning the Use of Art in Science*, edited by Brian S. Baigrie, 215–49. Toronto: University of Toronto Press, 1996.

Tufte, Edward R. *Beautiful Evidence*. Cheshire, CT: Graphics Press, 2006.

Tufte, Edward R. *Envisioning Information*. Cheshire, CT: Graphics Press, 1990.
Tufte, Edward R. *The Visual Display of Quantitative Information*. Cheshire, CT: Graphics Press, 1983.
Tufte, Edward R. *Visual Explanations: Images and Quantities; Evidence and Narrative*. Cheshire, CT: Graphics Press, 1997.
Vögtli, Alexander, and Beat Ernst, eds. *Wissenschaftliche Bilder: Eine kritische Betrachtung*. Basel: Schwabe, 2007.
Wechsler, Judith, ed. *On Aesthetics in Science*, Basel: Birkhäuser, 1988 [1978].
Wise, Norton. "Making Visible." *Isis* 97, no. 1 (2006): 75–82.
Wright, Richard. "Art and Science in Chaos: Contested Readings of Scientific Visualizations." In *FutureNatural: Nature, Science, Culture*, edited by George Robertson, 218–36. London: Routledge, 1996.

Selected Case Studies and Individual Issues Related to Imagery and the History of Science

Bedell, Rebecca B. *The Anatomy of Nature: Geology & American Landscape Painting, 1825–1875*. Princeton, NJ: Princeton University Press, 2001.
Bertin, Jacques. *Semiology of Graphics: Diagrams, Networks, Maps*. Madison: University of Wisconsin Press, 1983.
Bleichmar, Daniela. *Visible Empire: Botanical Expeditions and Visual Culture in the Hispanic Enlightenment*. Chicago: University of Chicago Press, 2012.
Brasseur, Lee E. *Visualizing Technical Information: A Cultural Critique*. Amityville, NY: Baywood, 2003.
Bredekamp, Horst. *Darwins Korallen: Frühe Evolutionsmodelle und die Tradition der Naturgeschichte*. Berlin: Wagenbach, 2005.
Bredekamp, Horst. *Die Fenster der Monade: Gottfried Wilhelm Leibniz' Theater der Natur und Kunst*. Berlin: Akademie Verlag, 2004.
Bredekamp, Horst. *Thomas Hobbes: Der Leviathan; Das Urbild des modernen Staates und seine Gegenbilder 1651–2001*. Berlin: Akademie Verlag, 2003.
Brons, Franziska, ed. "Imagination des Himmels." In *Bildwelten des Wissens: Kunsthistorisches Jahrbuch für Bildkritik*, vol. 5, no. 2. Edited by Horst Bredekamp, Matthias Bruhn, and Gabriele Werner. Berlin: Akademie Verlag, 2007.
Bruhn, Matthias. *Bildwirtschaft. Verwaltung und Verwertung der Sichtbarkeit*. Weimar: Verlag und Datenbank für Geisteswissenschaften /VDG, 2003.
Bruhn, Matthias, ed. "Ikonografie des Gehirns." In *Bildwelten des Wissens: Kunsthistorisches Jahrbuch für Bildkritik*, vol. 6, no. 1. Edited by Horst Bredekamp, Matthias Bruhn, and Gabriele Werner. Berlin: Akademie Verlag, 2008.
Campbell-Kelly, Martin, Mary Croarken, Raymond Flood, and Eleanor Robson, eds. *The History of Mathematical Tables: From Sumer to Spreadsheets*. Oxford: Oxford University Press, 2003.
Chadarevian, Soraya de. *Designs for Life: Molecular Biology after World War II*. Cambridge: Cambridge University Press, 2002.
DeLue, Rachael Z. *George Inness and the Science of Landscape*. Chicago: University of Chicago Press, 2004.
Dommann, Monika. *Durchsicht, Einsicht, Vorsicht: Eine Geschichte der Röntgenstrahlen 1896–1963*. Zurich: Chronos, 2003.
Elkins, James. "Harold Edgerton's Rapatronic Photographs of Atomic Tests." *History of Photography* 28, no. 1 (2004): 74–81.
Findlen, Paula. *Possessing Nature: Museums, Collecting, and Scientific Culture in Early Modern Italy*. Berkeley: University of California Press, 1996.
Fischel, Angela, ed. "Grenzbilder." In *Bildwelten des Wissens: Kunsthistorisches Jahrbuch für Bildkritik*, vol. 6, no. 2. Edited by Horst Bredekamp, Matthias Bruhn, and Gabriele Werner. Berlin: Akademie Verlag, 2009.
Freedberg, David. *The Eye of the Lynx: Galileo, His Friends, and the Beginnings of Modern Natural History*. Chicago: University of Chicago Press, 2003.
Fukuoka, Maki. *The Premise of Fidelity: Science, Visuality, and Representing the Real in Nineteenth-Century Japan*. Stanford, CA: Stanford University Press, 2012.
Galison, Peter. "Aufbau/Bauhaus: Logical Positivism and Architectural Modernism." *Critical Inquiry* 16, no. 4 (1990): 709–52.
Heck, Karsten, ed. *Bildendes Sehen. In Bildwelten des Wissens: Kunsthistorisches Jahrbuch für Bildkritik*, vol. 7, no. 1. Edited by Horst Bredekamp, Matthias Bruhn, and Gabriele Werner. Berlin: Akademie Verlag, 2009.
Heesen, Anke Te. *The World in a Box: The Story of an Eighteenth-Century Picture Encyclopedia*. Chicago and London: University of Chicago Press, 2002.
Helbig, Jutta, ed. *Präparate*. In *Bildwelten des Wissens: Kunsthistorisches Jahrbuch für Bildkritik*, vol. 9, no. 1. Edited by Horst Bredekamp, Matthias Bruhn, and Gabriele Werner. Berlin: Akademie Verlag, 2012.
Helden, Albert van, and Mary G. Winkler. "Representing the Heavens: Galileo and Visual Astronomy." *Isis* 83 (1992): 195–217.
Hentschel, Klaus. *Mapping the Spectrum: Techniques of Visual Representation in Research and Teaching*. Oxford: Oxford University Press, 2002.
Hentschel, Klaus, and Axel Wittmann, eds. *The Role of Visual Representations in Astronomy: History and Research Practice*, Series Acta Historica Astronomiae, vol. 9. Frankfurt a.M.: Verlag Harri Deutsch, 2000.
Hess, Volker, ed. *Normierung der Gesundheit: Messende Verfahren der Medizin als kulturelle Praktik um 1900*. Husum: Matthiesen, 1997.
Hunter, Matthew C. *Wicked Intelligence: Visual Art and the Science of Experiment in Restoration London*. Chicago: University of Chicago Press, 2013.
Klonk, Charlotte. *Science and the Perception of Nature: British Landscape Art in the Late Eighteenth and Early Nineteenth Centuries*. New Haven, CT: Yale University Press, 1996.
Krämer, Sybille. *Symbolische Maschinen: Die Idee der Formalisierung in geschichtlichem Abriß*. Darmstadt: Wissenschaftliche Buchgesellschaft, 1988.
Kusukawa, Sachiko. *From Counterfeit to Canon: Picturing the Human Body, Especially by Andreas Vesalius*. Berlin: Max Planck Institute for the History of Science, 2004.
Lawrence, Christopher, and Steven Shapin, eds. *Science Incarnate: Historical Embodiments of Natural Knowledge*. Chicago: University of Chicago Press, 1998.
Lefèvre, Wolfgang, ed. *Picturing Machines, 1400–1700*. Cambridge, MA: MIT Press, 2004.
Lepp, Nicola, Martin Roth, and Klaus Vogel, eds. *Der neue Mensch: Obsessionen des 20. Jahrhunderts*. Exhibition catalogue, Deutsches Hygiene Museum, Dresden, Germany. Ostfildern: Hatje Cantz, 1999.

Lüthy, Christoph. "The Invention of Atomist Iconography." In *The Power of Images in Early Modern Science*, edited by Wolfgang Lefèvre, Jürgen Renn, and Urs Schoepflin, 117–38. Basel: Birkhäuser, 2003.

MacGregor, Arthur. *Curiosity and Enlightenment: Collectors and Collections from the Sixteenth to Nineteenth Century*. New Haven, CT: Yale University Press, 2007.

Mayer-Deutsch, Angela. *Das Musaeum Kircherianum: Kontemplative Momente, historische Rekonstruktion, Bildrhetorik*. Zurich: diaphanes, 2010.

Meinel, Christoph. "Molecules and Croquet Balls." In *Models: The Third Dimension of Science*, edited by Soraya de Chadarevian and Nick Hopwood, 243–75. Palo Alto, CA: Stanford University Press, 2004.

Nasim, Omar W. "On Seeing an Image of a Spiral Nebula: From Whewell to Flammarion." *Nuncius: Journal for the History of Science* 24, no. 2 (2009): 393–414.

Neri, Janice. *The Insect and the Image: Visualizing Nature in Early Modern Europe, 1500–1700*. Minneapolis: University of Minnesota Press, 2011.

Newman, William R. *Promethean Ambitions: Alchemy and the Quest to Perfect Nature*. Chicago: University of Chicago Press, 2004.

Panofsky, Erwin. *Galileo as a Critic of the Arts*. The Hague: Nijhoff, 1954.

Parshall, Peter. "Imago Contrafacta: Images and Facts in the Northern Renaissance." *Art History* 16, no. 4 (December 1993): 554–79.

Putscher, Marie-Helene. *Geschichte der medizinischen Abbildung*. Munich: Moos, 1972.

Reichle, Ingeborg, Steffen Siegel, and Achim Spelten, eds. *Visuelle Modelle*. Munich: Fink, 2008.

Rifkin, Benjamin. *Human Anatomy: Depicting the Body from the Renaissance to Today*. London: Thames and Hudson, 2006.

Rosenberg, Daniel, and Anthony Grafton, eds. *Cartographies of Time: A History of the Timeline*. New York: Princeton Architectural Press, 2010.

Rudwick, Martin. *The Emergence of Visual Language for Geological Science, 1760–1840. History of Science*, 14 (1976): 149–95.

Samuel, Nina, ed. *The Islands of Benoît Mandelbrot: Fractals, Chaos, and the Materiality of Thinking*. New Haven, CT: Yale University Press, 2012.

Schmidt-Burkhardt, Astrit. *Sehende Bilder: Die Geschichte des Augenmotivs seit dem 19. Jahrhundert*. Berlin: Akademie Verlag, 1992.

Schmidt-Burkhardt, Astrit. *Stammbäume der Kunst: Zur Genealogie der Avantgarde*. Berlin: Akademie Verlag, 2005.

Schneider, Birgit. "Climate Model Simulation Visualization from a Visual Studies Perspective." *WIREs Climate Change* 3, no. 2 (March/April 2012): 185–93.

Schneider, Birgit, ed. *Diagramme und bildtextile Ordnungen*. In *Bildwelten des Wissens: Kunsthistorisches Jahrbuch für Bildkritik*, vol. 3, no. 1. Edited by Horst Bredekamp, Matthias Bruhn, and Gabriele Werner. Berlin: Akademie Verlag, 2005.

Schneider, Birgit. *Textiles Prozessieren: Eine Mediengeschichte der Lochkartenweberei*. Berlin and Zurich: diaphanes, 2007.

Schoonover, Carl. *Portraits of the Mind: Visualizing the Brain from Antiquity to the 21st Century*. New York: Harry N. Abrams, 2010.

Screech, Timon. *The Western Scientific Gaze and Popular Imagery in Later Edo Japan: The Lens within the Heart*. Cambridge: Cambridge University Press, 1996.

Shapin, Steven, and Simon Schaffer. *Leviathan and the Air-Pump: Hobbes, Boyle, and the Experimental Life*. Princeton, NJ: Princeton University Press, 1985.

Smith, Pamela H., and Paula Findlen. *Merchants and Marvels: Commerce, Science, and Art in Early Modern Europe*. New York: Routledge, 2002.

Swan, Claudia. "Ad Vivum, Naer het Leven, From the Life: Defining a Mode of Representation." *Word and Image* 11 (1995): 353–72.

Tufte, Edward R. *Beautiful Evidence*. Cheshire, CT: Graphics Press, 2006.

Velminski, Wladimir, and Gabriele Werner, eds. *Mathematische Forme(l)n*. In *Bildwelten des Wissens: Kunsthistorisches Jahrbuch für Bildkritik*, vol. 7, no. 2. Edited by Horst Bredekamp, Matthias Bruhn, and Gabriele Werner. Berlin: Akademie Verlag, 2010.

Vismann, Cornelia. *Medien der Rechtssprechung*, edited by Alexandra Kemmerer. Frankfurt a.M.: Fischer, 2011.

Voss, Julia. *Darwins Bilder: Ansichten der Evolutionstheorie 1837–1874*. Frankfurt a.M.: Fischer, 2007.

Wosk, Julie. *Women and the Machine: Representations from the Spinning Wheel to the Electronic Age*. Baltimore: Johns Hopkins University Press, 2001.

IMAGE TECHNIQUES / VISUAL MEDIA

(Manual) Drawing, Notation, and Illustration

Booker, Peter Jeffrey. *A History of Engineering Drawing*. London: Chatto and Windus, 1963.

Bredekamp, Horst. "Denkende Hände: Überlegungen zur Bildkunst der Naturwissenschaften." In *Räume der Zeichnung*, edited by Angela Lammert, Carolin Meister, Jan-Philipp Frühsorge, and Andreas Schalhorn, 10–24. Nuremberg: Verlag für moderne Kunst, 2007.

Bredekamp, Horst. *Galilei, der Künstler: Der Mond. Die Sonne. Die Hand*. Berlin: Akademie Verlag, 2007.

Bredekamp, Horst. "Gazing Hands and Blind Spots: Galileo as Draftsman." In *Galileo in Context*, edited by Jürgen Renn, 153–92. Cambridge: Cambridge University Press, 2001.

Bredekamp, Horst. "Die zeichnende Denkkraft. Überlegungen zur Bildkunst der Naturwissenschaften." In *Einbildungen* (Interventionen 14), edited by Jörg Huber, 155–71. Zurich: Springer, 2005.

Busch, Werner, Oliver Jehle, and Carolin Meister, eds. *Randgänge der Zeichnung*. Munich: Fink, 2006.

Elkins, James. "On Visual Desperation and the Bodies of Protozoa." *Representations* (1992): 33–56.

Emmons, Paul. "Size Matters: Virtual Scale and Bodily Imagination in Architectural Drawing." *Architectural Research Quarterly* 9, nos. 3–4 (2005): 227–35.

Fischel, Angela. *Natur im Bild: Zeichnung und Naturerkenntnis bei Conrad Gessner und Ulisse Aldrovandi*. Berlin: Mann, 2009.

Galison, Peter. "The Suppressed Drawing: Paul Dirac's Hidden Geometry." *Representations* 72 (Fall 2000): 145–66.

Haupt, Herbert, Thea Vignau-Wilberg, Eva Irblich, and Manfred Staudinger. *Le bestiaire de Rodolphe II*. Paris: Citadelle and Mazenod, 1990.

Hodges, Elaine R. S., ed. *The Guild Handbook of Scientific Illustration*. Hoboken, NJ: Wiley, 1989.

Holmes, Frederic Lawrence, Jürgen Renn, and Hans-Jörg

Rheinberger, eds. *Reworking the Bench: Research Notebooks in the History of Science*. Dordrecht: Kluwer, 2003.
Hunter, Matthew C. "I resolved to throw aside all manner of hypotheses . . . and to attend wholly to what the appearances themselves would teach me." In *Wicked Intelligence: Visual Art and the Science of Experiment in Restoration London*, 28–67. Chicago: University of Chicago Press, 2013.
Ingold, Tim. *Lines: A Brief History*. London: Routledge, 2007.
Krämer, Sybille. "'Epistemology of the Line': Reflections on the Diagrammatical Mind." In *Studies in Diagrammatology and Diagram Praxis*, edited by Olga Pombo and Alexander Gerner, 13–39. London: College Publications, 2010.
Mainberger, Sabine. *Experiment Linie: Künste und ihre Wissenschaften um 1900*. Berlin: Kulturverlag Kadmos, 2010.
Nasim, Omar W. "Extending the Gaze: The Temporality of Astronomical Paperwork." *Science in Context* 26, no. 2 (2013): 247–77.
Nasim, Omar W. "The 'Landmark' and 'Groundwork' of Stars: John Herschel, Photography and the Drawing of Nebulae." *Studies in History and Philosophy of Science* 42, no.1 (2011): 67–84.
Nasim, Omar W. "Observation, Working Images and Procedure: The 'Great Spiral' in Lord Rosse's Astronomical Record Books and Beyond." *The British Journal for the History of Science* 43, no. 3 (2010): 353–89.
Nedoluha, Alois. *Kulturgeschichte des technischen Zeichnens*. Vienna: Springer, 1960.
Nickelsen, Karin. *Draughtsmen, Botanists, and Nature: The Construction of Eighteenth-Century Botanical Illustrations*. New York: Springer, 2006.
Nissen, Claus. *Die botanische Buchillustration, ihre Geschichte und Bibliographie*. Stuttgart: A. Hirsemann, 1951.
Nissen, Claus. *Die zoologische Buchillustration, ihre Bibliographie und Geschichte*. 2 vols. Stuttgart: A. Hirsemann, 1969–1978.
Nuti, Lucia. "'To Make the Whole Progress a Lineall Visible Demonstration': The Journal of Edward Dummer." *Word & Image* 15, no. 3 (1999): 292–305.
Olmi, Giuseppe. *L'inventario del mondo: Catalogazione della natura e luoghi del sapere nella prima età moderna*. Bologna: Il Mulino, 1992.
Petherbridge, Deanna. *The Primacy of Drawing: Histories and Theories of Practice*. New Haven, CT: Yale University Press, 2010.
Rheinberger, Hans-Jörg. "Scrips and Scribbles." *Modern Language Notes* 118, no. 3 (2003): 622–36.
Schaffer, Simon. "On Astronomical Drawing." In *Picturing Science, Producing Art*, ed. Caroline A. Jones, Peter Galison, and Amy E. Slaton, 441–74. New York: Routledge, 1998.
Swan, Claudia. "Ad vivum, naer het leven, from the life: Defining a Mode of Representation." *Word and Image* 11 (1995): 353–72.
Wittmann, Barbara, ed. *Spuren erzeugen: Zeichnen und Schreiben als Verfahren der Selbstaufzeichnung*. Berlin: diaphanes, 2009.

Optical Instruments and Media of Seeing (Optical Toys, the Zograscope, the Microscope, and the Telescope)

Breidbach, Olaf. "Representations of the Microcosm: The Claim for Objectivity in 19th Century Scientific Microphotography." *Journal of the History of Biology* 35, no. 2 (2002): 221–50.
Breidbach, Olaf, Kerrin Klinger, and Matthias Müller, eds. *Camera Obscura*. Wiesbaden: Steiner Franz, 2013.
Brusati, Celeste. *Artifice and Illusion: The Art and Writing of Samuel Van Hoogstraten*. Chicago: University of Chicago Press, 1995.
Dewitz, Bodo, et al., eds. *Ich sehe was, was du nicht siehst: Sehmaschinen und Bilderwelten; Die Sammlung Werner Nekes*. Exhibition catalogue, Museum Ludwig Cologne, September 27–November 24, 2002. Göttingen: Steidl, 2002.
Ditzen, Stefan. *Kunstformen instrumenteller Sicherheit: Etappen einer Bildgeschichte des Mikroskops*. Aachen: Shaker, 2008.
Dupré, Sven, ed. "Optics, Instruments and Painting, 1420–1720: Reflections on the Hockney-Falco Thesis." *Early Science and Medicine* 10, no. 2 (2005): 125–339.
Fiorentini, Erna, ed. *Observing Nature—Representing Experience: The Osmotic Dynamics of Romanticism, 1880–1850*. Berlin: Reimer, 2007.
Fischel, Angela, ed. *Instrumente des Sehens*. In *Bildwelten des Wissens: Kunsthistorisches Jahrbuch für Bildkritik*, vol. 2, no. 2. Edited by Horst Bredekamp, Matthias Bruhn, and Gabriele Werner. Berlin: Akademie Verlag, 2002.
Gronemeyer, Nicole. *Optische Magie: Zur Geschichte der visuellen Medien in der Frühen Neuzeit*. Bielefeld: Transcript, 2004.
Hall, Edward T. *The Hidden Dimension*. Garden City, NY: Doubleday, 1966.
Hankins, Thomas L., and Robert J. Silverman. *Instruments and the Imagination*. Princeton, NJ: Princeton University Press, 1995.
Heering, Peter. "The Enlightened Microscope: Re-Enactment and Analysis of Projections with Eighteenth-Century Solar Microscopes." *British Journal for the History of Science* 41, no. 3 (2008): 345–67.
Hick, Ulrike. *Geschichte der optischen Medien*. Munich: Fink, 1999.
Hockney, David. *Secret Knowledge: Rediscovering the Lost Techniques of the Old Masters*. London: Thames & Hudson, 2006.
Kittler, Friedrich A. *Optical Media: Berlin Lectures 1999*. Cambridge: Polity, 2010.
Kusukawa, Sachiko, and Ian Maclean. *Transmitting Knowledge: Words, Images, and Instruments in Early Modern Europe*. Oxford: Oxford University Press, 2006.
Lynch, Michael. "The Externalized Retina: Selection and Mathematization in the Visual Documentation of Objects in the Life Sciences." *Human Studies* 11 (1988): 201–34.
McLuhan, Marshall. *Understanding Media: The Extensions of Man*. New York: McGraw-Hill, 1964.
Payne, Alina A. *The Telescope and the Compass: Teofilo Gallaccini and the Dialogue between Architecture and Science in the Age of Galileo*. Florence: L. S. Olschki, 2012.
Reuleaux, Franz. *The Kinematics of Machinery: Outlines of a Theory of Machines* [1875]. Trans. and ed. Alexander Blackie William Kennedy. London: Taylor and Sons, 1876.
Roos, Anna Marie. "A Speculum of Chymical Practice: Isaac Newton, Martin Lister (1639–1712), and the Making of Telescopic Mirrors." *Notes and Records of the Royal Society of London* 64, no. 2 (June 2010): 105–20.
Ruestow, Edward G. *The Microscope in the Dutch Republic: The Shaping of Discovery*. Cambridge: Cambridge University Press, 1996.
Schickore, Jutta. *The Microscope and the Eye: A History of Reflections, 1740–1870*. Chicago: University of Chicago Press, 2007.
Schramm, Helmar, Ludger Schwarte, and Jan Lazardzig, eds. *Instrumente in Kunst und Wissenschaft: Zur Architektonik kultureller Grenzen im 17. Jahrhundert*. Berlin: Walter de Gruyter, 2006.

Stafford, Barbara Maria. "'Magnifications': The Eighteenth-Century Fortunes of a Primitive and Universal 'Imagistic.'" *Prospettiva* 57–60 (April 1989–October 1990): 308–15.

Stafford, Barbara Maria, and Frances Terpak, eds. *Devices of Wonder: From the World in a Box to Images on a Screen*. Exhibition catalogue, J. Paul Getty Museum, Los Angeles, November 13, 2001–February 3, 2002. Los Angeles: Getty Publications, 2001.

Steadman, Philip. *Vermeer's Camera: Uncovering the Truth behind the Masterpieces*. Oxford: Oxford University Press, 2001.

Szekely, Francis. "Unreliable Observers, Flawed Instruments, 'Disciplined Viewings': Handling Specimens in Early Modern Microscopy." *Parergon* 28, no. 1 (2011): 155–76.

Wilson, Catherine. *The Invisible World: Early Modern Philosophy and the Invention of the Microscope*. Princeton, NJ: Princeton University Press, 1995.

Printing Techniques

Armstrong, Carol, and Catherine de Zegher, eds. *Ocean Flowers*. Princeton, NJ: Princeton University Press, 2004.

Bann, Stephen. *Parallel Lines: Printmakers, Painters, and Photographers in Nineteenth-Century France*. New Haven, CT: Yale University Press, 2001.

Dackerman, Susan, ed. *Prints and the Pursuit of Knowledge in Early Modern Europe*. Harvard Art Museum. Cambridge, MA: Harvard Art Museums, 2011.

Darnton, Robert. *The Business of Enlightenment: A Publishing History of the "Encyclopédie," 1775–1800*. Cambridge, MA: Belknap Press, 1979.

Dünkel, Vera, ed. *Kontaktbilder*. In *Bildwelten des Wissens: Kunsthistorisches Jahrbuch für Bildkritik*, vol. 8, no. 1. Edited by Horst Bredekamp, Matthias Bruhn, and Gabriele Werner. Berlin: Akademie Verlag, 2010.

Eisenstein, Elizabeth. *The Printing Revolution in Early Modern Europe*. Cambridge: Cambridge University Press, 1983.

Felfe, Robert, and Karin Leonhard. *Lochmuster und Linienspiel: Überlegungen zur Druckgrafik im 17. Jahrhundert*. Freiburg: Rombach, 2006.

Frieß, Peter. *Kunst und Maschine: 500 Jahre Maschinenlinien in Bild und Skulptur*. Munich: Deutscher Kunstverlag, 1993.

Giedion, Siegfried. *Mechanization Takes Command, a Contribution to Anonymous History*. New York: Oxford University Press, 1948.

Grasselli, Margaret Morgan, ed. *Colorful Impressions: The Printmaking Revolution in 18th-Century France*. Exhibition catalogue, National Gallery of Art, Washington, DC, October 26, 2003–February 16, 2004. Washington, DC: National Gallery of Art, 2003.

Hunter, Matthew C. "The Theory of the Impression According to Robert Hooke." In *Printed Images in Early Modern Britain: Essays in Interpretation*, edited by Michael Hunter, 167–90. Aldershot, UK: Ashgate, 2010.

Innis, Harold A. *The Bias of Communication*. With an introduction by Marshall McLuhan. Toronto: University of Toronto Press, 1964.

Innis, Harold A., and Mary Q. Innis. *Empire and Communications*. Toronto: University of Toronto Press, 1972.

Ivins, William M. *Prints and Visual Communication*. London: Routledge, 1953.

Johns, Adrian. *The Nature of the Book: Print and Knowledge in the Making*. Chicago: University of Chicago Press, 1998.

Koschatzky, Walter. *Die Kunst der Graphik: Technik, Geschichte, Meisterwerke*. Salzburg: DTV, 1972.

Kristeller, Paul. *Kupferstich und Holzschnitt in vier Jahrhunderten*. Berlin: Cassirer, 1921.

Landau, David, and Peter Parshall. *The Renaissance Print, 1470–1550*. New Haven, CT: Yale University Press, 1994.

MacGregor, William B. "The Authority of Prints: An Early Modern Perspective." *Art History* 22, no. 3 (September 1999): 389–420.

McLuhan, Marshall. *The Gutenberg Galaxy: The Making of Typographic Man*. Toronto: University of Toronto Press, 1962.

McLuhan, Marshall. *Understanding Media: The Extensions of Man*. New York: McGraw-Hill, 1964.

Mumford, Lewis. *Art and Technics*. London: Oxford University Press, 1952.

Mumford, Lewis. *Technics and Civilization*. New York: Harcourt, Brace and Co, 1934.

Prinz, Felix, ed. *Graustufen*. In *Bildwelten des Wissens: Kunsthistorisches Jahrbuch für Bildkritik*, vol. 8, no. 1. Edited by Horst Bredekamp, Matthias Bruhn, and Gabriele Werner. Berlin: Akademie Verlag, 2010.

Graphical Recording

Brain, Robert M., and M. Norton Wise. "Muscles and Engines: Indicator Diagrams in Helmholtz's Physiology." In *Universalgenie Helmholtz: Ruckblick nach 100 Jahren*, ed. Lorenz Krüger, 124–45. Akademie Verlag, 1994.

Chadarevian, Soraya de. "Die 'Methode der Kurven' in der Physiologie zwischen 1850 und 1900." In *Die Experimentalisierung des Lebens*, edited by Hans-Jörg Rheinberger and Michael Hagner. Berlin: Akademie Verlag, 1993.

Dagognet, François. *Étienne-Jules Marey: A Passion for the Trace* [1987], translated by Robert Galeta with Jeanine Herman. New York: Zone Books, 1992.

Holl Hebbel, E., and L. A. Geddes. "The Beginning of Graphic Recording." *Isis* 53 (1962): 289–324.

Holmes, Frederic L., and Kathryn M. Olesko. "Precision's Images: Helmholtz and Graphical Methods in Physiology." In *The Values of Precision*, ed. M. Norton Wise, 198–221. Princeton, NJ: Princeton University Press, 1995.

Mainberger, Sabine. *Experiment Linie. Künste und ihre Wissenschaften um 1900*. Berlin: Kulturverlag Kadmos, 2010.

Schmidgen, Henning. *Die Helmholtz-Kurven: Auf der Spur der verlorenen Zeit*. Berlin: Merve, 2009.

Photography and Film in the Sciences

Braun, Marta. *Picturing Time: The Work of Étienne-Jules Marey (1830–1904)*. Chicago: University of Chicago Press, 1992.

Bud, Robert, Bernard Finn, and Helmuth Trischler, eds. *Manifesting Medicine: Bodies and Machines (Artefacts: Studies in the History of Science and Technology)*. Vol. 1. Amsterdam: Harwood Academic Publishers, 1999.

Canales, Jimena. *A Tenth of a Second: A History*. Chicago: University of Chicago Press, 2009.

Cartwright, Lisa. *Screening the Body: Tracing Medicine's Visual Culture*. Minneapolis: University of Minnesota Press, 1995.

Derenthal, Ludger, and Christiane Stahl, eds. *Mikrofotografie: Schönheit jenseits des Sichtbaren*. Exhibition catalogue, Staatliche

Museen zu Berlin. Ostfildern: Hatje Cantz, 2010.
Didi-Huberman, Georges. *Invention of Hysteria: Charcot and the Photographic Iconography of the Salpêtrière*. Cambridge, MA: MIT Press, 2004 [first edition Paris, 1982].
Elkins, James, ed. *Six Stories from the End of Representation: Images in Painting, Photography, Astronomy, Microscopy, Particle Physics, and Quantum Mechanics, 1980–2000*. Chicago: University of Chicago Press, 2008.
Ellenbogen, Josh. *Reasoned and Unreasoned Images: The Photography of Bertillon, Galton, and Marey*. University Park: Pennsylvania State University Press, 2012.
Geimer, Peter. *Bilder aus Versehen. Eine Geschichte fotografischer Erscheinungen*. Hamburg: Philo Fine Arts, 2010.
Geimer, Peter, ed. *Ordnungen der Sichtbarkeit. Fotografie in Wissenschaft, Kunst und Technologie*. 3rd ed. Frankfurt a.M.: Suhrkamp, 2002.
Gunning, Tom. "To Scan a Ghost: The Ontology of Mediated Vision." *Grey Room* (2007): 94–127.
Hume, Naomi. "The Nature Print and Photography in the 1850s." *History of Photography* 35, no. 1 (2011): 44–58.
Keller, Corey, ed. *Brought to Light: Photography and the Invisible*. San Francisco Museum of Modern Art and Graphische Sammlung Albertina. New Haven, CT, and London: San Francisco Museum of Modern Art, 2008.
Krase, Andreas, and Agnes Matthias, eds. *Wahr-Zeichen. Fotografie und Wissenschaft*. Exhibition catalogue, Technische Sammlungen der Stadt Dresden, November 11, 2006–February 18, 2007. 2 vols. Dresden: Technische Universität Dresden, 2006.
Mnookin, Jennifer. "The Image of Truth: Photographic Evidence and the Power of Analogy." *Yale Journal of Law and the Humanities*, no. 10 (Winter 1997/98): 1–74.
Reichert, Ramón. *Im Kino der Humanwissenschaften: Studien zur Medialisierung wissenschaftlichen Wissens*. Bielefeld: Transcript, 2007.
Sekula, Allan. "The Body and the Archive." *October* 39 (1986): 3–64.
Thomas, Ann, ed. *Beauty of Another Order: Photography in Science*. Exhibition catalogue, National Gallery of Canada, Ottawa, 1997. New Haven, CT, and London: Yale University Press, 1997.
Tucker, Jennifer. *Nature Exposed: Photography as Eyewitness in Victorian Science*. Baltimore: Johns Hopkins University Press, 2006.

Novel Imaging Techniques and Media (MRI, PET Scanning, Oscilloscopy, X-Ray Imaging, CAT scanning)

Adelmann, Ralf, Jan Frercks, Martina Hessler, and Jochen Hennig. *Datenbilder. Zur digitalen Bildpraxis in den Naturwissenschaften*, Bielefeld: Transcript, 2009.
Burri, Regula Valérie. *Doing Images. Zur Praxis medizinischer Bilder*. Bielefeld: Transcript, 2008.
Dumit, Joseph. *Picturing Personhood: Brain Scans and Biomedical Identity*. Princeton, NJ: Princeton University Press, 2004.
Hennig, Jochen. *Bildpraxis: Visuelle Strategien in der frühen Nanotechnologie*. Bielefeld: Transcript, 2011.
Hennig, Jochen. "Changes in the Design of Scanning Tunneling Microscopic Images from 1980 to 1990." *Techné* 8, no. 2 (2004): 36–55.
Hennig, Jochen. "The Instrument in the Image: Revealing and Concealing the Condition of the Probing Tip in Scanning Tunneling Microscopic Image Design." In *Instruments in Art and Science—On the Architectonics of Cultural Boundaries in the 17th Century*, edited by Jan Lazardzig, Helmar Schramm, and Ludger Schwarte, 348–61. Berlin: de Gruyter, 2008.
Hinterwaldner, Inge. *Das systemische Bild: Ikonizität im Rahmen computerbasierter Echtzeitsimulationen*. Munich: Fink, 2010.
Kelty, Christopher, and Hannah Landecker. "A Theory of Animation: Cells, L-Systems, and Film." *Grey Room* 1, no.17 (2004): 30–63.
Lab. Jahrbuch für Künste und Apparate, edited by Kunsthochschule für Medien and Verein der Freunde der KHM, 1994–2006. Cologne: Walter König, 1994–2006.
Lübeck, Dieter. *Das Bild der Exakten—Objekt: Der Mensch; Zur Kultur der maschinellen Abbildungstechnik*. Munich: H. Moos, 1974.
Pasveer, Bernike. "Depiction in Medicine as a Two-Way Affair." In *Medicine and Change: Historical and Sociological Studies of Medical Innovation*, edited by Ilana Löwy, 85–105. Paris: Montrouge, 1993.
Pratschke, Margarete, ed. *Digitale Form*. In *Bildwelten des Wissens: Kunst-historisches Jahrbuch für Bildkritik*, vol. 3, no. 2. Edited by Horst Bredekamp, Matthias Bruhn, and Gabriele Werner. Berlin: Akademie Verlag, 2005.
Rasmussen, Nicolas. "Facts, Artifacts, and Mesosomes: Practicing Epistemology with the Electron Microscope." *Studies in History and Philosophy of Science* (Part A) 24, no. 2 (1993): 227–65.
Rasmussen, Nicolas. *Picture Control: The Electron Microscope and the Transformation of Biology in America, 1940–1960*. Stanford, CA: Stanford University Press, 1997.
Schwartzman, Madeline. *See Yourself Sensing: Redefining Human Perception*. London: Black Dog, 2011.
Tufte, Edward R. *The Cognitive Style of Powerpoint*. Cheshire, CT: Graphics Press, 2003.
Wilson, Stephen. *Information Arts: Intersections of Art, Science, and Technology*. Cambridge, MA: MIT Press, 2002.

JOURNALS

Berichte zur Wissenschaftsgeschichte: Organ der Gesellschaft für Wissenschaftsgeschichte. Weinheim: Wiley-VCH Verlag, since 1978.
Bildwelten des Wissens: Kunsthistorisches Jahrbuch für Bildkritik. Berlin: Akademie Verlag, since 2003.
The British Journal for the History of Science. Cambridge: Published by Cambridge University Press for the British Society for the History of Science, since 1962.
The British Journal for the Philosophy of Science. Oxford: Published by Oxford University Press for the British Society for the Philosophy of Science, since 1950.
Critical Inquiry. Chicago: University of Chicago Press, since 1974.
Fotogeschichte: Beiträge zur Geschichte und Ästhetik der Fotografie. Marburg: Jonas Verlag für Kunst und Literatur, since 1981.
Gegenworte: Hefte für den Disput über Wissen. Berlin: Akademie-Verlag, since 1998.
Grey Room. Cambridge, MA: MIT Press, since 2000.
History of Science. London: Sage, since 1962.
Isis. Chicago: Published by the University of Chicago Press for the History of Science Society, since 1913.

Journal of Visual Culture. London: Sage, since 2002.
Leonardo: Journal of the International Society for the Arts, Sciences and Technology. Cambridge, MA: MIT Press, since 1968.
Osiris. Chicago: Published by the University of Chicago Press for the History of Science Society, since 1936.
Representations. Berkeley: University of California Press, since 1983.
Rheinsprung 11: Zeitschrift für Bildkritik. Basel: Eikones, since 2011.
Science in Context. Cambridge: Published by Cambridge University Press for the Cohn Institute for the History and Philosophy of Science and Ideas, Tel Aviv University, since 1987.
Social Studies of Science: An International Review of Research in the Social Dimensions of Science and Technology. London: Sage, since 1975.
Studies in History and Philosophy of Science. Oxford: Elsevier Science, since 1970.
Technikgeschichte: Beiträge über die geschichtliche Entwicklung der Technik in ihren wissenschaftlichen, gesellschaftlichen, wirtschaftlichen und politischen Zusammenhängen. Berlin: Edition Sigma, since 1933.
Visible Language. Cincinnati, OH: University of Cincinnati, since 1967.

THE AUTHORS

Jana August

studied art history, cultural history and theory, and museology in Leipzig and Berlin. From 2004 to 2009, she was a scholarly assistant in the "Das Technische Bild" research project at the Hermann von Helmholtz-Zentrum für Kulturtechnik, Humboldt University of Berlin. In 2008–2009, she received a graduation fellowship from the FAZIT Foundation; in 2009–2010, she was a Scholar Associate at the Getty Research Institute, The Getty Center in Los Angeles. She has also contributed to several exhibition projects as a freelancer, including "World Knowledge: 300 Years of Science in Berlin" (2010–2011). Since 2011, August has been a fellow in the doctoral program at the Center for Literary and Cultural Research in Berlin, where she is working on her dissertation, which explores Alfred H. Barr's collection concept at the Museum of Modern Art in New York between 1929 and 1958.

Horst Bredekamp

studied art history, archaeology, philosophy, and sociology in Kiel, Munich, Berlin, and Marburg, where he received his PhD in 1974. After an internship at the Liebieghaus, Frankfurt am Main, Germany (1974–1976), he was appointed assistant professor at the University of Hamburg, where he received tenure as a professor of art history in 1982. Since 1993, he has been a tenured professor of art history at the Humboldt University of Berlin and a permanent fellow of the Institute for Advanced Study in Berlin. Among the awards he has received are the Sigmund Freud Award of the German Academy for Language and Poetry in Darmstadt, Germany (2001), the Aby M. Warburg Award of the City of Hamburg (2005), and the Max Planck Research Award of the Max Planck Society and the Alexander von Humboldt Foundation (2006). He has been a member of the Berlin-Brandenburg Academy of the Sciences and Humanities since 1995, the National Academy Leopoldina in Halle since 2004, and the European Academy in London since 2010.

Franziska Brons

studied art history, cultural history and theory, and musicology at the Free University of Berlin, the University of York (UK), and the Humboldt University of Berlin. She was a fellow of the Gerda Henkel Foundation, Düsseldorf; From 2006 to 2013, she was a research associate in the "Das Technische Bild" project at the Hermann von Helmholtz-Zentrum für Kulturtechnik, and a lecturer at the Institute of Art and Visual History, Humboldt University. In 2012, she completed her dissertation on the exposition of the medium of photography at the *International Photographic Exhibition Dresden 1909* at the Department of Art and Visual History, Humboldt University. Her teaching and research focus on issues of exhibiting and collecting photography and the visual history of the medium's scientific application. Since 2013 she has been a postdoctoral research associate at the chair of art history, Leuphana University, Lüneburg. Her current research focuses on the history and theory of the installation shot as well as on the aesthetics and exploration of underwater life in 19th century photography, early film and contemporary art.

Matthias Bruhn

studied art history and philosophy at the University of Hamburg. His dissertation on French 17th century painter Nicolas Poussin was published in 2000. From 1997 to 2001, he was associate of the research center on Political Iconography at Hamburg University's *Warburg-Haus*, subsequently Clark Art Institute fellow in 2000 and a postdoctoral fellow of the Getty Grant Program in 2001. Since 2005, he has been director of the research project "Das Technische Bild" at Humboldt University of Berlin, as well as coeditor of the yearbook *Bildwelten des Wissens. Kunsthistorisches Jahrbuch für Bildkritik.*

Stefan Ditzen

studied art history, psychology, and philosophy at Heidelberg University and art history, the history of the exact sciences and technology, and Spanish studies at the Humboldt University of Berlin, the Technical University of Berlin, the University of Granada, and the University of Florence. In 2003, he was a research associate on the "Das Technische Bild" project at the Hermann von Helmholtz–Zentrum für Kulturtechnik, Humboldt University. From 2003 to 2006, he was a fellow at the German Research Foundation's "Bild—Körper—Medium. Eine anthropologische Perspektive" Research Training Group at the Karlsruhe University of Arts and Design. In 2007, he obtained his Ph.D. in art history with a dissertation on the visual history of the microscope. He is currently a managing director at Meltwater Group, a software development company.

Vera Dünkel

studied visual arts, aesthetics, and art history at the Université de Paris I Panthéon-Sorbonne and the Humboldt University of Berlin. She is a research affiliate with the "Das Technische Bild" project at the Hermann von Helmholtz–Zentrum für Kulturtechnik, Humboldt University. In late 2012, she completed her doctoral thesis on early X-ray imaging.

Angela Fischel

studied cultural history and theory and art history at the Humboldt University of Berlin and at Middlesex University, London. From 2000 to 2008, she was a research associate in the "Das Technische Bild" project. In 2008, she obtained her Ph.D. with a dissertation on drawing and natural science in Conrad Gessner and Ulisse Aldrovandi. In 2009 she published her dissertation: *Natur im Bild. Zeichnung und Naturerkenntnis bei Conrad Gessner und Ulisse Aldrovandi* (Berlin: Gebrüder Mann). Her research focuses on the function of drawing collections in the context of early modern natural history, the art history of technical imagery, and the history of optical instruments in seventeenth- and eighteenth-century science and art.

Jochen Hennig

studied physics in Münster, Edinburgh, Granada, and Oldenburg and graduated with a thesis in the history of science. After a scholarly traineeship at the Deutsches Museum, Munich, he was a research associate in the "Das Technische Bild" research project at the Hermann von Helmholtz–Zentrum für Kulturtechnik, Humboldt University of Berlin, from 2003 to 2008. He obtained his PhD with a dissertation on the imagery of early nanotechnology that was published in 2011 with the title *Bildpraxis: Visuelle Strategien in der frühen Nanotechnologie.* He then directed the "World Knowledge: 300 Years of Science in Berlin" exhibition project on the occasion of the anniversaries of Humboldt University, the Charité Hospital, the Berlin-Brandenburg Academy of Sciences and Humanities, and the Max Planck Society (2010–2011). Since 2012, he has been coordinator of the scientific collections at Humboldt University.

Angela Mayer-Deutsch

studied art history and psychology at the Universities of Fribourg, Vienna, Frankfurt am Main, and the Humboldt University of Berlin. From 2003 to 2008, she was a research associate in the "Das Technische Bild" research project at the Hermann von Helmholtz–Zentrum für Kulturtechnik, Humboldt University. Her dissertation on Athanasius Kircher was published in 2010 under the title *Das Musaeum Kircherianum: Kontemplative Momente, historische Rekonstruktion, Bildrhetorik.* She currently works as a freelance art historian in Berlin.

Margarete Pratschke

studied art history, history, and communication studies at the Humboldt University of Berlin and the Free University of Berlin. From 2004 to 2010, she was a research associate in the "Das Technische Bild" research project at the Hermann von Helmholtz–Zentrum für Kulturtechnik, Humboldt University. In 2010, she completed her dissertation on the Windows operating system as a tableau, which traces the visual history of graphical user interfaces. From 2007 to 2009, she taught at the School of Design Thinking of the Hasso Plattner Institute, University of Potsdam. Since 2011, she has been a postdoctoral fellow at the chair for science studies at the Swiss Federal Institute of Technology, Zurich, and a research associate in the "Perception, Implicit Pictorial Knowledge and Cognition" module at the National Center of Competence in Research "Iconic Criticism—The Power and Meaning of Images" (eikones) at the University of Basel. For her habilitation, she is working on a new book on art history and experimental psychology of perception in the twentieth century.

Violeta Sánchez Lorbach

studied philosophy, sociology, law, and art history at the Humboldt University of Berlin, the Free University of Berlin, and the University of Potsdam and graduated in 2009 with a thesis on Aby Warburg's influence on the thinking of Giorgio Agamben. In 2004, she joined the "Das Technische Bild" research project at the Hermann von Helmholtz–Zentrum für Kulturtechnik, Humboldt University, as a scholarly assistant; in 2010 she was a research associate in the project as well as for the "World Knowledge: 300 Years of Science in Berlin" exhibition project. She currently works at the Berlin-based publishing house Vorwerk 8.

Birgit Schneider

studied art history and media studies, philosophy, and media art and multimedia at the Karlsruhe University of Arts and Design, Goldsmiths College, London, and the Humboldt University of Berlin. From 1998 to 2003, she worked as a graphic designer. From 2000 to 2007, she was a research associate in the "Das Technische Bild" project at the Hermann von Helmholtz–Zentrum für Kulturtechnik, Humboldt University, where she wrote her dissertation on the history of punched-card weaving, which was published in 2007 under the title *Textiles Prozessieren: Eine Geschichte der Lochkartenweberei.* Since 2008, she has been Dilthey Fellow of the Fritz Thyssen Foundation at the Institute for Arts and Media, University of Potsdam. In 2009, she was a substitute professor at the Bauhaus-Universität, Weimar. Her current research project is tentatively entitled "Images of the Climate: A Typology of Climate Visualization and Its Changes since 1800."

Heike Weber

studied history of science and technology at the Technical University of Berlin and media studies at the Free University of Berlin. She obtained her PhD at the Technical University of Munich with a study on the development and adoption of portable electronic devices that was published in 2008 under the title *Das Versprechen mobiler Freiheit: Zur Kultur- und Technikgeschichte von Kofferradio, Walkman und Handy.* After working at the Research Institute for the History of Science and Technology, Deutsches Museum, Munich, the "Das Technische Bild" research project at the Hermann von Helmholtz–Zentrum für Kulturtechnik, Humboldt University of Berlin, and the "Topologie der Technik" Research Training Group at the Technical University of Darmstadt, she has been a research associate in the history of technology at the Technical University of Berlin since 2008.

Reinhard Wendler

studied art history, musicology, and philosophy at the Technical University of Berlin and the Humboldt University of Berlin. He was a fellow in the "Das Technische Bild" research project and a research associate in the "Bild Schrift Zahl" project at the Hermann von Helmholtz–Zentrum für Kulturtechnik, Humboldt University. In 2008, he completed his dissertation in art history on the role of models in creative and scientific processes. From 2008 to 2010, he worked in the Fritz Thyssen Foundation–funded "Modelle als Akteure" research project at the Technical University of Berlin and the National Center of Competence in Research "Iconic Criticism—The Power and Meaning of Images" (eikones) at the University of Basel. From 2011 to 2012, he was a substitute junior professor of applied arts and design at the Department of Art and Visual History, Humboldt University. Since 2012, he has taught at the Zurich University of the Arts. His book *Das Modell zwischen Kunst und Wissenschaft* was published in 2012.

Gabriele Werner

studied art history, literature, German philology, and philosophy at the University of Hamburg and obtained her PhD in 1995 with a dissertation on mathematics in Surrealism, which was published in 2002 under the title *Mathematik im Surrealismus: Man Ray, Max Ernst, Dorothea Tanning.* From 2002 to 2003, she directed the "Das Technische Bild" research project at the Hermann von Helmholtz–Zentrum für Kulturtechnik, Humboldt University of Berlin; she has also been coeditor of its yearbook, *Bildwelten des Wissens: Kunsthistorisches Jahrbuch für Bildkritik.* From 2003 to 2008, she was a professor of art history at the University of Applied Arts Vienna. In the winter semester of 2008–2009, she was the Käthe Leichter visiting professor of gender studies in the Department of Art History at the University of Vienna; in the summer of 2009, she was a professor of recent and contemporary art history in the same department. Since 2011, she has been a professor at the Weißensee School of Art, Berlin.

INDEX

Page numbers in *italics* refer to illustrations and/or captions.

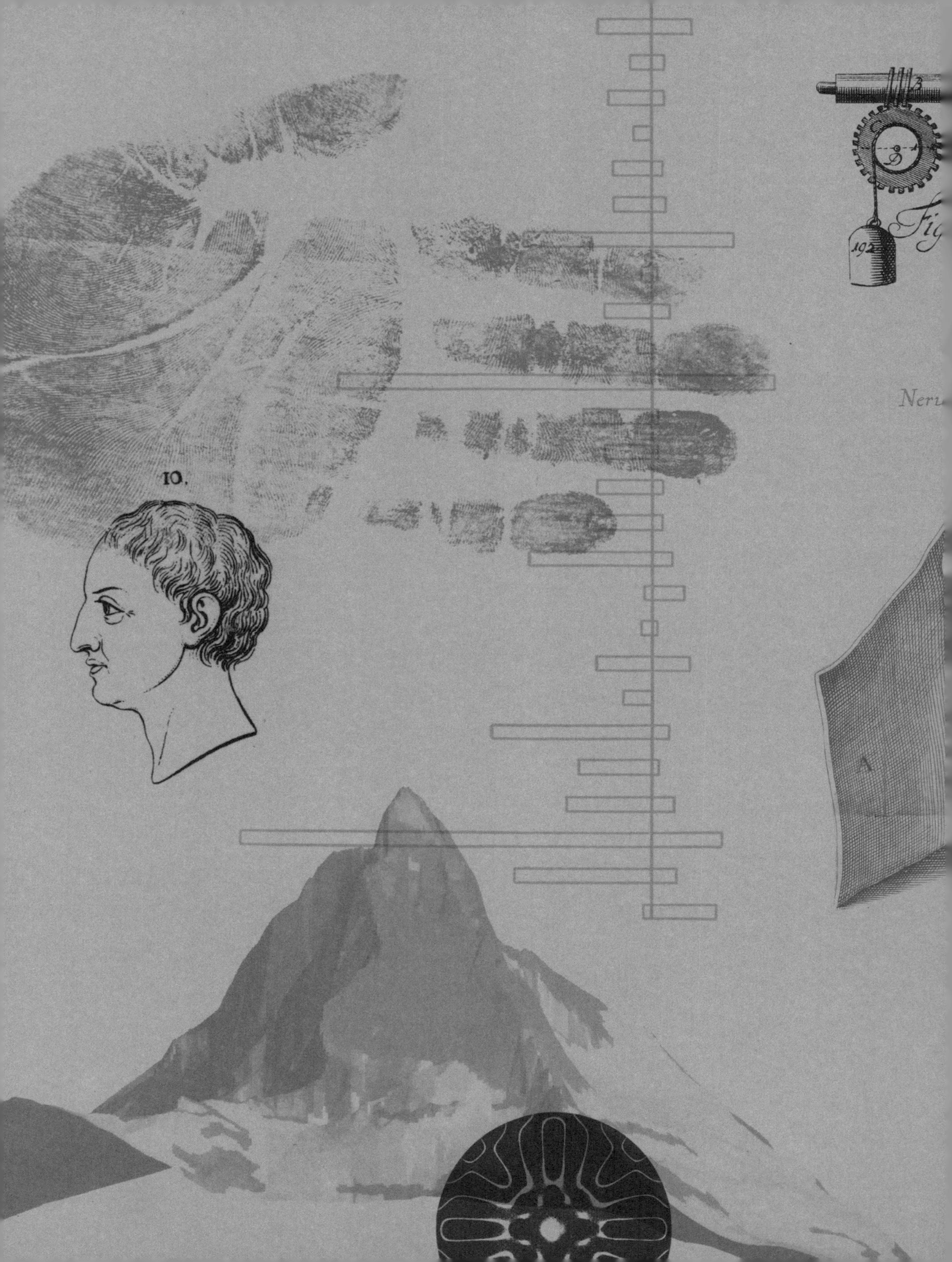
10.
192